REACTION MECHANISMS
IN SULPHURIC ACID

ORGANIC CHEMISTRY
A SERIES OF MONOGRAPHS

ALFRED T. BLOMQUIST, Editor
Department of Chemistry, Cornell University, Ithaca, New York

1. Wolfgang Kirmse. CARBENE CHEMISTRY, 1964; 2nd Edition, 1971
2. Brandes H. Smith. BRIDGED AROMATIC COMPOUNDS, 1964
3. Michael Hanack. CONFORMATION THEORY, 1965
4. Donald J. Cram. FUNDAMENTALS OF CARBANION CHEMISTRY, 1965
5. Kenneth B. Wiberg (Editor). OXIDATION IN ORGANIC CHEMISTRY, PART A, 1965; PART B, *In preparation*
6. R. F. Hudson. STRUCTURE AND MECHANISM IN ORGANO-PHOSPHORUS CHEMISTRY, 1965
7. A. William Johnson. YLID CHEMISTRY, 1966
8. Jan Hamer (Editor). 1,4-CYCLOADDITION REACTIONS, 1967
9. Henri Ulrich. CYCLOADDITION REACTIONS OF HETEROCUMULENES, 1967
10. M. P. Cava and M. J. Mitchell. CYCLOBUTADIENE AND RELATED COMPOUNDS, 1967
11. Reinhard W. Hoffman. DEHYDROBENZENE AND CYCLOALKYNES, 1967
12. Stanley R. Sandler and Wolf Karo. ORGANIC FUNCTIONAL GROUP PREPARATIONS, VOLUME I, 1968; VOLUME II, 1971
13. Robert J. Cotter and Markus Matzner. RING-FORMING POLYMERIZATIONS, PART A, 1969; PART B, *In preparation*
14. R. H. DeWolfe. CARBOXYLIC ORTHO ACID DERIVATIVES, 1970
15. R. Foster. ORGANIC CHARGE-TRANSFER COMPLEXES, 1969
16. James P. Snyder (Editor). NONBENZENOID AROMATICS, I, 1969; II, *In preparation*
17. C. H. Rochester. ACIDITY FUNCTIONS, 1970
18. Richard J. Sundberg. THE CHEMISTRY OF INDOLES, 1970
19. A. R. Katritzky and J. M. Lagowski. CHEMISTRY OF THE HETEROCYCLIC N-OXIDES, 1970
20. Ivar Ugi (Editor). ISONITRILE CHEMISTRY, 1971
21. G. Chiurdoglu (Editor). CONFORMATIONAL ANALYSIS, 1971
22. Gottfried Schill. CATENANES, ROTAXANES, AND KNOTS, 1971
23. M. Liler. REACTION MECHANISMS IN SULPHURIC ACID AND OTHER STRONG ACID SOLUTIONS, 1971

In preparation
J. B. Stothers. CARBON-13 NMR SPECTROSCOPY
Maurice Shamma. THE ISOQUINOLINE ALKALOIDS: CHEMISTRY AND PHARMACOLOGY

Reaction Mechanisms in Sulphuric Acid

and other Strong Acid Solutions

M. LILER

Lecturer in Chemistry
The University of Newcastle upon Tyne, England

1971
ACADEMIC PRESS
London and New York

ACADEMIC PRESS INC. (LONDON) LTD
Berkeley Square House
Berkeley Square
London, W1X 6BA

U.S. Edition published by
ACADEMIC PRESS INC.
111 Fifth Avenue
New York, New York 10003

Copyright © 1971 by ACADEMIC PRESS INC. (LONDON) LTD

All Rights Reserved

No part of this book may be reproduced in any form by photostat, microfilm, or any other means, without written permission from the publishers

Library of Congress Catalog Card Number: 70-149701
ISBN: 0-12-450050-1

PRINTED IN GREAT BRITAIN BY
WILLIAM CLOWES AND SONS LIMITED,
LONDON, COLCHESTER AND BECCLES

Preface

When it was suggested to me that I should write a monograph under the present title, I thought that the scope of the work would be too narrow if it were restricted to concentrated sulphuric acid; I have therefore adopted a wider interpretation of the title as covering reactivity in strongly acid solutions in general. Sulphuric acid is certainly the most important strongly acid medium and may be regarded as the prototype of such media, with some specificities of its own in the concentrated acid region. Some emphasis has been placed on these aspects, but I consider that some of the most valuable contributions to chemistry that have emerged from studies of reactivity in sulphuric acid solutions are due to the very wide range of acidities available in sulphuric acid/water mixtures, and have therefore covered all aspects of reactivity over the whole of this acidity range.

An attempt was made to take account of the most recent developments in this field, including work published in the course of 1969, while the writing of this book was in progress. The coverage of the literature for that year, however, is not complete. As the manuscript was completed, I became aware of the recent publication of "Acidity Functions" by C. H. Rochester (Academic Press, June 1970), which covers in part similar ground, although with different emphasis, and may therefore be regarded as complementary to this volume.

I wish to acknowledge useful discussions with, and help from, many of my colleagues. My sincere thanks are also due to Miss Doreen Moor, who typed some of the most difficult parts of the manuscript.

I thank Professor J. T. Edward and the Royal Society of Canada for permission to reproduce from one of their publications Fig. 2.2, and the Chemical Society (London) for permission to reproduce from some of my own publications Figs. 3.7, 3.8. and 3.13.

I would also like to thank the staff of Academic Press Inc. (London) Ltd., and in particular the production department, for their highly competent help, to which many improvements of the manuscript and the high standard of production of the text are due.

M. LILER

June 1971

Contents

Preface v

General Introduction xi

CHAPTER 1 Properties of sulphuric acid/water mixtures as solvents
1.1. Introduction 1
    1.1.1. Concentration units 2
    1.1.2. Preparation and analysis of solutions 3
1.2. Physical properties 4
    1.2.1. Mechanical properties 4
    1.2.2. Electrical properties 8
1.3. Thermodynamic properties 12
    1.3.1. Phase changes 13
    1.3.2. Thermochemical quantities 16
1.4. Spectroscopic properties 17
    1.4.1. Rayleigh scattering 18
    1.4.2. Raman and infrared spectra 18
    1.4.3. Ultraviolet absorption spectra 21
    1.4.4. X-Ray diffraction 22
    1.4.5. Proton magnetic resonance spectra 23
1.5. Isotopically substituted sulphuric acids 24
    1.5.1. Dideuterosulphuric acid 24
    1.5.2. ^{35}S-labelled sulphuric acid 25

CHAPTER 2 Acidity functions
2.1. Introduction 26
2.2. Definitions 27
    2.2.1. Acidity functions for simple protonation . . . 27
    2.2.2. Acidity function for complex ionizations . . . 29
    2.2.3. Acidity function for olefin protonation . . . 30
2.3. Evaluation of acidity functions 31
    2.3.1. Indicator measurements 31
    2.3.2. Electrochemical measurements 34
2.4. Various acidity function scales 35
    2.4.1. The H_0 or H_0' acidity function 35
    2.4.2. The H_0''' acidity function 40
    2.4.3. The H_A acidity function 41
    2.4.4. The H_I acidity function 42
    2.4.5. The H_+ acidity function 43

	2.4.6.	The H_- acidity function	44
	2.4.7.	The H_R and H_R' acidity function	45
2.5.	Verification of the activity coefficient postulate		47
	2.5.1.	Measurements of activity coefficients of neutral indicator bases	47
	2.5.2.	Relative activity coefficients of some indicator conjugate acids	49
	2.5.3.	Relative activity coefficient ratios	51
2.6.	Physical significance of acidity functions		52
	2.6.1.	A chemical hydration treatment of acidity functions	52
	2.6.2.	The acidity function H_0 in the vicinity of 100 per cent sulphuric acid	56

CHAPTER 3   Protonation of very weak bases

3.1.	Introduction		59
3.2.	Methods of determination of basicities in sulphuric acid media		61
	3.2.1.	The spectrophotometric method	61
	3.2.2.	The n.m.r. spectroscopic method	63
	3.2.3.	The Raman spectrometric method	65
	3.2.4.	The distribution or extraction method	66
	3.2.5.	The cryoscopic method	67
	3.2.6.	The conductometric method	71
	3.2.7.	The conductometric titration method	83
	3.2.8.	Correlation of basicity determinations in pure sulphuric acid with H_0	85
3.3.	Basicities of very weak bases		86
	3.3.1.	Introduction	86
	3.3.2.	Bases involving N^{III} and P^{III} as the basic centre	95
	3.3.3.	Bases involving more complex groups containing N, P and As	112
	3.3.4.	Bases involving O and S as the basic centre	118
	3.3.5.	Hydrocarbon bases	139
	3.3.6.	Inorganic bases	143

CHAPTER 4   Complex ionizations

4.1.	Introduction		145
4.2.	Methods of study of complex ionizations		146
	4.2.1.	The cryoscopic method	146
	4.2.2	The conductometric method	149
	4.2.3.	The conductometric titration method	149
	4.2.4.	The spectrophotometric method	150
4.3.	Survey of complex ionizations		150
	4.3.1.	Organic solutes	150
	4.3.2.	Inorganic solutes	160
	4.3.3.	Organometallic solutes	165

CHAPTER 5   Reaction mechanisms in sulphuric acid solutions

5.1.	Introduction		167
5.2.	Mechanisms and mechanistic criteria in strong acid solutions		168
	5.2.1.	Main types of reaction mechanism in strong acid solutions	168
	5.2.2.	Mechanistic criteria	170

5.3.	Methods of kinetic investigation	181
5.4.	Mechanisms of reactions in sulphuric acid solutions	183
	5.4.1. Hydrolyses	183
	5.4.2. Hydration and dehydration reactions	210
	5.4.3. Isomerizations and rearrangements	232
	5.4.4. Decarbonylations and decarboxylations	253
	5.4.5. Electrophilic aromatic substitutions by the hydrogen ion	262
	5.4.6. Electrophilic aromatic substitutions by other electrophiles	276
	5.4.7. Miscellaneous reactions involving carbonium ions.	297

References 307
Author Index 329
Subject Index 343

General Introduction

1. *Historical*

Sulphuric acid solutions and concentrated sulphuric acid have been used as catalysts and as reaction media for a number of reactions throughout the 19th century and have played a considerable part in the development of organic chemistry in general. The first decade of the 20th century marks an important point in the development of our understanding of these reactions. By that time basic ideas about electrolyte solutions in aqueous media had been well established through the work of Arrhenius and Van't Hoff, and interest in other media as solvents began to develop with the pioneering work of Walden (1901 and 1902). Concentrated sulphuric acid as solvent, already studied by Walden, attracted the attention of Hantzch (1908 and 1909), who carried out extensive exploratory investigations on the behaviour of a large number of solutes in sulphuric acid using the cryoscopic and conductometric methods. These have remained two of the most important methods of investigation of solutions in concentrated sulphuric acid up to the present time. Hantzch was primarily concerned with the state of solutes in this solvent, i.e. with their modes of ionization, and made the important discovery that carbonium ions are obtained from triphenyl carbinols in this medium. He also made some observations on unstable solutes, which could not be recovered unchanged by diluting their solutions in sulphuric acid with water. The first kinetic studies of reactions in concentrated sulphuric acid also date from the first decade of this century (Martinsen, 1905 and 1907; Lichty, 1907). This rising interest in concentrated sulphuric acid as solvent and reaction medium was a reflexion of the great advances made in the production of sulphuric acid at that time due to the introduction of the contact process, a detailed account of which was given by Knietsch (1901). The properties and potential uses of this substance were clearly a matter of considerable interest to the chemist and industrial chemist, and a better understanding of its solvent properties was imperative. During the first two decades of the century a great deal of information on the physical properties of the water/sulphur trioxide system was accumulated. The phase relations of concentrated sulphuric acid with a number of organic and inorganic substances were also studied. The third decade of the century saw an intensified interest in the kinetics of reactions in concentrated sulphuric acid and sulphuric acid/water mixtures, mainly decompositions with gas evolution, which were amenable to study by the methods available at the time.

The next major advance and stimulus for further work came in the nineteen thirties with Hammett's introduction of the acidity function concept (Hammett and Deyrup, 1932). Since then there has been steady progress in our understanding of the changes that organic substances undergo in strongly acid media.

One of the very important contributory factors to this has been the development of new and better instrumental methods for the study of reactions of organic compounds, in particular of the spectrophotometric method. The classical methods of cryoscopy and conductance have, however, continued to be used and have been further developed to yield results often unobtainable by the spectrophotometric method. Major leading centres in the developments in this field in the course of the last 25 years have been the school of Professor C. K. Ingold in Britain, with a transfer to Canada through the leading figure of Professor R. J. Gillespie, and the schools of Professor M. S. Newman and N. C. Deno in the U.S.A. Many notable contributors to this field have come from these schools, but there have also been many contributions from other centres. Further historical references will be made in the introductions to the chapters in this book.

2. Scope and Layout

The scope of this work may be defined as covering reactivity in strongly acid solutions to which the concepts of dilute aqueous chemistry do not apply. This refers particularly to the measure of acidity, which in these solutions is not the pH but its analogue, the acidity function. Sulphuric acid/water mixtures take pride of place amongst such acid media. In the moderately concentrated range, perchloric acid/water mixtures compete with sulphuric acid solutions as media for acid-catalysed reactions, but concentrated sulphuric acid as a reaction medium has certain specificities which are unparalleled *in toto* by any other acid medium. These arise from a combination of high acidity with low water activity. Pure fluorosulphuric acid has some properties similar to 100 per cent sulphuric acid, but some other acidic media (hydrogen fluoride/ boron trifluoride, fluorosulphuric acid/antimony pentafluoride) are considerably more acidic and contain weaker nucleophiles. For this reason they are better solvents for stabilizing carbonium ions. Quite strong nucleophiles, water and hydrogen sulphate ions, are present in sulphuric acid solutions, and are responsible for a number of reactions in these media. While reactions in sulphuric acid solutions are the main topic of this monograph, frequent reference to other acidic media will also be made, in order to provide a connected picture of reactivity in strongly acid solutions.

One of the most valuable features of sulphuric acid/water mixtures is the very wide acidity range that they offer, extending continuously from dilute aqueous acid to strongly acidic oleum solutions. At these acidities numerous compounds, normally regarded as neutral, become protonated. With some compounds protonation leads to unstable species, which undergo further reaction. Therefore, knowledge of the protonation behaviour of substances is an essential preliminary to an understanding of reaction mechanisms in sulphuric acid solutions. In view of this, Chapters 1 and 2, which discuss general properties of sulphuric acid/water mixtures and acidity functions respectively, are followed by Chapters 3 and 4, which discuss protonation and more complex modes of ionization of compounds in these media. Reaction mechanisms are then discussed in Chapter 5.

CHAPTER 1

Properties of Sulphuric Acid/Water Mixtures as Solvents

1.1. Introduction

Two of the most important characteristics of sulphuric acid/water mixtures and dilute oleums as reaction media have long been recognized to be their protonating ability and their affinity for water, i.e. their dehydrating power. A number of other properties of these media are, however, also important, either as general characteristics or as reflections of some special features of the mixtures, which can be used in studying reactions in these media and which may affect the stability of solutes in solution by solute/solvent interactions. In these respects the intermolecular structure of the concentrated acid is of particular importance. Since components of the solvent medium take part in some reactions, a knowledge of the nature and concentrations of the species present in the solvent mixtures as functions of their stoichiometric composition is essential for the unravelling of the mechanisms of such reactions. Information about this has largely been obtained from thermodynamic and spectroscopic studies of the mixtures.

A separate chapter will be devoted to acidity functions (Chapter 2), both as measures of the protonating ability of the solvent media and as reflections of solute/solvent interactions in them. The present chapter will be devoted to all other properties of sulphuric acid/water mixtures, the 100 per cent sulphuric acid region and dilute oleums. It will cover their physical, thermodynamic and spectroscopic properties.

The properties of isotopically substituted sulphuric acids are also of interest, especially those of deuterosulphuric acid, because of their usefulness in studies of reaction mechanism. One section of this chapter therefore gives a brief account of their preparation and properties.

Before giving an account of the properties of the mixtures, it is appropriate to define the various concentration scales which are in common use in dealing with sulphuric acid/water mixtures and dilute oleums. Also a description of the methods of preparation and analysis of the solutions is most conveniently given at this point.

1.1.1. CONCENTRATION UNITS

The most commonly used concentration units for sulphuric acid/water mixtures, as for mixtures of other acids with water, are weight per cent, molality, and molarity of the acid. Another method of expressing their concentration in terms of $A = $ mol H_2O/mol H_2SO_4 was introduced by Kunzler and Giauque (1952b), because of its suitability in the study of the properties of the solid hydrates of sulphuric acid. This scale offers no advantage in the study of the liquid phase. Apart from this mole ratio scale, the mole fraction scale is also used. The simple weight per cent scale, always used in studies of acidity functions, will be used in this chapter.

For solutions in 100 per cent sulphuric acid a new and unusual concentration unit has been proposed by Gillespie and Solomons (1960) called a "molon" and defined as mol kg^{-1} of solution. This unit is particularly suitable for self-dissociated solvents and solvents in which solutes dissolve by involving some solvent molecules in a reaction, for example, as in the protonation of a base:

$$B + H_2SO_4 \rightleftharpoons BH^+ + HSO_4^- \tag{1.1}$$

In such a case molality can be ambiguous if it is not made clear whether the solvent taking part in this reaction is regarded as part of the solute $BH^+HSO_4^-$ or not. Also the concentrations of the solvent self-dissociation species, whose estimation necessarily depends on certain assumptions, are best expressed per kilogram of the solvent as such, which is regarded as a solution of the self-dissociation species. The suggested symbol for the molon unit is w, to indicate that it is a weight concentration. Like the molality, this concentration unit is independent of temperature. It can easily be converted into molarity, if the density of the solution is known, by means of the relationship $c = \rho w$. This also makes it convenient for use in conductance work.

There are several concentration scales for oleum solutions. Three of these may be described as traditional ways of expressing concentrations of oleums. Probably the most common way is in terms of the percentage of free sulphur trioxide, which is the weight per cent of sulphur trioxide in the binary mixture of sulphuric acid and sulphur trioxide. The most direct way is in terms of the weight per cent of sulphur trioxide in the binary mixture sulphur trioxide/water. The third customary way is in terms of more than 100 per cent sulphuric acid, i.e. the weight per cent of sulphuric acid is taken to mean the total weight of sulphuric acid that may be obtained from the total sulphur trioxide present.

$$\text{Weight \% } H_2SO_4 = \frac{\text{Weight } SO_3 \cdot 98 \cdot 082/80 \cdot 066}{\text{Weight } SO_3 + \text{Weight } H_2O} \times 100 \tag{1.2}$$

Apart from this, the concentrations of dilute oleums can be expressed as molality of SO_3 or as molality of $H_2S_2O_7$ (which are not the same), or in terms of molon units (which are the same whether the solute is regarded as SO_3 or as $H_2S_2O_7$). Kunzler and Giauque (1952b) have also defined the unit A as (mol H_2O − mol SO_3)/mol SO_3, the values for oleum solutions being negative (from 0 to −1). Also the mole fraction of sulphur trioxide in the sulphuric acid/sulphur trioxide binary mixtures is often used.

1.1.2. PREPARATION AND ANALYSIS OF SOLUTIONS

Solutions of sulphuric acid were normally standardized against alkali in the usual way until the work of Kunzler (1953), which showed that sulphuric acid itself could be a highly accurate primary standard. It should be distilled first for the removal of impurities, and then distilled under constant known pressure to obtain the constant boiling azeotrope (b.p. approx. 330°C at 1 atm pressure). Kunzler (1953) showed that the composition of the constant boiling sulphuric acid is much less dependent on pressure than the composition of the constant boiling hydrochloric acid. The following concentrations were found for the pressure range close to normal atmospheric pressure:

mmHg	% H_2SO_4
700	98·495
750	98·482
800	98·469

Constant boiling sulphuric acid can be used as a standard for the preparation of more dilute solutions.

An equally accurate primary standard for the preparation of solutions of any acid content up to 100 per cent is the 100 per cent acid itself, which can be prepared by mixing concentrated sulphuric acid and dilute oleum until the composition of the maximum freezing point has been adjusted. If the freezing temperature is within 0·001°C of the freezing point of the 100 per cent acid (+10·371°C), the acid composition will be within 0·002 per cent of the 100 per cent acid. If it is within 0·010°C, the composition will be within 0·012 per cent of 100 per cent.

There are several rapid methods of preparing sulphuric acid of approximately 100 per cent composition. The "water titration" method is the oldest and easiest. It was proposed by Setlik as long ago as 1889. It consists in adding water to oleum solutions until they no longer give fumes. Brand (1946) used this method at temperatures below 10°C and was able to reach an end-point within 0·02 per cent of the absolute acid. A variation of this method (Kunzler, 1953) is to add concentrated sulphuric acid to slightly fuming oleum as long as a "fog" is formed above the surface when moist air is passed over it. The addition of concentrated acid is preferable to that of water because the heat evolved is much less and consequently less sulphur trioxide escapes from the solution to form fog. Also a more accurate variation in composition can be achieved. With some experience a rapid preparation by this method in an open beaker can yield acid of composition within 0·1 per cent of the absolute acid. A "thermometric titration" method, based on the enormous difference between the partial molar heat content of sulphur trioxide in fuming and in concentrated acid, was suggested by Somiya (1927). This method was reassessed by Kunzler (1953), who found that the break in the temperature against composition curve was rather gradual in the 100 per cent region, but the required 100 per cent composition could be obtained within $\pm 0\cdot05$ per cent. A conductometric titration of oleum with concentrated acid until the specific conductance of the 100 per cent acid is reached (see Section 1.2.2.2) can also yield acid as accurately adjusted to 100 per cent as by the freezing-point method.

All the methods described in the preceding paragraph are at the same time methods of determining the sulphur trioxide content of oleums. However, for this purpose water should preferably be used as titrant. A direct potentiometric titration of sulphur trioxide in oleum using water as titrant, a chloranil indicator electrode, and a Hg/Hg_2SO_4 reference electrode, has also been described (van der Heijde, 1955).

The storage of 100 per cent sulphuric acid and transfer to weighing bottles have also been described by Kunzler (1953). A dry box should preferably be used in the transfer operation, but rapid transfers in the open atmosphere need not introduce an error of more than 0·01 per cent. In preparing the solutions the weighing bottles are set into water, and this is allowed to cover the acid immediately, before complete mixing is achieved. The advantage of this procedure is that no loss of sulphur trioxide occurs.

1.2. Physical Properties

Many physical properties of pure sulphuric acid and its mixtures with water and sulphur trioxide were measured with remarkable accuracy at the turn of this century. Later work has brought little improvement on these early results. The density of sulphuric acid solutions is one of their most widely used properties, closely followed by their conductance. Other mechanical and electrical properties have also been studied with considerable care and have shed a great deal of light on the structure of these solutions. Such properties will be dealt with in the following two sections.

1.2.1. MECHANICAL PROPERTIES

1.2.1.1. *Densities*

(a) *Sulphuric acid/water mixtures.* Numerous determinations of the densities of sulphuric acid/water mixtures were carried out in the course of the nineteenth

1. PROPERTIES OF SULPHURIC ACID/WATER MIXTURES AS SOLVENTS

and early twentieth century, and extensive tabulations of data are available for a range of temperatures in *International Critical Tables* and *Landolt-Börnstein Tabellen*. Selected values for 25°C are given in Table 1.1. Densities

TABLE 1.1

Physical properties of sulphuric acid/water mixtures at 25°C

Weight % H_2SO_4	Density† (g cm^{-3})	−Excess volume (10^{-2} cm^3 g^{-1})	Viscosity† (10^{-7} N s cm^{-2})	Surface tension‡ (10^{-4} N cm^{-1})	Specific conductance§ (Ω^{-1} cm^{-1})
0	0·9971	0	0·891	7·19	
5	1·0300	0·74		7·20	0·230
10	1·0640	1·76	1·21	7·21	0·426
15	1·0994	2·51		7·24	0·598
20	1·1365	3·20	1·53	7·28	0·716
25	1·1750	3·81		7·34	0·792
30	1·2150	4·33	2·02	7·39	0·827
35	1·2563	4·76		7·44	0·815
40	1·2991	5·10	2·6$_8$	7·475	0·769
45	1·3437	5·39		7·485	0·702
50	1·3911	5·63	3·8$_3$	7·47	0·618
55	1·4412	5·86		7·43	0·527
60	1·4940	6·03	5·9	7·37	0·437
65	1·5490	6·13		7·29	0·346
70	1·6059	6·14	10·0	7·19	0·254
75	1·6644	6·06	14·3	7·05	0·186
80	1·7221	5·78	20·3	6·86	0·142
85	1·7732	5·18	24·6	6·56	0·1250
90	1·8091	4·01	23·5	6·05	0·1321
95	1·8286	2·33	21·9	5·43	0·1210
98	1·8310	1·05	23·2	5·11	0·0712
100	1·8269	0	24·54	4·95	0·0104

† *International Critical Tables* (see text).
‡ Morgan and Davis (1916).
§ Haase *et al.* (1966).

of sulphuric acid/water mixtures cover a wide span of values and, since they can be determined easily with an accuracy of 1 in 10^5, they have been used extensively for the determination of acid concentration in the mixtures. The densities of the mixtures are greater than the additive values and reflect a strong interaction between the two components. The volume contractions that occur on mixing have been calculated for 25°C and are also shown in Table 1.1. The maximum volume contraction occurs at 65–70 per cent sulphuric acid.

This is less than the composition of the 1:1 mole ratio, which corresponds to the main interaction in the system

$$H_2SO_4 + H_2O \rightleftharpoons H_3O^+ + HSO_4^- \qquad (1.3)$$

This can be explained in terms of solvation of the hydronium ion to give aggregates $H_9O_4^+$, which are at present believed to be the predominant form of the hydrogen ion in aqueous solution. The electrostriction that occurs in the formation of the hydrogen-bonded solvate (1.1) results in a reduction of

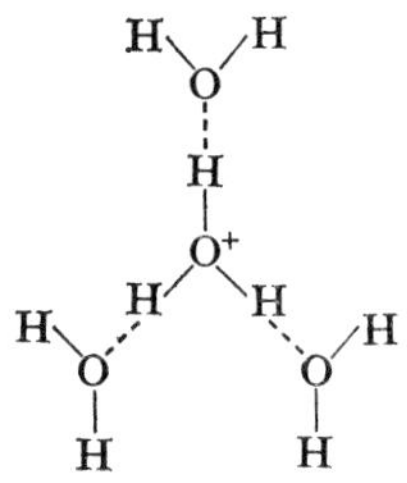

(1.1)

volume. Both components of the binary system are associated by hydrogen bonding in the pure state, as many of their properties show. The presence of the hydronium ions enhances the structure in the water-rich mixtures more than in the concentrated acid region, and this results in a maximum of volume contraction at concentrations 65–70 per cent acid.

TABLE 1.2

The properties of 100 per cent sulphuric acid over a temperature range

t (°C)	d_4^t (g cm^{-3})†	η (10^{-7} N s cm^{-2})†	γ(10^{-4} N cm^{-1})‡
10	1·8416	44·88	
20	1·8318	29·56	
30	1·8212	20·62	4·962
40	1·8117	14·67	
50	1·8017	10·93	4·906
60	1·7921	8·50	
70	1·7816		

† Greenwood and Thompson (1959).
‡ Morgan and Davis (1916).

(b) *The 100 per cent sulphuric acid region.* The densities of 100 per cent sulphuric acid and the densities of very dilute solutions of water and sulphur trioxide in it have been determined accurately by Gillespie and Wasif (1953b). Their value for the density of the 100 per cent acid at 25°C, $d_4^{25} = 1·8269$ g cm^{-3}, has been closely confirmed by Greenwood and Thompson (1959), who found a value of 1·8267 g cm^{-3}. Greenwood and Thompson also determined densities at other temperatures, and some of their values are reproduced in Table 1.2. This temperature dependence of the density can be represented by the equation

$$d_4^t = 1·8516 - 1·000 \times 10^{-3} t \text{ (g cm}^{-3}\text{)} \tag{1.4}$$

also due to Greenwood and Thompson (1959). Most earlier determinations gave lower values for the pure acid and *International Critical Tables* give 1·8255 g cm^{-3} at 25°C, but a higher value of 1·8278 g cm^{-3} has also been reported (Tutundžić and Liler, 1953).

(c) *Dilute oleum solutions.* Solutions of sulphur trioxide in 100 per cent sulphuric acid show increasing densities with increasing concentration. Recent accurate determinations are again due to Gillespie and Wasif (1953b) and are reproduced in Table 1.3.

TABLE 1.3

The properties of dilute oleum solutions at 25°C (Gillespie and Wasif, 1953b)

$m_{H_2S_2O_7}$	d_4^{25} (g cm^{-3})	η (10^{-7} N s cm^{-2})
0·0190	1·8270	24·54
0·0470	1·8280	24·54
0·2250	1·8330	24·57
0·3550	1·8360	24·66
0·5360	1·8407	24·74
0·6920	1·8439	24·78
0·8350	1·8480	24·82

Accurate density data on the more concentrated oleum solutions became available only recently, when they were measured by Popiel (1964) in connection with the work on Raman spectra of oleums. He confirmed some of the values earlier determined by Knietsch (1901). Popiel's density data for dilute oleum solutions are consistently lower than those of Gillespie and Wasif (1953b) by about 0·2 per cent, which represents only moderately good agreement.

1.2.1.2. *Viscosities*

(a) *Sulphuric acid/water mixtures.* Accurate measurements of the viscosities of sulphuric acid/water mixtures were carried out by Dunstan and Wilson (1907) at 25°C and by Bingham and Stone (1923) at 10, 20 and 40°C. On the basis of both sets of data, *International Critical Tables* give the best values for 25°C, which are higher than the original data of Dunstan and Wilson, as corrected by Dunstan (1914). Since, however, the value reported by Dunstan for the 100 per cent acid is lower by about 4 per cent than the best more recent determinations (Gillespie and Wasif, 1953b; Greenwood and Thompson, 1959), the values reported in *International Critical Tables* are more consistent with these newer data for the 100 per cent acid, and are therefore reproduced in Table 1.1.

As the figures show, there is an almost exponential rise in viscosity with increasing sulphuric acid concentration up to a sharp maximum at a concentration of 86 per cent. This is slightly higher than the composition of the 1:1 hydrate (84·5 per cent). Shifts in viscosity maxima corresponding to compound formation towards the more viscous component are commonly observed in binary liquid mixtures and suggest that the compound is partially dissociated into its components. In this particular case the dissociation of the compound corresponds to equilibrium in reaction (1.3).

(b) *Dilute oleum solutions.* Viscosities of oleum solutions have also been measured by Dunstan and Wilson (1908) at 60°C for up to 70 per cent free sulphur trioxide. Their value for the viscosity of the 100 per cent acid at 60°C ($8·32 \times 10^{-7}$ N s cm^{-2}) is in good agreement with the value in Table 1.2. They found a viscosity maximum at 40·6 per cent free sulphur trioxide and concluded that this was evidence for the formation of molecular aggregates of the composition of disulphuric acid ($H_2SO_4 . SO_3$). They also inferred that sulphuric acid itself was strongly associated. Highly accurate measurements of the viscosity of very dilute oleums at 25°C by Gillespie and Wasif (1953b) are reported in Table 1.3.

1.2.1.3. *Surface Tension*

The surface tension of sulphuric acid/water mixtures was measured by Morgan and Davis (1916) at 0, 30 and 50°C. The values reported in Table 1.1 were obtained by interpolation from plots of their data. This property also shows a distinct maximum at concentrations of 40–50 per cent acid, i.e. like the excess volume again on the aqueous side of the 1:1 hydrate. Surface tension is also a measure of the cohesive forces in the liquid and the maximum shows that these are stronger in the solutions of the acid than in either pure component.

1.2.2. ELECTRICAL PROPERTIES

1.2.2.1. *Dielectric Constant*

Walden's interest in the factors which determine the ionizing power of solvents, expressed in his paper published in 1902, led him to attempt to measure the dielectric constant of absolute sulphuric acid alongside many other solvents

(1903). However, 100 per cent sulphuric acid is a highly conducting liquid (it has a specific conductance comparable to that of 0·1 N KCl) and measurement of the dielectric constant of highly conducting liquids presents considerable problems. Walden used the method of Drude (1897 and 1902) and was able only to estimate that the dielectric constant of absolute sulphuric acid at 20°C was >84, i.e. greater than that of water. The first measurements were made by Brand et al. (1953), who studied dielectric dispersion at high frequencies (> 100 MHz) and obtained the static dielectric constant by extrapolation, a procedure which inevitably involves some uncertainty. Gillespie and Cole (1956) worked at lower frequencies, using a bridge circuit and a condenser with a variable electrode separation in order to eliminate the electrode polarization capacity. Gillespie and White (1958) applied yet another method, the Fürth "force" method, which had to be modified for viscous liquids. The two methods used by Gillespie and his collaborators gave concordant results, showing that the dielectric constant decreases with increasing temperature. The most recent determinations by Borovikov and Fialkov (1965) made use of resonance in a high frequency (15–18 MHz) oscillating circuit. While Gillespie and White (1958) believe their method to be accurate to no better

TABLE 1.4

The dielectric constant of 100 per cent sulphuric acid

t (°C)	ϵ	References†
8	122	a
9·5 ± 1·5	120 ± 8	b
16·5 ± 1·5	112 ± 8	b
20	110	c
25	101 ± 6	a
25	100·5 ± 2	d

† Reference: a, Gillespie and Cole (1956); b, Gillespie and White (1958); c, Brand et al. (1953); d, Borovikov and Fialkov (1965).

than ±5 per cent, Borovikov and Fialkov claim an accuracy of ±2 per cent. Table 1.4 gives the values obtained by all these authors. No measurements were carried out on the highly conducting sulphuric acid/water mixtures.

1.2.2.2. *Electrical Conductivity*

(a) *Sulphuric acid/water mixtures.* For many decades the most extensive and the most accurate data for the specific conductance of sulphuric acid/water mixtures were those of Kohlrausch (1876) for 18°C. In the last two decades new accurate measurements have become available for a range of temperatures, partly in response to industrial needs. Data of moderate accuracy for 25, 60 and 95°C were obtained by Roughton (1951), and Campbell *et al.* (1953) provided accurate data from 5–96 per cent acid for 50 and 75°C. The most accurate data for the temperature range 0–50°C (at 10 deg intervals) were provided by Haase *et al.* (1966) but only up to 86 per cent acid. Equally accurate extensive tabulations of data were published by Darling (1964) for 0·5–99 per cent acid at 1 per cent intervals of acid composition, covering the temperature range 0–240°F (at 10 deg intervals). The values reported in Table 1.1 for 25°C are based mainly on the data of Haase *et al.* (1966) and are supplemented at high concentrations by some data based on Darling's tables. The two sets of data appear to be in very good agreement.

The outstanding feature of the conductance isotherms is a broad maximum at 30–35 per cent acid, present at all temperatures, and a lower maximum at 90–95 per cent acid, present only at temperatures below 75°C. Between the two maxima there is a minimum corresponding closely to the composition of the 1:1 hydrate of sulphuric acid and hence also to the viscosity maximum. While proton-jump conduction accounts for the high conductivity in the water-rich mixtures, and likewise for the lower maximum in the sulphuric acid region (see Chapter 3), the conduction in mixtures close to the composition of the 1:1 hydrate, i.e. hydronium hydrogen sulphate, is more akin to that occurring in molten salts. The three-dimensional network of hydrogen-bonded bridges, essential for proton-jump conduction, cannot be present in this composition region. The conductance minimum shifts with increasing temperature towards higher concentrations of sulphuric acid, in close correspondence with the shifts of the viscosity maximum with temperature, suggesting that viscosity is a determining factor for conductance in this region. This also points to ion migration as the mechanism of conduction in this region. The two conductance maxima shift with increasing temperature in the same direction as the conductance minimum.

(b) *The* 100 *per cent sulphuric acid region.* Unlike the conductance of sulphuric acid/water mixtures, which seems to have been little used for analytical purposes, the conductance in the 100 per cent sulphuric acid region, owing to a distinct minimum long believed to correspond to the 100 per cent acid, was used from the earliest days of study of this solvent as a means of adjusting the concentration of the acid to 100 per cent (Section 1.1.2). According to a survey of the early work on conductance of sulphuric acid in the region of the conductance minimum, carried out by Hetherington *et al.* (1955b), the most accurate early data were obtained by Lichty (1908), who studied the 90–100 per cent region. His value for the minimum conductance, $1·041–1·043 \times 10^{-2}$ Ω^{-1} cm^{-1} at 25°C, is in excellent agreement with the currently accepted value of $1·0432 \times 10^{-2}$ Ω^{-1} cm^{-1} (Bass *et al.*, 1960b). However

Knietsch (1901) had already shown that the conductance minimum corresponds to less than 100 per cent acid. Kunzler and Giauque (1952a) established that at the freezing point of the acid (10·37°C) the conductance minimum corresponds to 99·996 ± 0·001 per cent acid. The composition of minimum conductance changes by less than 0·003 per cent over the temperature range 5–20°C (Kunzler, 1953).

The most accurate currently available conductance measurements for the region of 100 per cent sulphuric acid have been reported by Bass *et al.* (1960b) as conductances of hydronium hydrogen sulphate and of disulphuric acid in the solvent 100 per cent sulphuric acid. They represent only a slight revision of the highly accurate earlier measurements of Gillespie *et al.* (1957). The values given in Table 1.5 are due partly to this earlier source (for $m \leqslant 0·005$ mol kg^{-1}). Measurements covering the same concentration range have also been reported for 9·66 and 40°C (Gillespie *et al.*, 1957).

Two points of view have been put forward to explain the very high conductance of pure sulphuric acid. According to Hammett and Lowenheim (1934) and Gillespie and Wasif (1953a) the high conductance of the acid is due to highly conducting ions produced by the two types of self-ionization of the pure acid, i.e. autoprotolysis

$$2H_2SO_4 \rightleftharpoons H_3SO_4^+ + HSO_4^- \tag{1.5}$$

TABLE 1.5

Specific conductances for the 100 per cent sulphuric acid region at 25°C (Bass *et al.*, 1960b, and Gillespie *et al.*, 1957)

m (mol kg^{-1})	Specific conductance (10^{-2} Ω^{-1} cm^{-1})	
	$H_3O \cdot HSO_4$	$H_2S_2O_7$
0·000	1·0439	1·0439
0·001	1·0436	1·0447
0·002	1·0432	1·0459
0·003	1·0437	1·0473
0·004	1·0440	1·0489
0·005	1·0453	1·0506
0·01	1·0584	1·0635
0·02	1·1172	1·1047
0·04	1·329	1·204
0·06	1·601	1·307
0·08	1·879	1·407
0·10	2·162	1·502
0·14	2·686	1·676
0·18	3·174	1·830
0·24	3·822	2·033
0·32	4·562	2·260
0·40	5·198	2·45

and dissociation into water and sulphur trioxide

$$H_2SO_4 \rightleftharpoons H_2O + SO_3 \tag{1.6}$$

followed by the ionization of both to give highly conducting ions

$$H_2O + H_2SO_4 \rightleftharpoons H_3O^+ + HSO_4^- \tag{1.7}$$

$$SO_3 + 2H_2SO_4 \rightleftharpoons H_3SO_4^+ + HS_2O_7^- \tag{1.8}$$

Evidence for the very high mobility of the HSO_4^- and $H_3SO_4^+$ ions will be discussed later. Both ions conduct by proton jumps in the three-dimensional network of the associated solvent (Gillespie and Wasif, 1953a). An estimate of the extents of self-dissociation by Bass *et al.* (1960a) and by Gillespie *et al.* (1960) has led to values of equivalent conductances of these ions which decrease rapidly with increasing concentrations of electrolytes (Flowers *et al.*, 1960a). In order to account for this rapid decrease Wyatt (1961) suggested that a different mechanism of conductance, which he called "asymmetric dissociation", made a substantial contribution to the conductance of the pure acid. Both points of view will be discussed further in Section 3.2.6 (p. 71).

(c) *Oleum solutions.* The resistances of oleum solutions were measured by Kohlrausch (1882) and by Knietsch (1901), who, however, did not determine their cell constants. The only accurate measurements for dilute oleum solutions are therefore those of Gillespie *et al.* (1957), which were slightly revised by Bass *et al.* (1960b) (see Table 1.5). Recently, extensive data of 1 per cent accuracy were reported by Popiel (1964), who covered the whole oleum range up to 100 per cent sulphur trioxide. His conductance values for dilute oleums are lower than those of Bass *et al.* (1960b) reported in Table 1.5. The conductance of oleum solutions reaches a maximum at 12·4 per cent free sulphur trioxide, and then falls off steadily towards pure sulphur trioxide.

1.3. Thermodynamic Properties

The temperatures of phase changes are of importance with sulphuric acid/water mixtures as with any other solvent, because they determine the useful liquid range of the mixtures as solvents. With 100 per cent sulphuric acid the solid/liquid phase change assumes particular importance, however, because pure sulphuric acid has proved to be a solvent with a high cryoscopic constant and, in spite of the complications due to self-dissociation, has been used extensively as a solvent for cryoscopic studies. A vast volume of information on the behaviour of solutes in this solvent has been obtained by this means and the method has also been used in kinetic studies (see Section 3.2.5, p. 67). Freezing point studies in the region of pure sulphuric acid and its mono- and dihydrates have also been used by Giauque *et al.* (1956) to

1. PROPERTIES OF SULPHURIC ACID/WATER MIXTURES AS SOLVENTS

determine the variation of the free energy of water in the concentrated acid. The vapour/liquid equilibria are also an important source of information from which the activities of water and of sulphuric acid in the mixtures can be calculated. These are necessary in the analysis of some acidity functions and in the kinetic analysis of some reactions taking place in these media. Thermochemical data are needed for the recalculation of the activities of the two components in the mixtures to temperatures other than the standard temperature of 25°C for which activity data are available. All these thermodynamic properties will be discussed in the following two sections.

1.3.1. PHASE CHANGES

1.3.1.1. *The Solid/Liquid Equilibria*

(a) *Freezing points of 100 per cent sulphuric acid and its hydrates.* There has been a good deal of disagreement in the earlier literature about the exact temperature of the maximum of the freezing point curve at the composition of 100 per cent sulphuric acid. This appears to have been due mainly to the use of inaccurate thermometers and probably also to the presence of impurities in the acid. According to a survey of the older literature by Gillespie *et al.* (1950a), values between +10·0 and +10·6°C were commonly reported. After repeated investigations of the freezing-point curve in the vicinity of the 100 per cent composition, agreement was reached in recent years when the precise determination of Kunzler and Giauque (1952c), which gave a value of +10·371°C, was confirmed by Bass and Gillespie (1960). However, in their summarizing paper Giauque *et al.* (1960) give a lower figure of +10·31°C. As the monohydrate of sulphuric acid freezes at +8·48°C (Giauque *et al.*, 1960), the practical lower limit of the liquid range in the concentrated sulphuric acid region may be taken to be +10°C. In the more dilute sulphuric acid work at lower temperatures is possible, as the freezing points of the higher hydrates are considerably lower.

(b) *The cryoscopic constant of sulphuric acid.* The determination of the cryoscopic constant of sulphuric acid presented considerable difficulties, owing to the fact that the freezing-point maximum is not sharp due to the self-dissociation of the pure acid. This difficulty was often overcome by carrying out cryoscopic measurements in the slightly aqueous acid, in which water as solute represses the self-dissociation of the solvent. Using a large number of solutes, both fully dissociated as bases and undissociated, Gillespie *et al.* (1950a) obtained as the best experimental mean value $k_f = 5·98$ K mol^{-1} kg. Later, when the latent heat of fusion ($\Delta H_f°$) was determined (Rubin and Giauque, 1952), the cryoscopic constant could be calculated from its definition $k_f = RT_0^2/m_1 \Delta H_f°$, where m_1 is the molality of the solvent and T_0 its freezing point. The value so obtained, $6·12 \pm 0·02$ K mol^{-1} kg (Gillespie, 1954), is regarded as the best and is currently accepted.

1.3.1.2. The Liquid/Vapour Equilibria

(a) *Boiling points of sulphuric acid/water mixtures.* Tabulations of the boiling points of sulphuric acid/water mixtures available in standard works of reference (e.g. *International Critical Tables*) show that there is a gradual increase in the boiling points with increasing weight per cent of sulphuric acid up to 50 per cent acid, which has a boiling point of 123°C, but above this concentration boiling points rise sharply to a maximum corresponding to an azeotrope in the binary system water/sulphur trioxide. A recent estimate of the best values of the boiling point and composition at this point under a pressure of 1 atm by Gmitro and Vermeulen (1964) sets the boiling point at $326 \pm 5°C$ and the composition at 98·48 per cent sulphuric acid. Thus, in concentrated sulphuric acid of more than 80 per cent kinetic studies at temperatures above 200°C are possible and some have been reported.

(b) *Vapour pressures and activities of water and sulphuric acid in the mixtures.* The vapour pressure above sulphuric acid/water mixtures is due almost entirely to water, owing to the low volatility of sulphuric acid. Numerous earlier measurements of this property were assembled by Greenewalt (1925) and his final figures are quoted in reference works. The more recent measurements of Shankman and Gordon (1939), who used a static method and studied the concentration range 16–70 per cent acid at 25°C, are at present the best available. They were confirmed by Hornung and Giauque (1955), who made vapour pressure measurements at three selected concentrations at higher temperatures and calculated the values at 25°C using thermal data obtained in their laboratory. At higher concentrations of sulphuric acid direct vapour pressure measurements become difficult and therefore Giauque *et al.* (1956) used a freezing-point method to obtain activities of water at concentrations > 70 per cent. The activities at 25°C were again calculated using thermal data. Activities of water at 25°C based on all these measurements, as well as the activity coefficients of sulphuric acid calculated from them by means of the Gibbs–Duhem equation, are reported by Giauque *et al.* (1960) in their summarizing paper. The standard state for sulphuric acid was taken to be the hypothetical ideal 1 molal solution and the data are given for round values of molality. These data have been recalculated here to the weight per cent concentration scale and to pure sulphuric acid as the standard state, and are given together with water activities at 5 per cent intervals of concentration in Table 1.6. Interpolation was carried out using large-scale plots of data.

Liquid phase thermodynamic data of Giauque *et al.* (1960) have also been used as a basis for the calculation of partial pressures of water, sulphuric acid and sulphur trioxide by Gmitro and Vermeulen (1964), together with the information on the vapour phase equilibrium

$$H_2SO_4(g) \rightleftharpoons H_2O(g) + SO_3(g) \tag{1.9}$$

TABLE 1.6

Activities of water and sulphuric acid in sulphuric acid/water mixtures at 25°C (based on data of Giauque et al., 1960)

Weight % H_2SO_4	$-\log a_{H_2O}$	$-\log a_{H_2SO_4}$
0	0	
5	0·009	
10	0·019	
15	0·034	
20	0·060	
25	0·084	
30	0·124	
35	0·178	
40	0·251	8·000
45	0·341	7·325
50	0·453	6·627
55	0·602	5·900
60	0·788	5·180
65	1·032	4·374
70	1·352	3·553
75	1·750	2·733
80	2·276	1·914
85	2·906	1·152
90	3·638	0·602
95	4·469	0·237
98	5·284	0·084
100	5·807	0

The results of their calculations show that the vapour pressure of sulphur trioxide becomes noticeable only at acid concentrations > 98 per cent. Knietsch (1901) reported that 100 per cent sulphuric acid showed a measurable vapour pressure of sulphur trioxide and that 98·33 per cent sulphuric acid was found to be the best medium for the absorption of sulphur trioxide. These findings led Sackur (1902) to suggest the existence of the equilibrium

$$2H_2SO_4 \rightleftharpoons H_2SO_4 \cdot H_2O + SO_3 \qquad (1.10)$$

in the liquid 100 per cent acid. This was confirmed by Gillespie et al. (1950a) by cryoscopic studies in which water was found to suppress the self-dissociation of the solvent more effectively than metal hydrogen sulphates, which can suppress only the autoprotolysis. Therefore it was inferred that water suppresses also another kind of self-dissociation apart from autoprotolysis, i.e. the ionic self-dehydration (equations 1.6–1.8, p. 12), which is the current representation of the above equilibrium. The vapour pressures of oleums were measured by Miles et al. (1940) for up to 65 per cent free sulphur trioxide over the temperature range 17–62°C.

1.3.2. THERMOCHEMICAL QUANTITIES

1.3.2.1. *Relative Partial Molal Enthalpies and Heat Capacities*

Highly accurate values of relative partial molal enthalpies and heat capacities at 25°C, obtained mainly in the calorimetric studies of sulphuric acid/water mixtures by Kunzler and Giauque (1952b), have been tabulated for closely spaced concentrations by Giauque *et al.* (1960), together with absolute partial

TABLE 1.7

Relative partial molal enthalpies and partial molal heat capacities for sulphuric acid/water mixtures at 25°C (based on data of Giauque *et al.*, 1960)

Weight % H_2SO_4	$-\bar{L}_1$ (J mol^{-1})	$-\bar{L}_2$ (J mol^{-1})	$\bar{C}_{p_1}$ (J mol^{-1} K^{-1})	$\bar{C}_{p_2}$ (J mol^{-1} K^{-1})
0	0		75·295	
10	26·3	71 404	74·693	92·72
20	143·5	68 107	74·371	107·28
25	304·6	65 145	74·425	106·73
30	570·7	61 346	75·776	88·03
35	956·5	52 689	77·626	66·53
40	1 468·6	52 400	78·073	62·43
45	2 068·2	48 053	77·458	66·69
50	2 771·9	43 710	74·199	86·11
55	3 630·0	39 451	70·998	102·22
60	4 711·2	35 196	68·333	112·76
65	6 100·3	30 656	63·488	128·45
70	7 962·2	25 782	56·049	147·82
75	10 602	20 418	47·601	165·35
80	14 602	14 108	40·88	176·06
85	20 866	6 937	68·91	146·11
90	26 832	2 196	95·23	123·60
95	30 501	539·7	75·52	131·75
98	32 300	189·1	55·27	135·65
100	67 467	0		138·91

molal entropies and the temperature coefficients of heat capacities. Only the relative partial molal enthalpies and partial molal heat capacities are essential for the recalculation of the activities of sulphuric acid and water to other temperatures, and therefore only these quantities are reported in Table 1.7. The figures for round concentrations have been obtained by numerical interpolation. Activities of water and sulphuric acid at 45°C and 65°C have already been calculated from Giauque's data by Kort and Cerfontain (1968) for acid concentrations of 70–98 per cent.

1.3.2.2. Latent Heats of Fusion of Sulphuric Acid and its Monohydrate

These quantities are of interest in the present context because of their relationship to the cryoscopic constants of the two solids. So far only pure sulphuric acid has been used as a solvent in cryoscopic studies, but in principle there is no reason why the monohydrate should not be used in the same way. It also has a conveniently high freezing point of $+8.48°C$. The most reliable values for the latent heats of fusion of sulphuric acid and its monohydrate are due to Rubin and Giauque (1952) and are 10.711 kJ mol^{-1} for the pure acid and 19.439 kJ mol^{-1} for its monohydrate, both at the melting points of the solids.

1.3.2.3. Specific Heat of 100 per cent Sulphuric Acid

There is a sharp rise in the specific heat of sulphuric acid/water mixtures near absolute sulphuric acid to 1.416 J K^{-1} g^{-1} and then again a sharp drop with increasing concentration of sulphur trioxide (Kunzler and Giauque, 1952b; Giauque et al., 1960). This indicates that a small part of the heat capacity at the composition of absolute sulphuric acid is due to the temperature coefficients of the heats of self-dissociation.

1.3.2.4. The Heats of Autoprotolysis and of Ionic Self-dehydration

Measurements of partial molal heat contents at 25°C of some metal hydrogen sulphates, which suppress the autoprotolysis of 100 per cent sulphuric acid (equation 1.5), have led to estimates of the heat of autoprotolysis of sulphuric acid between 19.25 and 21.34 kJ mol^{-1} (Kirkbride and Wyatt, 1958). The same type of measurement involving water as solute, which in addition suppresses also the ionic self-dehydration (equations 1.6–1.8, p. 12), yielded an estimate of the heat of that reaction as 25.94 kJ mol^{-1}. Dacre and Wyatt (1961) also obtained the heat of autoprotolysis from measurements of the heat of a strong acid/strong base reaction in 100 per cent sulphuric acid and again arrived at a figure of about 21 kJ mol^{-1}, in agreement with the above estimate.

1.4. Spectroscopic Properties

Numerous spectroscopic studies of pure sulphuric acid, sulphuric acid/water mixtures and dilute oleums have been described in the literature. Two aspects of these studies are of interest in connection with the use of these liquids as reaction media: firstly, information on the molecular structure of the species in solution and on their concentrations, and secondly information on the intermolecular structure of the liquids, i.e. on association, which is responsible, for example, for a special conductance mechanism. The main results of spectroscopic studies will be presented from these points of view in the following five sections.

1.4.1. RAYLEIGH SCATTERING

Before discussing Raman scattering, some recent observations on the Rayleigh scattering by concentrated sulphuric acid, as compared with that by water, nitric acid and acetic acid, deserve mention. Rao and Ramanaiah (1966) have studied the intensity distribution of the wings accompanying the Rayleigh line in the spectra of these liquids. They have found that the wings from sulphuric acid and water are highly polarized, whereas those from nitric acid and acetic acid are highly depolarized. They suggest that the origin of these wings from sulphuric acid and water is not the rotation of molecules, but rather intermolecular forces (hydrogen bonding) and consider that this represents evidence for symmetric-type lattice oscillations in these liquids, i.e. for the almost crystalline short-range order. Hydrogen bonds are of course also present in acetic acid, but their distribution is unsymmetrical and the extent of hydrogen bonding less extensive. The wing from nitric acid is thought to be due to the rotation of free molecules between shells of water molecules attached to ions.

1.4.2. RAMAN AND INFRARED SPECTRA

The nature and concentrations of molecular and ionic species in sulphuric acid/water mixtures, in pure sulphuric acid and in dilute oleums are of considerable interest in kinetic studies in these media, because all the species present are potential reactants or catalysts. Most of the information on such species has been obtained from Raman spectra, and photoelectric recording of the intensities of Raman bands has made estimation of their concentrations possible. Studies of Raman spectra are carried out using glass cells, and so strong acid solutions present no difficulty. The study of infrared spectra of such solutions became possible, however, only by using sodium chloride cells protected with polythene or by working with silver chloride plates (Walrafen and Dodd, 1961). Recently the use of pure silicon windows was suggested by Stopperka (1966a).

(a) *Pure sulphuric acid.* A number of lines have been observed in both the Raman and the infrared spectrum of pure sulphuric acid. These spectra are virtually identical in the 600 to 1500 cm^{-1} region. In a recent review Gillespie and Robinson (1965) quoted all the frequencies and their assignments, given in their own earlier work (Gillespie and Robinson, 1962a), the work of Giguère and Savoie (1960), and the work of Walrafen and Dodd (1961). An extensive survey of earlier literature is also given by Stopperka (1966a), who has revised some assignments. Most authors agree that the presence of the self-dissociation species in the pure acid is not detectable by these methods.

(b) *Sulphuric acid/water mixtures.* The strongest bands of undissociated sulphuric acid which disappear on dilution are those at 1370 cm^{-1}, 970 cm^{-1} and 910 cm^{-1}. Observations of these changes in the infrared spectra were reported by Fabbri and Roffia (1959). The most extensive studies of changes in the intensities of these lines in the Raman spectra with dilution are due to Young, whose work was reviewed by Young et al. (1959), and to Zarakhani and Vinnik (1963). Young et al. (1959) report only a few measurements of the intensity of the 910 cm^{-1} line in the concentrated acid and conclude that

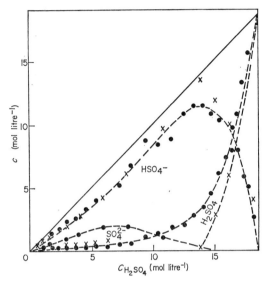

FIG. 1.1. Concentrations of species in sulphuric acid/water mixtures from Raman spectra, (x) Young et al. (1959); (•) Zarakhani and Vinnik (1963).

the concentration of the undissociated acid becomes negligible below 14 M (a concentration of 15·25 M corresponds to the 1:1 hydrate). Zarakhani and Vinnik (1963) have studied the same line more thoroughly in the concentrated acid and conclude first of all that its frequency changes as follows:

% acid	98	90	80	70	40
ν cm^{-1}	913	911	909	892	884

The line is still observable in 15 per cent sulphuric acid. Since its frequency changes, Zarakhani and Vinnik (1963) infer that there is a structural change in the species producing this Raman shift and postulate the existence of complexes $H_2O \cdot H_2SO_4$ or ion pairs $H_3O^+ \cdot HSO_4^-$ to account for this, giving preference to the latter. Their estimates of the concentrations of the species producing this line (from peak heights to ± 5 per cent) are considerably higher than those of Young (see Fig. 1.1). A Raman frequency ascribed to the HSO_4^- ion has previously been reported at 895·5 cm^{-1} (Woodward and Horner, 1934). A similar frequency shift of the infrared line at 905 cm^{-1} of the pure acid to 885 cm^{-1} in 80 per cent acid has been observed by Walrafen and Dodd (1961), who believe it to be a result of the absence of H_2SO_4.

The most intense band of the HSO_4^- ion, which is used for its quantitative estimation from the Raman spectrum, is reported at frequencies of 1036–1053 cm^{-1} (most often taken as 1040 or 1050 cm^{-1}). Young et al. (1959) estimated the concentration of the HSO_4^- ions from the integrated intensities of this line over the whole sulphuric acid/water concentration range at temperatures of 0, 25 and 50°C. Zarakhani and Vinnik (1963) used the same line to estimate the concentrations of HSO_4^- ions from peak heights (to ± 5 per cent) at 25°C. Both sets of results are represented in Fig. 1.1, and it can be seen that the agreement is excellent up to about 8 M. Above this concentration the estimates of Young et al. are considerably higher, especially in the vicinity of the maximum. Young et al. had estimated from their data an equilibrium quotient for the reaction

$$H_2O + H_2SO_4 \rightleftharpoons H_3O^+ + HSO_4^- \qquad (1.11)$$

of 15 to 20. The data of Vinnik and Zarakhani (1963) lead to a value of only 4·2–6·2. Deno and Taft (1954) proposed a value of 50 for this quantity, and Gillespie (1950a) estimated a value of 1·2 from the cryoscopic behaviour of water in sulphuric acid. The suggestion of Wyatt (1957) of a value of 2000 appears to be definitely too high on the basis of Raman results.

More recently Young and Walrafen (1961) carried out a more detailed study of the concentrated sulphuric acid region and reported a frequency shift of the 1040 cm^{-1} line to 1055 cm^{-1} with increasing acid concentration. They suggested that the formation of the species $H_5SO_5^+$ or $H_3O^+(H_2SO_4)$ was responsible for this. This frequency shift was definitely not confirmed by Zarakhani and Vinnik (1963). A small shift of the corresponding infrared line from 1045 to 1052 cm^{-1} with increasing concentration of the acid was observed by Walrafen and Dodd (1961), but thought probably not to be significant. An absorption band of concentrated sulphuric acid at 1690 cm^{-1}, which shifts with increasing concentration towards 1668 cm^{-1} and disappears in 100 per cent sulphuric acid, could, according to Stopperka (1966a), be due to this species. It may be concluded that spectroscopic evidence for the existence of the species $H_5SO_5^+$ is inconclusive. A solvation number of 1 for the H_3O^+ ion in sulphuric acid has been assumed by both Gillespie (1950a) and Wyatt (1960) and it is probable that solvation occurs, but what change in the Raman spectrum this should produce is uncertain.

There is a wide disagreement between the estimates of the SO_4^{2-} ion concentration by Young et al. (1959) and by Zarakhani and Vinnik (1963), although both are based on the intensities of the line at 980 cm^{-1} (see Fig. 1.1). The reasons for the discrepancy are not clear.

An intense band at 1190 cm^{-1} appears in the infrared spectrum as the dilution of the acid is increased (Walrafen and Dodd, 1961). The same band is observed in the spectrum of aqueous hydrochloric acid and has therefore been ascribed to the H_3O^+ ion.

The disappearance of the aggregates of water $(H_2O)_5$ with increasing acid concentration was followed by Stopperka (1966a) in the infrared region of about 700 cm^{-1}.

Views on frequency assignments in these spectra are often divergent. Some guide to assignment has been obtained from a comparison of the spectra of concentrated D_2SO_4 with those of concentrated H_2SO_4 (Walrafen and Dodd, 1961). Infrared lines at $\sim$ 380, 562, 900, 985 (broad), 1195 and 1350 cm^{-1} of the deuterio acid are only slightly shifted in frequency from the corresponding lines of the protio acid, and are therefore probably produced by the SO vibrations. Other infrared lines of deuterosulphuric acid at $\sim$ 305, 475, 1820 and 2230 cm^{-1} apparently correspond to the lines at $\sim$ 420, 675, 2430 and 3000 cm^{-1} of the protio acid. These lines, which are shifted to lower frequency by approximately $\sqrt{2}$ by deuterium substitution, are apparently due to the vibrations of OH groups.

(c) *Oleum solutions*. There has been a good deal of controversy in recent years concerning the spectroscopic evidence for the presence or absence of polysulphuric acids and other species in oleum solutions. Walrafen and Young (1960) found a rising concentration of disulphuric acid and its anion with increasing sulphur trioxide content of oleums, with a maximum of 9·6 mol litre^{-1}, when the mole fraction of SO_3 reaches *ca.* 0·55. The appearance of molecular sulphur trioxide is first observed at a mole fraction of about 0·45. Some evidence for the presence of $HS_2O_7^-$ ions was also found and their concentration estimated as *ca.* 4 mol litre^{-1} near the maximum of specific conductance (at $X_{SO_3} = 0·3$), but no evidence for polysulphuric acids, $H_2S_3O_{10}$ and $H_2S_4O_{13}$. Gillespie and Robinson (1962b) have provided evidence for both these species, but no evidence for the $HS_2O_7^-$ ion. Stopperka (1966b) was also unable to detect any positive evidence for the presence of $HS_2O_7^-$ ions. Their concentration must therefore be below the limit of detection by this method. However, unlike Gillespie and Robinson (1962b), he found no evidence for the presence of $H_2S_4O_{13}$. Recently Walrafen (1964) conceded some evidence for $H_2S_3O_{10}$ at higher oleum concentrations. The conclusion from Raman spectra then is that in dilute oleums only disulphuric acid (and possibly also its anion) is present, but no polysulphuric acids.

The spectrum of the sulphuric acidium ion $H_3SO_4^+$ was also not observed in dilute oleum solutions because of its low concentration, but Gillespie and Robinson (1962c) were able to observe it in solutions of tetrahydrogenosulphatoboric acid, which behaves as a strong acid in pure sulphuric acid and produces an appreciable concentration of this ion in solution.

1.4.3. ULTRAVIOLET ABSORPTION SPECTRA

The ultraviolet method of estimating the concentrations of species is applicable only to oleum solutions, because sulphur trioxide absorbs in a region where the absorption of water and sulphuric acid is small. The absorption of sulphur trioxide increases with decreasing wavelength in the region of 220–300 nm, but the molar extinction coefficients are known only for gaseous sulphur trioxide (Fajans and Goodeve, 1936). In order to estimate the concentrations of species in the equilibrium

$$H_2SO_4 + SO_3 \rightleftharpoons H_2S_2O_7 \quad (1.12)$$

Cerfontain (1961a) made the assumption that the ratio of extinction coefficients of sulphur trioxide in oleum solution to that in the gaseous phase is a constant, and that disulphuric acid also does not absorb in the ultraviolet. By combining his measurements with the vapour pressure data of Miles *et al.* (1940), he was able to show that the activity coefficient of sulphur trioxide is independent of oleum composition and equal to unity if referred to 100 per cent sulphuric acid. Since Brand and Rutherford (1952) had provided some evidence that $f_{H_2S_2O_7}/f_{H_2SO_4}$ is also independent of oleum composition up to 34 per cent sulphur trioxide and also equal to unity if referred to 100 per cent sulphuric acid, the equilibrium constant for the above reaction may be written simply as $K = [H_2S_2O_7]/[H_2SO_4][SO_3]$. Sulphur trioxide concentrations calculated for various assumed values of the equilibrium constant were compared with the experimental concentrations from ultraviolet spectra. The closest fit was obtained for $K = 10$ mol^{-1} litre and $\alpha_\lambda = 0.16$ (this being the ratio of the extinction coefficient of sulphur trioxide in oleum to that in the gaseous phase). The low value of α_λ remains unexplained. Taking into account also an estimate of the degree of dissociation of $H_2S_2O_7$ from cryoscopic data of Brayford and Wyatt (1956) and Dacre and Wyatt (1960), Cerfontain decided on a most probable value of $K \geqslant 5$ mol^{-1} litre for the above reaction. This leads to an estimate of the concentration of sulphur trioxide in 100 per cent sulphuric acid of, at most, 8×10^{-5} mol litre^{-1}.

1.4.4. X-RAY DIFFRACTION

Accurate bond length data for the SO_4 tetrahedra of solid sulphuric acid became available only recently (Pascard-Billy, 1965). It was established that the tetrahedra possess a two-fold axis and the following SO distances were reported: SO(H) 1·535 Å and SO 1·426 Å, which can be correlated with the estimated $\bar{u}$-bond orders of these bonds of 0·38 and 0·66, respectively (Cruickshank, 1961). Solid sulphuric acid has a layer-type structure (Pascard, 1955), each sulphuric acid molecule being hydrogen-bonded to four others. The structure of the liquid is probably similar to that of the solid, analogously to the ice/water case.

The diffraction pattern of concentrated sulphuric acid shows clear, although diffuse, bands or "halos", characteristic of highly ordered, associated liquids. A study of these bands and their changes with dilution was carried out by Bose (1941). The main diffraction band of 99·2 per cent sulphuric acid corresponds to a Bragg spacing of 4·07 Å, and an inner weaker diffuse band to a spacing of 7·98 Å. These presumably correspond to the first and the second shell of nearest neighbours. The dilution of the acid to the molar ratio 1:1 (84·5 per cent) leads to a sharp drop in both these spacings to 3·6 and 7·32 Å respectively. Further dilution to molar ratios 2:1 and 3:1 does not lead to further large changes. The change becomes large again when diluting from a molar ratio 10:1 to 20:1 (21·4 per cent acid), where the bands become less diffuse and correspond to the spacings of 3·28 and 6·28 Å. At the limit the band for pure water is observed, corresponding to a spacing of 3·42 Å. These changes of spacings parallel the volume contractions on mixing, since this clearly affects the distance between neighbouring molecules. However, a calculation of changes in the spacings from volume changes would be difficult. The important point is that the diffraction pattern persists throughout the whole sulphuric acid/water concentration range, indicating a high degree of internal order in all the mixtures.

1.4.5. PROTON MAGNETIC RESONANCE SPECTRA

Sulphuric acid/water mixtures, in which there is rapid proton exchange between all the species involved in the main equilibrium at moderate and high acid concentrations

$$H_2O + H_2SO_4 \rightleftharpoons H_3O^+ + HSO_4^- \qquad (1.13)$$

show only a single line in the n.m.r. spectrum. The chemical shifts of that line relative to pure water as external reference have been measured by several authors (Gutowsky and Saika, 1953; Hood and Reilly, 1957; Gillespie and White, 1960). A down-field shift is observed with increasing acid concentration, which goes through a maximum at about 17 mol litre^{-1} and then decreases at still higher concentrations. Usually chemical shifts, uncorrected for the difference between the magnetic susceptibilities of the sample and the reference, have been reported, but the corrections are not large. For example, Gillespie and White (1960) report chemical shifts for the 100 per cent acid as $-5\cdot90$ ppm, uncorrected, and $-5\cdot95$ ppm corrected. Other authors find higher values for the pure acid: $-6\cdot1$ ppm (Gutowsky and Saika, 1953), $-6\cdot15$ ppm (Hood and Reilly, 1957) and $-6\cdot08$ ppm (Stopperka, 1966b) (all uncorrected). Attempts were made by Hood and Reilly (1957) and by Gillespie and White (1960) to estimate the position of the above equilibrium from the chemical shifts, on the assumption that the observed chemical shifts of the mixtures are averages (weighted according to proton fractions) of the chemical shifts of the species involved in the equilibrium. The reasonable agreement of the calculated degrees of dissociation with those obtained from Raman spectra is, however, probably fortuitous (Gillespie and White, 1960), because the above treatment implies that the ions do not affect the chemical shift of the remaining molecular species, which is probably not true. The solvation of ions, whether at the sulphuric acid end or at the water end of the range of mixtures, undoubtedly breaks up the hydrogen-bonded structure of the solvent. Thus ions affect the chemical shift of the solvent molecules by the effect of their charge in the solvation shell, and outside it, by disturbing the hydrogen-bonded structure of the remaining solvent. Hydrogen bonding is known to lead to large shifts to low field (see Pople et al., 1959) and the disruption of the structure therefore makes a positive contribution to the shielding of the protons. The chemical shift of the ions themselves is probably sensitive to the nature of the neighbouring molecules or ions (in ion pairs).

Chemical shifts of oleum solutions have also been measured by Gutowsky and Saika (1953), by Gillespie and White (1960) and by Stopperka (1966b). The effect of increasing concentrations of sulphur trioxide is almost nil up to 20 per cent free SO_3, but at higher concentrations there is a steady up-field shift, probably due to the break-up of hydrogen-bonded structure (Stopperka, 1966b).

1.5. Isotopically Substituted Sulphuric Acids

Isotopically substituted sulphuric acids are valuable alternatives to the normal protio sulphuric acid in some studies of the behaviour of solutes in concentrated sulphuric acid or in studies of reaction mechanism. Dideuterosulphuric acid is more generally useful, but ^{35}S-labelled sulphuric acid also has some more limited applications.

1.5.1. DIDEUTEROSULPHURIC ACID

1.5.1.1. *Preparation*

Equimolar amounts of heavy water and sulphur trioxide are the starting materials. Small amounts of sulphur trioxide for laboratory scale preparations are most easily prepared by distilling the oxide from a concentrated oleum solution, as described by Greenwood and Thompson (1960). Gillespie et al. (1957) recommend prior purification of commercial 65 per cent oleum by refluxing with chromium trioxide for 10 hours to decompose any oxidizable impurities. Direct bubbling of undiluted sulphur trioxide through heavy water is not a good procedure, because of the large heat evolution. Sulphur trioxide diluted with an inert gas (e.g. helium) may, however, be bubbled safely. Alternatively, sulphur trioxide and heavy water are distilled into two adjacent strongly cooled limbs on a vacuum line. The two limbs are then sealed off together, and sulphur trioxide is allowed to distil slowly into heavy water as the temperature rises (Greenwood and Thompson, 1960).

1.5.1.2. *Properties*

The highest value for the freezing point of dideuterosulphuric acid was reported by Flowers et al. (1958) as $+ 14\cdot35 \pm 0\cdot02°C$. Greenwood and Thompson (1959) report a lower value of $+14 \cdot 1°C$, but believe that this could be raised by fractional crystallization. Mountford and Wyatt (1966) report $+14 \cdot 305°C$.

The most extensive comparison of the properties of pure deuterosulphuric acid with those of pure sulphuric acid has been carried out by Greenwood and Thompson (1959) over the temperature range 7·5–70°C, but the most accurate values for its properties at 10, 25 and 40°C are due to Flowers et al. (1958). Some properties for 25°C are given in Table 1.8.

TABLE 1.8

Some properties of 100 per cent deuterosulphuric acid at 25°C (Flowers et al., 1958)

Freezing point	$+14 \cdot 35 \pm 0 \cdot 02°C$
Density	$1 \cdot 8573$ g cm^{-3}
Viscosity	$2 \cdot 488 \times 10^{-7}$ N s cm^{-2}
Specific conductance	$0 \cdot 2568 \pm 0 \cdot 0005 \times 10^{-2}$ Ω^{-1} cm^{-1}
Specific conductance "minimum"	$0 \cdot 2540 \pm 0 \cdot 0005 \times 10^{-2}$ Ω^{-1} cm^{-1}
Composition at minimum conductance	$0 \cdot 0045$ mol $D_2S_2O_7$ kg$_{soln.}^{-1}$

By comparing these properties with those of pure sulphuric acid (Section 1.2), it can be seen that the substitution of deuterium for hydrogen raises the melting point and the density of the compound, has little effect on the molar volume and viscosity, and lowers the specific electrical conductivity considerably. The position of the conductance minimum is also different, and is in fact on the oleum side of the 100 per cent composition.

No detailed studies of the properties of deuterium oxide/deuterosulphuric acid mixtures are available.

1.5.2. ^{35}S-LABELLED SULPHURIC ACID

The starting material for the preparation of this acid is neutron-irradiated potassium chloride. The nuclear reaction ^{35}Cl(n,p)^{35}S produces ^{35}S in the crystals in the $+6$ oxidation state, i.e. as sulphate. The possible processes that lead to this transformation have been discussed by Koski (1949). When irradiated crystals are dissolved in aqueous solution containing sulphide, sulphite and sulphate ions as carriers, upon separation almost all activity is found as sulphate. The sulphate is separated as barium sulphate. Its conversion into ^{35}S-labelled sulphuric acid has been described by Masters and Norris (1952). The dried barium sulphate is dissolved in 96 per cent sulphuric acid, and the solution is then equilibrated with sulphur dioxide at 350°C (Norris, 1950). Isotopic exchange takes place and ^{35}S-labelled sulphur dioxide is obtained. This can be used to label inactive sulphuric acid by the same equilibration procedure at higher temperature.

More recently two other methods of converting neutron-irradiated potassium chloride into labelled sulphuric acid have been described (Nikolov and Mikhailov, 1965; Suarez, 1966).

CHAPTER 2

Acidity Functions

2.1. Introduction

The most important property of sulphuric acid/water mixtures (and mixtures of strong acids with water in general) from the point of view of their application as reaction media is undoubtedly their acidity, which is measurable in terms of their acidity function. Since this concept was first introduced by Hammett and Deyrup in 1932 and its applications to kinetic studies reviewed by Hammett (1940), there has been a steady development of ideas on the subject up to the present time. While Hammett's acidity function was defined as a measure of the ability of the medium to protonate simple bases, other acidity functions have since been defined which measure the ability of the acid media to effect ionizations of other kinds. Apart from that, it has been realized that the Hammett acidity function is not generally applicable even to the protonation of all kinds of simple bases. This has stimulated further research into the validity of the basic assumptions underlying the concept of the "acidity of the medium" and has led to the development of some theoretical ideas on its physical significance.

The subject has often been reviewed. One of the best known older reviews, by Paul and Long (1957), has been followed more recently by equally excellent reviews by Arnett (1963), Edward (1964), Vinnik (1966) and Boyd (1969). Each of these reviews has been written with different emphasis and they are therefore all useful in highlighting different aspects of the subject.

The present chapter will deal with definitions of acidity functions, with the problems involved in their experimental determination, and with attempts to verify some of the assumptions involved. Finally, current theoretical ideas on the physical meaning of acidity functions and some attempts to correlate various acidity functions in terms of these ideas will be discussed. The applications of acidity functions to kinetic problems, which stem from the fact that substances undergoing acid-catalysed reactions are protonated in acid media prior to reaction, will be discussed in the introduction to Chapter 5.

2.2. Definitions

There are in fact only three basically different definitions of acidity functions for three different types of protonation equilibria. The first refers to the protonation of simple bases, the second to the protonation of bases which is followed by dehydration, and the third to the protonation of olefins to give carbonium ions. All three types of equilibria are very important for studies of reaction mechanisms, and will be dealt with in detail in the following three sections.

2.2.1. ACIDITY FUNCTIONS FOR SIMPLE PROTONATION

In moderately concentrated or concentrated acid media, in which the concept of pH cannot be defined with any degree of accuracy, a new measure of acidity or the activity of the proton is needed. Hammett and Deyrup (1932) proposed that this might be obtained from measurements of protonation equilibria of simple basic indicators, which ionize according to the equation:

$$B + H^+ \rightleftharpoons BH^+ \tag{2.1}$$

The equilibrium constant for this reaction is commonly given as the ionization constant of the conjugate acid, i.e.

$$K_{BH^+} = \frac{[B][H^+]f_B f_{H^+}}{[BH^+] f_{BH^+}} \tag{2.2}$$

where f_B, f_{BH^+} and f_{H^+} are the activity coefficients of the base, the conjugate acid and the hydrogen ion, respectively. (For simplicity, the hydrogen ion is represented here as H^+, but is in fact present in a hydrated form. The question of hydration will be taken up in Section 2.6.) The indicator ratio is measurable spectrophotometrically, but the activity coefficients in concentrated acid solutions are unknown. Therefore Hammett proposed that the acidity of the medium should be defined by

$$h_0 \equiv a_{H^+} \frac{f_B}{f_{BH^+}} = K_{BH^+} \frac{[BH^+]}{[B]} \tag{2.3}$$

assuming that the ratio of the activity coefficients of the base and its conjugate acid for the fairly large molecules of organic bases is independent of the nature of the base, i.e. approximately the same for all bases of the same charge type. The negative logarithm of the acidity of the medium

$$H_0 = -\log h_0 = pK_{BH^+} - \log \frac{[BH^+]}{[B]} \tag{2.4}$$

was termed the acidity function. Since the ionization ratio $I = [BH^+]/[B]$ for any indicator base is measurable from 30–10 to 0·1–0·03, which corresponds to 2–3 units of H_0, in order to extend the scale into concentrated acid solutions it is necessary to use a number of indicators, which ionize in overlapping ionization ratios over a range of acid concentrations. It we take another indicator equilibrium

$$C + H^+ \rightleftharpoons CH^+ \qquad (2.5)$$

and measure its ionization ratio $[CH^+]/[C]$ in the same acid concentration range as for B, then in the region of overlap the following relationship holds:

$$pK_{BH^+} - pK_{CH^+} = \log\frac{[BH^+]}{[B]} - \log\frac{[CH^+]}{[C]} \qquad (2.6)$$

Thus it is possible first to determine pK_{CH^+} with reference to pK_{BH^+} at lower acidities, and then to extend the acidity function scale to higher acidities using equation (2.4). The reference point for the acidity scale is chosen so that in very dilute acid solution $H_0 = -\log[H^+]$. This means that indicators are chosen which are appreciably ionized in very dilute solutions of strong acids, in which the hydrogen ion concentration may be taken to be equal to the stoichiometric concentration of the strong acid, and their pK_{BH^+} values are determined under these conditions. The base ionization constants of all other indicators which are protonated at higher acid concentrations are ultimately based upon these pK_{BH^+} values determined in very dilute acid, and therefore the reference point for the acidity scale is a state of infinite dilution in water. This means that the acidity function is a thermodynamic acidity scale and all pK values obtained by its use are thermodynamic pK values. This remains true, however, only if all the indicators used in defining the scale behave as "Hammett bases", i.e. if in the regions of overlap the $\log I$ values for pairs of indicators remain perfectly parallel when plotted against acid concentration. This will be so only if the Hammett assumption that

$$\frac{f_{BH^+}}{f_B} = \frac{f_{CH^+}}{f_C} = \frac{f_{DH^+}}{f_D} = \text{etc.} \qquad (2.7)$$

applies. While the measurements of Hammett and Deyrup (1932) suggested that this held approximately for substituted anilines and some other indicators, it has become clear in recent years that strict adherence to the above relationship is possible only in series of indicators of closely similar structure, and that even within series of bases containing the same basic centre deviating members frequently occur. This has led, on the one hand, to attempts to obtain direct information on the activity coefficients of the base and the conjugated acid, and on the other, to a proliferation of acidity function scales. These recent developments will be discussed in Sections 2.4 and 2.5.

It remains to mention here that acidity functions have been defined not only for electrically neutral bases but also for positively and negatively charged bases, which protonate according to

$$BH^+ + H^+ \rightleftharpoons BH_2^{2+} \qquad (2.8)$$

and
$$A^- + H^+ \rightleftharpoons HA \tag{2.9}$$

The acidity functions which measure the ability of the acid medium to effect these protonations are given by

$$H_+ = pK_{BH^{2+}} - \log \frac{[BH_2^{++}]}{[BH^+]} \tag{2.10}$$

for cation bases (Brand et al., 1952; Bonner and Lockhart, 1957), and by

$$H_- = pK_{HA} - \log \frac{[HA]}{[A^-]} \tag{2.11}$$

for anion bases in acid solution (Boyd, 1963b).

2.2.2. ACIDITY FUNCTION FOR COMPLEX IONIZATIONS

Certain basic molecules ionize in acid solutions in a more complex manner

$$ROH + H^+ \rightleftharpoons R^+ + H_2O \tag{2.12}$$

which may be regarded as protonation accompanied by dehydration. This kind of behaviour was found for aryl-carbinols in the first instance, and an acidity function based on their ionization was first proposed by Lowen et al. (1950) and by Gold and Hawes (1951), and was further developed by Deno et al. (1955) and by Gold (1955). The bases which behave in this manner are often termed secondary bases, and the water molecule may be regarded as their conjugate acid. Analogously to the ionization of simple bases, the equilibrium constant for the ionization of secondary bases is commonly given for the ionization of their conjugate acid, i.e.

$$K_{R^+} = \frac{[ROH] f_{ROH} a_{H^+}}{[R^+] f_{R^+} a_{H_2O}} \tag{2.13}$$

The ionization ratio $I = [ROH]/[R^+]$ is again measurable by spectrophotometric means, and a measure of the acidity of the medium with respect to secondary bases can be defined by

$$h_R \equiv \frac{a_{H^+} f_{ROH}}{a_{H_2O} f_{R^+}} = K_{R^+} \frac{[R^+]}{[ROH]} \tag{2.14}$$

assuming that the ratio of activity coefficients f_{ROH}/f_{R^+} will be approximately the same for a series of indicators of closely similar structure. This should be a reasonable assumption, especially since both R^+ and ROH are large organic structures, the charge in the carbonium ion R^+ being as a rule highly delocalized. The corresponding acidity function is again defined as the negative logarithm of the above function:

$$H_R = -\log h_R = pK_{R^+} - \log \frac{[R^+]}{[ROH]} \tag{2.15}$$

In the earlier literature symbols J_0 and C_0 were also used to denote this acidity function, but the symbol H_R is preferable, since it indicates its close relationship to the H_0 acidity function. The standard state for this acidity function is also an infinitely dilute solution in water, and for dilute acid solutions H_R becomes equal to pH. Also, as the activity of water approaches unity, H_R becomes equal to H_0. The procedure for determining H_R values is the same as for the H_0 function, namely the stepwise extension of the scale into more and more concentrated acid solutions by means of a series of overlapping indicators.

The definition of the J_0 acidity function, as originally suggested by Gold and Hawes (1951), is not strictly equivalent to that of the H_R acidity function. They have assumed that f_{R+}/f_{ROH_2+} may be taken as unity in aqueous acid solvents (in view of the large size and structural similarity of the two ions) and this has led them to define J_0 so that

$$J_0 = H_0 + \log a_{H_2O} \qquad (2.16)$$

The definition of H_R merely implies that activity coefficient ratios for pairs of secondary bases are equal to unity. There are instances of reactions in acid solution, whose rate depends not only on acidity, but also on water activity (certain dehydration reactions), and for which therefore the J_0 acidity function appears to be the relevant function to use in correlating rates.

2.2.3. ACIDITY FUNCTION FOR OLEFIN PROTONATION

Although the protonation of diaryl olefins according to the equation

$$\underset{Ar}{\overset{Ar}{\diagdown}}C=CH_2 + H^+ \rightleftharpoons \underset{Ar}{\overset{Ar}{\diagdown}}\overset{+}{C}-CH_3 \qquad (2.17)$$

is an equilibrium of the type $B + H^+ \rightleftharpoons BH^+$, it was found to deviate widely from the equation

$$H_0 = pK_{BH+} - \log [BH^+]/[B] \qquad (2.18)$$

This means that the activity coefficient postulate that f_{BH+}/f_B should be equal to f_{R+}/f_{olefin} does not hold for this type of base, and hence a new acidity function, closely related to the H_R acidity function, was proposed by Deno et al. (1959). The failure of the protonation of olefins to follow the H_R acidity function was also confirmed, and it meant that R^+ was in equilibrium with the olefin and not with the alcohol. The new acidity function is defined by

$$h'_R \equiv a_{H+} \frac{f_{olefin}}{f_{R+}} = K_{R+} \frac{[R^+]}{[olefin]} \qquad (2.19)$$

where K_{R^+} is the ionization constant for the ionization of a carbonium ion into a proton and the corresponding olefin given by

$$K_{R^+} = \frac{[\text{olefin}] f_{\text{olefin}} a_{H^+}}{[R^+] f_{R^+}} \quad (2.20)$$

Since it was found by Deno and Perizzolo (1957) that activity coefficients in sulphuric acid solution of olefins and the corresponding carbinols are approximately equal, equality in the activity coefficient ratios in the definitions of h'_R and h_R can be assumed, and hence a relationship between h_R and h'_R can be obtained directly:

$$h'_R = h_R a_{H_2O} \quad (2.21)$$

and

$$H'_R = H_R - \log a_{H_2O} \quad (2.22)$$

The symbol H'_R was suggested by Kresge and Chiang (1961a).

The H'_R acidity function could not be easily determined using olefin protonations, because the olefins studied so far do not afford an extensive enough series of overlapping equilibria. Therefore this acidity function is more easily obtained from the relationship (2.22), from the known H_R acidity function and the known data on the activity of water. The acidity function so obtained is successful in describing the protonation behaviour of arylolefins (Deno et al., 1959 and Deno et al., 1960).

2.3. Evaluation of Acidity Functions

The general principle of the evaluation of acidity functions by means of a series of overlapping indicator equilibria has already been described in the preceding section. The central problem is the accurate determination of the ionization ratios, and the methods used for this are the same regardless of the nature of the indicator. Also, similar problems arise in measurements with all kinds of indicators. These questions will be discussed in the present section.

2.3.1. INDICATOR MEASUREMENTS

2.3.1.1. *Principles and Methods*

The measurement of the ionization ratios of indicators is carried out as a rule spectrophotometrically, because the unionized and the ionized forms of the indicators differ considerably in their absorption spectra, both in the positions of maximum absorption and in the values of specific absorption coefficients. It is sufficient if only one form has a pronounced absorption maximum, but more accurate determination of ionization ratios is possible if both forms give well-defined peaks sufficiently far apart.

The procedure is as follows. The absorption spectra of each of the two forms are determined first, at such acidities of the medium at which there may be at most 1 per cent of the other form present, and then absorption is measured at the same wavelengths over a range of intermediate acid concentrations. At intermediate acidities both forms are present, and if Beer's law holds, the molar extinction coefficient is given by

$$\epsilon = \frac{c_u \epsilon_u - c_i \epsilon_i}{c_u - c_i} \quad (2.23)$$

where ϵ_u and ϵ_i are the molar extinction coefficients of the unionized and the ionized forms, respectively, and c_u and c_i are their concentrations. When this is solved for the ionization ratio, the following relationship is obtained:

$$I = \frac{c_i}{c_u} = \frac{\epsilon - \epsilon_u}{\epsilon_i - \epsilon} \quad (2.24)$$

Indicators used originally by Hammett and Deyrup (1932) had only one of the two forms coloured, and simple colorimetric measurements of the ionization ratio were therefore possible. When absorption spectra are recorded spectrophotometrically over a range of wavelengths equation (2.24) may, in principle, be applied at any wavelength, but maximum accuracy is achieved when it is applied at the wavelength where the molar extinction coefficients of the two forms differ most. This is the wavelength of maximum absorption of one or the other form, and the method is most commonly applied at such a wavelength. When both forms show clear absorption maxima sufficiently far apart, the superposed spectra over a range of ionization ratios should show a clear isobestic point. This is often not so, owing to medium effects on spectra, and some methods for overcoming this difficulty have been developed.

An accurate method of determining ionization ratios when medium effects are present, or in general when absorption spectra of the ionized and the unionized form are more complex, showing a number of maxima and no clear isobestic points, has been devised by Davis and Geissman (1954). Two wavelengths are chosen in this method, λ_i and λ_u, close to the points of maximum difference between the extinction coefficients of the completely ionized and the essentially unionized forms. Then if F is the mole fraction of the ionized form (i.e. $c_i/(c_i + c_u)$), the molar extinction coefficients at the two wavelengths will be given by

$$(\epsilon)_{\lambda_i} = (\epsilon_u)_{\lambda_i} + [(\epsilon_i)_{\lambda_i} - (\epsilon_u)_{\lambda_i}] F \quad (2.25)$$

$$(\epsilon)_{\lambda_u} = (\epsilon_i)_{\lambda_u} + [(\epsilon_u)_{\lambda_u} - (\epsilon_i)_{\lambda_u}] (1 - F) \quad (2.26)$$

Subtracting the two equations we obtain

$$D = (\epsilon)_{\lambda_i} - (\epsilon)_{\lambda_u} = [(\epsilon_u)_{\lambda_i} - (\epsilon_u)_{\lambda_u}] + [(\epsilon_i)_{\lambda_i} - (\epsilon_u)_{\lambda_i} + (\epsilon_u)_{\lambda_u} - (\epsilon_i)_{\lambda_u}] F \quad (2.27)$$

Both terms in square brackets are constants and if, following Davis and Geissman (1954), they are denoted by

$$c = 1/[(\epsilon_i)_{\lambda_i} - (\epsilon_u)_{\lambda_i} + (\epsilon_u)_{\lambda_u} - (\epsilon_i)_{\lambda_u}]$$

and $K = c[(\epsilon_u)_{\lambda_u} - (\epsilon_u)_{\lambda_i}]$, equation (2.27) becomes

$$D = -\frac{K}{c} + \frac{1}{c}F \qquad (2.28)$$

It can be seen that D is linearly related to the fraction of the compound in the ionized form. Since $F = K + cD$, the ionization ratio I is given by

$$I = \frac{F}{1-F} = \frac{K+cD}{1-K-cD} \qquad (2.29)$$

This method of determining ionization ratios minimizes errors due to the effect of changes of medium upon absorption and also errors due to inaccuracies in the concentration of the indicator, since it relies upon differences in absorption at two wavelengths.

2.3.1.2. *Medium Effects and Accuracy of Measurements*

The presence of medium effects upon absorption is deduced mainly from the failure of the superposed absorption spectra to intersect in an isobestic point, i.e. a point in which the molar extinction coefficients of the unionized and the ionized form are equal, as theoretically required. This is the result of a shift in the absorption of one or of both forms with changing acid concentration. Owing to this the ionization ratios estimated from measurements at different wavelengths often differ considerably. The problem was studied by Bascombe and Bell (1959), who report typical variations in the ionization ratios of about 5 per cent, depending on the wavelength chosen and the I values themselves, but with some indicators the variations are much greater. The oldest method of dealing with this difficulty was suggested by Flexser *et al.* (1935), who simply shifted the absorption curves along the wavelength axis until an isobestic point was obtained. This method was later found to be unsatisfactory (Yates *et al.*, 1964), because it is possible as a rule to obtain more than one isobestic point by this procedure and the choice is arbitrary. Errors due to medium effects are least when wavelengths close to the absorption maximum of one or the other form are chosen for the estimation of ionization ratios. The results of Bascombe and Bell (1959) show that ionization ratios cannot be obtained with an accuracy better than 2–5 per cent when measurements are carried out at a single wavelength. In favourable cases an accuracy of ±0·3 per cent in the ionization ratio can be attained when I is determined at several wavelengths (Vinnik and Ryabova, 1964). Ryabova *et al.* (1966), who also found that with more complex absorption spectra the estimated ionization ratios depended on the wavelength, recommended that an integral of the area under the absorption curve (the integral intensity) be used. This method is, however, laborious and is not always feasible. Davis and Geissman (1954) also state that areas under the absorption spectrum plots between two arbitrarily chosen wavelengths often give estimates of ionization ratios in good agreement with those obtained by their more elaborate method, described in the preceding section, which is the best method so far developed for the partial cancellation of medium effects. Even so, in cases in which complex absorption spectra are found, it is difficult to estimate the logarithms of ionization ratios of indicators, and hence their pK values, to better than 0·1 unit.

It is even more difficult to correct adequately for medium effects upon the values of molar extinction coefficients, which are sometimes observed independently of the changes produced by the change in the ionization ratio. It has been suggested (Bascombe and Bell, 1959; Edward and Wang, 1962, and Katritzky *et al.*, 1963) that one could assume this effect to be linearly dependent on acidity and thus obtain corrected ϵ values by an extrapolation procedure.

Taking all this into account, it may be concluded that the cumulative error in the evaluation of acidity functions arises mainly from errors in the determination of the pK of individual indicators used in the overlap procedure (Vinnik, 1966). The pK of the first indicator may be known to $\pm(0\cdot02-0\cdot04)$ units. If the logarithms of the ionization ratios of all subsequent indicators are determinable to $\pm(0\cdot02-0\cdot04)$ units, which is the usual accuracy obtainable, the error in the acidity function will be the sum of such errors for all indicators involved in its determination. This means that when 9–10 indicators are involved (as for 100 per cent sulphuric acid, for example), the error in the acidity function value may amount to $\pm0\cdot4$ units. Over narrower acidity function intervals of 3–5 units, which is the normal range used in kinetic studies, the relative error between the start and the end of the interval will probably be only $\pm(0\cdot08-0\cdot13)$ units. Any errors due to the uncertainty in the acid concentration should be negligible by comparison, in view of the possibility of treating 100 per cent sulphuric acid as a highly accurate primary standard (see Chapter 1).

2.3.2. ELECTROCHEMICAL MEASUREMENTS

Following the suggestion of Koepp *et al.* (1960), that the normal potential of the redox system ferrocene–ferricinium is virtually independent of the nature of the solvent, especially in solvents of high dielectric constant, Strehlow and Wendt (1961) made measurements of the electromotive force of the cell:

Pt, $H_2|H_2O-H_2SO_4$, Ferrocene–Ferricinium (1:1)|Pt,

containing sulphuric acid solutions of varying concentrations up to 100 per cent, and defined the function

$$R_0(\text{H}) = \frac{F}{2 \cdot 303 RT}(E_{(x)} - E_{(1)}) \qquad (2.30)$$

where $E_{(x)}$ and $E_{(1)}$ are electromotive forces of the cell at proton activities x and unity, respectively. This redox function should be a logarithmic measure of the proton activity, according to

$$\log a_\text{H} = -R_0(\text{H}) \qquad (2.31)$$

if the basic assumption applies, and should afford a simple means of obtaining this quantity at any acid concentration from two e.m.f. measurements, in contrast to the laborious indicator procedure.

The values of the $R_0(\text{H})$ function obtained are more negative up to 85 per cent sulphuric acid than H_0 values from indicator measurements, and less negative above this concentration. The independence of the standard potential of the ferrocene-ferricinium couple on the medium is in doubt for strongly acid media, however, owing to the possibility of protonation of ferrocene (Curphey et al., 1960). For other media the electrochemical method offers some advantages (Vedel, 1967).

According to Bates (1966), a combination of indicator H_0 determinations and $R_0(\text{H})$ measurements affords a means of determining the indicator activity coefficient ratio from

$$H_0 - R_0(\text{H}) = -\log(f_\text{B}/f_{\text{BH}^+}) \qquad (2.32)$$

In view of the unknown effect of protonation of ferrocene on the potential of the ferrocene/ferricinium electrode in concentrated sulphuric acid, it is doubtful whether meaningful values of the activity coefficient ratios could be obtained from this equation, at least for concentrated acid solutions.

The use of the glass electrode for the determination of the acidity function has also been suggested (Clerc et al., 1965), but only through the use of a calibration curve against the indicator H_0 function.

Electrochemically determined acidity functions are of only limited interest for kinetic applications, where acidity functions determined using indicators chemically similar to the substrates are more useful.

2.4. Various Acidity Function Scales

2.4.1. THE H_0 OR H_0' ACIDITY FUNCTION

2.4.1.1. Sulphuric Acid/Water Mixtures

Hammett's activity coefficient postulate, which was stated in Section 2.2, is most easily verifiable in terms of the equation

$$\log\frac{[\text{BH}^+]}{[\text{B}]} - \log\frac{[\text{CH}^+]}{[\text{C}]} = \text{p}K_{\text{BH}^+} - \text{p}K_{\text{CH}^+} + \log\frac{f_\text{B} f_{\text{CH}^+}}{f_{\text{BH}^+} f_\text{C}} \qquad (2.33)$$

according to which, if the postulate that $f_{BH^+}/f_B = f_{CH^+}/f_C$ holds, the last term should become zero. This means that in plots of the logarithms of ionization ratios for various overlapping indicators against acid concentration, the lines obtained should be parallel within a given concentration range. The interest thus centres on the slopes $d(\log I)/d(\%\text{ acid})$. The slopes were found to be parallel for most bases used by Hammett and Deyrup (1932), who chose to work with a series of substituted primary nitroanilines, but included also in some regions of acid concentration bases of a different type (p-nitroazobenzene, benzalacetophenone, β-benzoylnaphthalene, p-benzoyldiphenyl and anthraquinone). Benzalacetophenone is now known to undergo isomerization (Noyce et al., 1959), and anthraquinone shows a complex absorption spectrum, which is in addition subject to a medium shift. The most striking failure of parallelism was, however, observed with a tertiary aniline (N,N-dimethyl-2,4,6-trinitroaniline), which was also included in the series, and which showed a distinctly higher slope. Since it became clear more recently that a number of bases show slopes $d(\log I)/d(-H_0)$ different from unity, i.e. deviate from the scale defined by Hammett, the H_0 scale has been re-evaluated by Jörgenson and Hartter (1963) in terms of a series of indicators consisting of primary anilines only. Ionization ratios were determined from spectrophotometric measurements at a single wavelength (λ_{max}). Acidity function measurements using the same indicators were repeated more recently by Ryabova et al. (1966), who paid more attention to the evaluation of ionization ratios of some indicators showing complex absorption spectra (2-bromo-4,6-dinitroaniline and 2,4-dichloro-6-nitroaniline). They used integral intensities from λ_{max} towards higher wavelengths in evaluating ionization ratios for these indicators. Very recently a redetermination of the scale using the same set of indicators was carried out by Johnson et al. (1969). This is in good agreement with the two previous determinations, especially with the data of Ryabova et al. (1966), the discrepancy not exceeding 0·14 units at concentrations less than 90 per cent. At higher concentrations the values of Johnson et al. (1969) are more negative (up to 0·37 units at 99 per cent), but their tabulation of data for this region is less extensive. Therefore preference is here given to the data of Ryabova et al. (1966), which are reproduced in Table 2.1.

At the aqueous end, the scale is based upon the ionization of p-nitroaniline, which was studied carefully by Paul (1954), by Bascombe and Bell (1959), and again by Ryabova et al. (1966), who finally adopted for it a pK value of 1·00 ± 0·01. Johnson et al. (1969) report this figure to an even higher degree of accuracy (1·004 ± 0·001). Hydrochloric and perchloric acid have more commonly been used for the very dilute acid solutions in the studies of this indicator, rather than sulphuric acid, because of the difficulties in evaluating the hydrogen ion concentration in its solutions due to its second dissociation.

Sulphuric acid solutions have also been used by Bascombe and Bell (1959) and by Ryabova *et al.* (1966).

The temperature dependence of the H_0 acidity function has been repeatedly studied (Gel'bshtein *et al.*, 1956; Biggs, 1961; Boyd and Wang, 1965), but

TABLE 2.1

The H_0 acidity function of sulphuric acid/water mixtures at 25°C
(Ryabova *et al.*, 1966)

Weight % H_2SO_4	H_0	Weight % H_2SO_4	H_0
1	+0·84	67	−5·48
3	+0·31	70	−5·92
5	−0·02	72	−6·23
8	−0·28	75	−6·72
10	−0·43	77	−7·05
12	−0·58	80	−7·52
14	−0·73	82	−7·84
16	−0·85	85	−8·29
18	−0·97	87	−8·60
20	−1·10	90	−9·03
22	−1·25	92	−9·33
25	−1·47	94	−9·59
27	−1·61	96	−9·88
30	−1·82	98	−10·27
32	−1·96	99	−10·57
35	−2·19	99·1	−10·62
37	−2·34	99·2	−10·66
40	−2·54	99·3	−10·72
42	−2·69	99·4	−10·77
45	−2·95	99·5	−10·84
47	−3·13	99·6	−10·92
50	−3·41	99·7	−11·01
52	−3·60	99·8	−11·18
55	−3·91	99·85	−11·28
57	−4·15	99·90	−11·43
60	−4·51	99·95	−11·64
62	−4·82	100	−11·94
65	−5·18		

the widest temperature range (40, 60, 80 and 90°C) was covered by the recent accurate measurements of Johnson *et al.* (1969), in addition to the values for 25°C already mentioned. The acidity function values become progressively less negative with increasing temperature.

A number of simple bases follow the H_0 acidity function fairly closely, i.e. show unit slope in plots of log I *vs.* $-H_0$ (within the limits of experimental error of 0·05–0·1). Some carbonyl bases (substituted acetophenones and benzaldehydes) behave in this way, but substituted benzophenones show deviations at acid concentrations above 60 per cent, corresponding to less negative acidity function values at a given acid concentration (Bonner and Phillips, 1966). Tertiary aromatic amines, primary amides and indoles show very large deviations, so that new acidity functions have been evaluated in terms of the protonation of these bases, and new symbols have been suggested for them. They will be discussed in Sections 2.4.2, 2.4.3, and 2.4.4. Since the H_0 acidity function is now defined exclusively in terms of primary aniline indicators, it is often denoted by H_0', to distinguish it from the acidity function defined in terms of tertiary aniline indicators, for which the symbol H_0''' has been suggested (Arnett and Mach, 1964).

It may be mentioned at this point that there is a linear correlation between the chemical shift of the proton magnetic resonance of sulphuric acid/water mixtures relative to water and the acidity function H_0 up to about 75 per cent acid, but above this concentration the relationship breaks down (Gillespie and White, 1960; Zarakhani and Vinnik, 1962).

2.4.1.2. *Dilute Oleum Solutions*

In determining the acidity function of pure sulphuric acid, Vinnik and Ryabova (1964) used *p*-nitrotoluene as indicator. Jörgenson and Hartter (1963) proposed the use of 3-bromo-2,4,6-trinitroaniline and 3-chloro-2,4,6-trinitroaniline for this purpose, but Vinnik and Ryabova (1964) found that the ionization ratio for these indicators increases more sharply with rising acidity than that for 3-methyl-2,4,6-trinitroaniline and for 2,4,6-trinitroaniline. The reason for this breakdown of the activity coefficient postulate for these indicators is not known (Vinnik, 1966). The use of nitrocompounds as indicators for the determination of the acidity function of 100 per cent sulphuric acid and dilute oleum solutions was first suggested by Brand (1950) and measurements were carried out by Brand *et al.* (1952). More recently these measurements were repeated by Vinnik and Ryabova (1964) for dilute oleum solutions using *p*-chloronitrobenzene up to 11 per cent free sulphur trioxide. On the basis of these measurements and the measurements of Brand *et al.* (1952) Table 2.2 was compiled (Vinnik, 1966). The values in this Table are consistent with the acidity function values for sulphuric acid/water mixtures in Table 2.1.

As can be seen from the Table the acidity rises sharply in dilute oleum solutions and there is no doubt that this rise is due to the increasing protonating ability of the medium. At higher concentrations, however, interactions of solutes with sulphur trioxide as an aprotic acid become increasingly probable (Vinnik, 1966). Work on acidity in more concentrated oleum solutions has been done by Lewis and Bigeleisen (1943) and by Coryell and Fix (1955).

Substituted nitrobenzenes have also been used to determine the acidity function of mixtures of sulphuric acid with chlorosulphuric acid (Pal'm, 1956). In this system the acidity increases with the concentration of chlorosulphuric acid and reaches a value of $-1\cdot 89$ H_0 units in the 100 per cent acid relative to 100 per cent sulphuric acid. The acidity functions of mixtures of sulphuric acid with fluorosulphuric acid and of solutions of tetrahydrogenosulphatoboric acid in pure sulphuric acid have also been measured (Gillespie, 1963).

TABLE 2.2

The H_0 acidity function for oleum solutions at 25°C (Vinnik, 1966)

Weight % free SO_3	$-H_0$	Weight % free SO_3	$-H_0$
0.15	12.10	8.0	13.10
0.2	12.13	9.0	13.15
0.4	12.26	10.0	13.21
0.5	12.30	11	13.27
0.6	12.33	12	13.31
0.8	12.38	14	13.39
1.0	12.43	16	13.47
1.5	12.54	18	13.55
2.0	12.62	20	13.62
2.5	12.68	22	13.68
3.0	12.74	25	13.76
3.5	12.78	27	13.81
4.0	12.83	30	13.91
5.0	12.90	32	13.98
6.0	12.98	36	14.13
7.0	13.04		

2.4.1.3. *Deuterium Oxide/Deuterosulphuric Acid Mixtures*

Högfeldt and Bigeleisen (1960) suggested the symbol D_0 for acidity functions in deuterium oxide and measured the acidity function for solutions of deuterosulphuric acid up to 11 M using primary nitroanilines as indicators. The scale was extended up to 12 M using N,N-dimethyl-2,4,6-trinitroaniline and benzalacetophenone. It was found that the values of the acidity functions H_0 and D_0 were identical within the limits of error at all concentrations, except in the very dilute range ($H_0 = 3$–$0\cdot 5$), where the deuterio solutions were up to $0\cdot 1$ H_0 units less acidic. This difference between H_0 and D_0 is consistent with the fact that HSO_4^- is a stronger acid than DSO_4^- (Drucker, 1937). A difference was also found in the pK values of indicators, the deuterio conjugate acids being weaker than the protio conjugate acids by 0.4–0.6 pK units, depending on the pK values themselves.

2.4.1.4. *Sulphuric Acid Mixtures with Partially Aqueous and Non-aqueous Solvents*

Sulphuric acid/water mixtures are rather poor solvents for many organic compounds. Therefore introduction of an organic component (ethanol, dioxane, etc.) is often considered as a means of improving the solubility of substrates in kinetic work, or else glacial acetic acid is substituted for water. Acidity functions for some such media have been evaluated, although difficulties arise if the dielectric constant of the medium becomes too low. Only a brief reference to such measurements will be made here.

Measurements of the acidity functions in sulphuric acid/water mixtures containing 20 volume per cent of ethanol have been reported by Yeh and Jaffé (1959a), who used azobenzenes as indicators, and by Dolman and Stewart (1967), who used substituted diphenylamines up to 11·2 M sulphuric acid. The agreement between the two sets of measurements is not very good at concentrations above 5 M.

An H_0 acidity function scale was established also in 5 per cent dioxane/aqueous sulphuric acid by Noyce and Jörgenson (1961 and 1962) for up to 75 per cent acid. In 40 per cent and 60 per cent aqueous dioxane, perchloric acid has been studied up to 4·5 M (Bunton et al., 1957).

The acidity function for mixtures of acetic acid with sulphuric acid was measured by Hall and Spengeman (1940) for up to 52·5 per cent sulphuric acid ($H_0 = -6$). In spite of the low dielectric constant of these mixtures, this function has been used successfully in correlating kinetic data (see Chapter 5).

2.4.1.5. *Perchloric Acid/Water Mixtures*

In the moderately concentrated acid range perchloric acid solutions are often used interchangeably with sulphuric acid solutions as media for certain reactions. For the sake of completeness and in order to provide a connected picture of acid-catalyzed reactions in moderately concentrated acid, some such studies will be mentioned in Chapter 5, alongside studies in sulphuric acid/water mixtures. It is therefore appropriate to refer at this point to acidity function measurements in such media. The most recent determinations, with extension up to 79 per cent acid, are due to Yates and Wai (1964). The H_0 scale in this acid becomes rapidly more negative than that in aqueous sulphuric acid, 79 per cent perchloric acid having an indicator acidity equal to that of 98 per cent sulphuric acid. This is the upper concentration limit accessible at 25°C, because of the crystallization of the monohydrate (84·8 per cent acid).

2.4.2. THE H_0''' ACIDITY FUNCTION

The deviation of N,N-dimethyl-2,4,6-trinitroaniline from the behaviour of other Hammett bases in the original set of Hammett indicators (Hammett and Deyrup, 1932) was found to be a general feature of the behaviour of tertiary aromatic amines. These indicators therefore generate their own acidity function, which decreases more steeply with acid concentration than the H_0' acidity function. Current views on the cause of this discrepancy will be

TABLE 2.3

The H_0''' acidity function of sulphuric acid/water mixtures at 25°C
(Arnett and Mach, 1964)

Weight % H_2SO_4	H_0'''	Weight % H_2SO_4	H_0'''
1	+0·84	54	−4·97
4	+0·14	58	−5·58
8	−0·34	62	−6·22
10	−0·53	66	−6·91
12	−0·69	68	−7·28
14	−0·88	70	−7·65
16	−1·09	72	−8·01
18	−1·28	74	−8·35
20	−1·47	76	−8·71
22	−1·66	78	−9·07
24	−1·86	80	−9·44
26	−2·06	82	−9·80
28	−2·25	84	−10·13
30	−2·44	86	−10·47
34	−2·83	88	−10·81
38	−3·25	90	−11·14
42	−3·67	92	−11·44
46	−4·12	94	−11·74
50	−4·54		

discussed in Sections 2.5 and 2.6. At this point merely the values of this new acidity function, as determined by Arnett and Mach (1964), are reported in Table 2.3. This acidity function scale is based on the thermodynamic pK of N,N-dimethyl-4-nitroaniline in very dilute aqueous hydrochloric acid.

2.4.3. THE H_A ACIDITY FUNCTION

A striking deviation from the Hammett acidity function was also observed in the protonation of amides, but in the opposite direction to that of tertiary aromatic amines, i.e. slopes less than unity in plots of log$I$ vs. $-H_0$ were consistently found (Yates et al., 1964). Therefore a new acidity function for the protonation of primary amides was determined and designated by H_A. The values of this acidity function coincide for up to about 15 per cent sulphuric acid with H_0 values, but diverge from it increasingly at higher concentrations. The function has been evaluated for up to 81·8 per cent sulphuric acid (Yates et al., 1964). Its values at round concentrations are reported in Table 2.4.

TABLE 2.4

The H_A acidity function of sulphuric acid/water mixtures at 25°C
(Yates et al., 1964)

Weight % H_2SO_4	H_A	Weight % H_2SO_4	H_A
15	−0·73	50	−2·50
20	−0·98	55	−2·77
25	−1·24	60	−3·08
30	−1·59	65	−3.39
35	−1·74	70	−3·74
40	−2·00	75	−4·16
45	−2·24	80	−4·62

This acidity function is based, like the H_0 acidity function, on the pK of p-nitroaniline in very dilute acid, because no amide sufficiently basic to ionize appreciably in dilute acid could be found. The most basic amide studied, 2-pyrrole-carboxamide, however, seems to have an activity coefficient ratio equal to that of p-nitroaniline in acid solutions of different concentration, and therefore the H_A acidity scale may be regarded as fairly firmly based on dilute aqueous solution as the standard state (Edward, 1964). The $H_A - H_0$ difference in hydrochloric solutions is less than in sulphuric acid solutions (Yates and Riordan, 1965).

2.4.4. THE H_I ACIDITY FUNCTION

Indoles have proved to be a class of simple bases which bring out particularly clearly the limitations of the Hammett activity coefficient postulate, because in a group of 25 substituted indoles, in spite of structural similarity, a variety of slopes of log$I$ vs. $-H_0$ plots were observed (Hinman and Lang, 1964). The slopes fell into three groups, each representative of a class of indicator bases. The largest group are indoles having one or more alkyl groups in the hetero ring, ranging from methyl and ethyl at the 1- and 2-positions, to t-butyl, carboxymethyl and β-aminoethyl at the 3-position. The average slope of log$I$ vs. $-H_0$ plots for this group of indoles is about 1·3, and the acidity function

H_I, defined in terms of their protonation equilibria, thus decreases more steeply with acid concentration than the H_0 acidity function. For 1,3-disubstituted indoles the slopes are even higher, although the 1,2,3-trimethyl derivative falls into the first class. Indole itself and indoles unsubstituted in the hetero ring show, on the contrary, lower slopes. The acidity function for indoles was defined in terms of the largest class, and since the strongest base 1,2-dimethylindole could be studied in very dilute acid solution, the scale is based on dilute aqueous solution as the standard state. The values are listed in Table 2.5.

TABLE 2.5

The H_I acidity function of sulphuric acid/water mixtures at 25°C
(Hinman and Lang, 1964)

Weight % H_2SO_4	H_I	Weight % H_2SO_4	H_I
1	+0·9	40	−3·38
5	+0·1	45	−4·01
10	−0·53	50	−4·65
15	−0·93	55	−5·32
20	−1·39	60	−6·01
25	−1·85	65	−6·83
30	−2·34	70	−7·88
35	−2·84		

2.4.5. THE H_+ ACIDITY FUNCTION

The H_+ acidity function was first defined by Brand et al. (1952) and measured in oleum solutions using m-nitroanilinium ion as indicator. In the limited acid concentration ranges of 30–55 per cent, where 4-nitro-1,2-phenylenediamine was used as indicator, and of 75–95 per cent, where 4-aminoacetophenone was used, parallelism between the logarithms of the ionization ratios of Hammett neutral bases and these cation bases was observed (Bonner and Lockhart, 1957). The conclusion was reached that H_0 and H_+ are either identical or differ by a small constant value. More recently Vetešnik et al. (1968) have used a series of substituted 1,2-phenylenediamines and 2,3-dimethylquinoxalines to determine the H_+ acidity function for up to 80 per cent sulphuric acid at 25°C. The thermodynamic $pK_{BH_2^{++}}$ value for 4-methyl-1,2-phenylenediamine, determined spectrophotometrically in aqueous buffer solutions, was used as the basis for the H_+ scale. The H_+ scale follows the H_0 scale closely up to approximately −5 (65 per cent acid). The values are reported in Table 2.6.

TABLE 2.6

The H_+ acidity function of sulphuric acid/water mixtures at 25°C
(Vetešnik et al., 1968)

Weight % H_2SO_4	H_+	Weight % H_2SO_4	H_+
5	−0·02	45	−2·97
10	−0·51	50	−3·38
15	−0·93	55	−3·80
20	−1·24	60	−4·25
25	−1·59	65	−4·72
30	−1·87	70	−5·13
35	−2·12	75	−5·36
40	−2·59	80	−5·53

2.4.6. THE H_- ACIDITY FUNCTION

An H_- acidity scale could only be established by means of strongly acidic indicators. Such indicators were found recently in a class of cyanocarbon acids (Boyd, 1961). The anions of these acids show strong electronic absorptions which disappear on protonation, and so their ionization ratios could be established by ultraviolet and visible spectrophotometry. Salts of the following cyanocarbon acids were used: 1,1,2,3,3-pentacyanopropene, 2-dicyanomethylene-1,1,3,3-tetracyanopropene, bis(tricyanovinyl)amine, tricyanomethane, 1,1,2,6,7,7-hexacyano-1,3,5-heptatriene, methyl dicyanoacetate and p-(tricyanovinyl)-phenyldicyanomethane, and were found to fulfil the requirement

TABLE 2.7

The H_- acidity function of sulphuric acid/water mixtures at 25°C (Boyd, 1961)

Weight % H_2SO_4	H_-	Weight % H_2SO_4	H_-
10	−0·09	50	−3·91
15	−0·55	55	−4·39
20	−0·99	60	−4·90
25	−1·47	65	−5·57
30	−1·96	70	−6·21
35	−2·44	75	−6·79
40	−2·93	80	−7·28
45	−3·42		

of parallelism in the plots of log$I$ against acid concentration. An acidity function has hence been evaluated, as reported in Table 2.7. The pK_a of p-(tricyanovinyl)-phenyl-dicyanomethane could be obtained by measurements in dilute acid and the scale is therefore based on the standard state of infinite dilution in water.

TABLE 2.8

The H_R acidity function of sulphuric acid/water mixtures at 25°C (Deno et al., 1955)

Weight % H_2SO_4	H_R	Weight % H_2SO_4	H_R
0·5	+1·25	52	−7·01
1·0	+0·92	54	−7·44
3·0	+0·37	56	−7·90
5·0	−0·07	58	−8·40
10	−0·72	60	−8·92
15	−1·32	70	−11·52
20	−1·92	80	−14·12
25	−2·55	90	−16·72
30	−3·22	92	−17·24
35	−4·00	94	−17·78
40	−4·80	96	−18·45
45	−5·65	98	−19·64
50	−6·60		

2.4.7. THE H_R AND H_R' ACIDITY FUNCTIONS

The H_R acidity function has been determined by Deno et al. (1955) by means of a series of eighteen substituted arylmethanols, and its applicability was tested with another ten. 4,4′,4″-Trimethoxytriphenylmethanol ionizes sufficiently in very dilute acid to provide a reference for the acidity scale, which is thus based on infinitely dilute aqueous solution as the standard state. The values of the H_R function are reported in Table 2.8.

The H_R scale does not change much with temperature up to 65 per cent acid, but becomes progressively more negative with increasing temperature at higher concentrations (Arnett and Bushik, 1964).

Some measurements on the H_R acidity function in dideuterosulphuric acid (the D_R function) have also been reported (de Fabrizio, 1966). The same numerical values have been found as for the protio acid.

As has been mentioned in Section 2.2, the H_R' acidity function can most easily be evaluated from the H_R function using equation (2.22), p. 31, and the activity of water (Table 1.6, p. 15). Some earlier data on the protonation of arylolefins conformed to the acidity function so obtained (Deno *et al.*, 1959; Deno *et al.*, 1960), but a recent actual determination of the scale by means of a series of carbon bases (substituted azulenes, diarylethylenes and aromatic

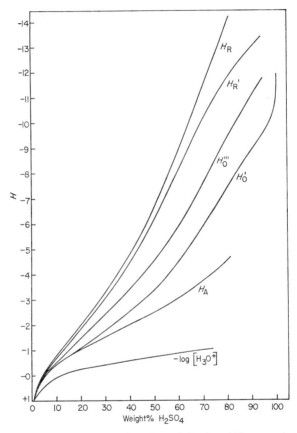

FIG. 2.1. Some acidity functions for sulphuric acid/water mixtures.

polyethers) gave values of the acidity function deviating from those calculated from equation (2.22) by 0·2 to 0·9 units towards lower acidity (Reagan, 1969). The measured function and H'_R change with solvent composition at approximately the same rate over most of the range of acid concentration, which means that they are equally useful in correlating kinetic data. The symbol H_C was suggested for this measured function, based on carbon bases (Reagan, 1969). The discrepancy between H'_R and H_C is due to the inadequacy of the basic assumption in deriving equation (2.22) that $f_{ROH}/f_{R^+} = f_{olefin}/f_{R^+}$, from which it follows that $f_{ROH} = f_{olefin}$. This is not strictly true, probably owing to the hydrogen-bonding ability of the carbinol.

Figure 2.1 gives a plot of most acidity functions discussed in this Section against weight per cent of sulphuric acid in sulphuric acid/water mixtures. The acidity function for oleum solutions is not shown.

2.5. Verification of the Activity Coefficient Postulate

The numerous acidity functions for the protonation of neutral bases, reported in the preceding Section, could only arise if Hammett's postulate of the equality of activity coefficient ratios (equation 2.7, p. 28) does not hold strictly for all neutral bases. The realization that so many bases deviate in their protonation behaviour from the behaviour of Hammett bases has intensified efforts in recent years to obtain independent information on the activity coefficients of the indicators and their conjugate acids. The activity coefficients of the neutral indicators are measurable quantities. The activity coefficients of the conjugate acids are not obtainable by any kind of measurement, but relative values were obtained by Boyd (1963a) on the basis of certain assumptions. These studies of activity coefficients are described in the present Section.

2.5.1. MEASUREMENTS OF ACTIVITY COEFFICIENTS OF NEUTRAL INDICATOR BASES

Measurements of solubilities of indicators in sulphuric acid/water mixtures or of distribution of indicators between these solvent mixtures and an inert solvent have been the source of all the information on activity coefficients of neutral bases so far obtained. The calculation of activity coefficients from solubilities is based on the equation

$$f = f'(c'/c) \qquad (2.34)$$

where c and c' are the concentrations and f and f' the activity coefficients of the substance in saturated solution in two different solvent media. Since a dilute solution of the substance in pure water is taken as the reference state for which $f' = 1$, equation (2.34) becomes

$$f = c_0/c \qquad (2.35)$$

where c_0 is the solubility in pure water and f and c now refer to a saturated solution in a sulphuric acid/water mixture. This method was first used by Hammett and Chapman (1934). The concentrations of the indicators are normally determined spectrophotometrically. An increase of solubility with increasing acid concentration, the so-called "salting-in" of the neutral solute by the acid, corresponds to a decrease in its activity coefficient. The activity coefficient was found to vary in different ways with increasing acid concentration even within the class of substituted nitroanilines. To take an extreme example, while the activity coefficient of *p*-nitroaniline increases up to about 25 per cent sulphuric acid to a value of 2 (Librovich and Vinnik, 1966), that of 2,4,6-trinitroaniline falls off steadily with increasing concentration down to 10^{-3} in 88·5 per cent acid (Boyd, 1963a). The average behaviour of some other members of this series is less extreme (e.g. 2,6-dichloro-4-nitroaniline, 2,4-dichloro-6-nitroaniline and 2,4-dinitroaniline) and activity coefficients close to unity are observed up to about 30 per cent acid, with a steady decrease at higher concentrations. At higher acid concentrations, where the indicators become partly protonated, the activity coefficient of the base can still be obtained by taking into account that the total solubility is now due to the neutral base and the protonated base:

$$c = c_B + c_{BH^+} = c_B(1 + h_0/K_{BH^+}) \qquad (2.36)$$

The activity coefficient is then calculated from the equation

$$f = \frac{c_0}{c}(1 + h_0/K_{BH^+}) \qquad (2.37)$$

the h_0 of the medium and the K_{BH^+} of the indicator being known. As the solubility in most instances increases with increasing acid concentration, at higher acid concentration, where the solubility becomes very high, distribution measurements between the acid solvents and an inert solvent (carbon tetrachloride, 1,2-dichloroethylene) may be used to obtain the activity coefficients of the free base. If the distribution coefficients between pure water and the inert solvent and between the acid mixture and the inert solvent are denoted by γ_0 and γ respectively, the activity coefficient is given by

$$f = \frac{\gamma_0}{\gamma}(1 + h_0/K_{BH^+}) \qquad (2.38)$$

Using this method, activity coefficient measurements for 2,4-dinitroaniline were extended right up to 100 per cent sulphuric acid, and for 2,4,6-trinitroaniline up to 80 per cent (Grabovskaya *et al.*, 1967). The activity coefficients from distribution measurements for the last mentioned compound are not in good agreement with those of Boyd (1963a) from solubility measurements and suggest that the behaviour of this indicator is closely similar to that of other nitroanilines.

Other indicators apart from nitroanilines have also been studied, in particular a number of nitrocompounds. Apart from nitrobenzene, which was studied over a very wide acid concentration range (Hammett and Chapman,

1934; Arnett et al., 1962), most measurements on these compounds refer to concentrated acid (Brand et al., 1959; Grabovskaya and Vinnik, 1966). The large increase in the solubility of nitrobenzene in >90 per cent acid, in which protonation still does not occur, were interpreted in terms of the formation of hydrogen-bonded complexes by Deno and Perizzolo (1957). This is a reasonable suggestion, since mononitrocompounds are known to form solid 1:1 addition compounds with sulphuric acid (Masson, 1931; Liler, 1959). The behaviour of carboxylic acids is analogous (Hammett and Chapman, 1934).

Of particular interest are measurements on some benzamides (Sweeting and Yates, 1966), in view of the deviation of the protonation of amides from the behaviour of primary aniline indicators. The activity coefficients of neutral benzamides show a closely similar acidity dependence to that of primary nitroanilines, which means that the difference in the protonation behaviour of these two classes of compound must be primarily due to a difference in the activity coefficients of the conjugate acids.

Some activity coefficient measurements are also available on other neutral indicators. Activity coefficients of some arylmethanols were measured by Deno and Perizzolo (1957), as well as those of some hydrocarbons. The results indicated little variation in the activity coefficients of these compounds from unity up to 40–50 per cent sulphuric acid. The more recent measurements of Reagan (1969), however, show that there is a difference in the activity coefficients of triarylmethanols and naphthalene (as a typical hydrocarbon) which accounts for the difference between the H'_R acidity function and the measured carbon base acidity function H_C.

2.5.2. RELATIVE ACTIVITY COEFFICIENTS OF SOME INDICATOR CONJUGATE ACIDS

Since single ion activity coefficients are not obtainable from any kind of thermodynamic measurement, Boyd (1963a) devised an ingenious method of obtaining relative activity coefficients referred to a chosen standard ion from solubility measurements on the cyanocarbon salts of indicators. The pentacyanopropenide anion (PCP$^-$), which is a sufficiently weak base not to be protonated in acid solution, was used in combination with the cations of the H_0 and H_R indicators. The salts so obtained, consisting of a large cation and a large anion, both singly charged, are sparingly soluble. The tetraethylammonium ion (TEA$^+$) was chosen as the standard ion. The solubilities of these salts were studied over a range of concentrations of sulphuric acid, and all salts having a common anion, it was assumed that the following relationship should hold:

$$\frac{f_{BH^+}}{f_{TEA^+}} = \frac{f_{BH^+} f_{PCP^-}}{f_{TEA^+} f_{PCP^-}} = \frac{f_{\pm}^2(BH^+, PCP^-)}{f_{\pm}^2(TEA^+, PCP^-)} \qquad (2.39)$$

The activity coefficient ratio was termed the relative activity coefficient and was denoted by $f^*_{BH^+}$. The results show that the relative activity coefficients of most anilinium cations increase with acid concentration, but at strikingly different rates for primary, secondary, tertiary and quaternary anilinium cations. The relative activity coefficient for the trimethylphenylammonium ion remains close to unity right up to 60 per cent sulphuric acid, as expected in view of its similarity with the standard ion (TEA$^+$). The relative activity coefficients of tertiary (N,N-dimethyl-), secondary (N-ethyl-), and primary anilinium ions increase with increasing acid concentration the more rapidly the greater the number of NH protons in the cation. The f_+^* of the N,N-dimethylanilinium ion increases about twentyfold from 0–70 per cent acid, whereas primary anilinium ions show on the average a thousandfold increase over the same acid concentration range. This may be explained in terms of hydrogen bonding of the anilinium cations to water molecules in the solvent mixtures (Taft, 1960b). For example, the anilinium ion may form hydrogen bonds with three water molecules (**2.1**), whereas the N,N-dimethylanilinium ion will bond with only one (**2.2**).

$$\begin{array}{cc}
\begin{array}{c} \text{OH}_2 \\ \vdots \\ \text{H}^+ \\ | \\ \text{Ar—N—H}\ldots\text{OH}_2 \\ | \\ \text{H} \\ \vdots \\ \text{OH}_2 \end{array} & \qquad
\begin{array}{c} \text{CH}_3 \\ | \\ \text{Ar—N}^+\text{—H}\ldots\text{OH}_2 \\ | \\ \text{CH}_3 \end{array} \\
(\mathbf{2.1}) & (\mathbf{2.2})
\end{array}$$

Since the activity of water in 70 per cent sulphuric acid falls to one twentieth of that in pure water, the reduced solvation destabilizes primary anilinium ions more than the secondary or the tertiary ions, and this leads to more steeply rising activity coefficients with increasing acidity.

The application of this method of determining relative activity coefficients of indicator conjugate acids is unfortunately not applicable in a straightforward way to amide cations, because they are fully dissociated in dilute aqueous solution, and it is therefore impossible to refer their activity coefficients to infinite dilution in water as the standard state. An attempt has been made to overcome this difficulty for benzamide by taking a reference state in 52·8 per cent sulphuric acid (Sweeting and Yates, 1966), but the amide is less than 90 per cent protonated in this acid. In view of this the conclusion drawn that the relative activity coefficient of the benzamidinium cation rises more steeply with acid concentration than that of the anilinium ion is not

2. ACIDITY FUNCTIONS

entirely certain, but it is consistent with the views on the reasons for different acidity function behaviour of amides compared with amines, to be discussed further in Section 2.6.1.

2.5.3. RELATIVE ACTIVITY COEFFICIENT RATIOS

On the basis of the results discussed in Sections 2.5.1 and 2.5.2, Yates (1964) and Edward (1964) have calculated ratios of relative activity coefficients of

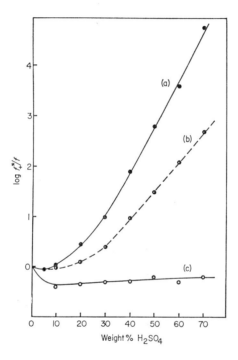

FIG. 2.2. Variation of the relative activity coefficient ratio with acid concentration: (a) primary anilines; (b) tertiary anilines; (c) olefins. (After Edward, 1964.)

cations to those of neutral indicator bases, in order to verify Hammett's activity coefficient postulate. The plots of these ratios for primary anilines (a), for tertiary aromatic amines (b) and for carbonium ions and olefins (c) are given in Fig. 2.2. In plotting curve (b) an average activity coefficient for primary anilines was used, since values for N,N-dimethylaniline are not available. Bearing in mind a possible uncertainty here, it can be seen from Fig. 2.2 that activity coefficient ratios for different types of indicator bases are indeed different, and should lead to different acidity functions, as observed.

2.6. Physical Significance of Acidity Functions

Before the experimental approach described in the preceding Section was made to the question of why different indicator bases define different acidity functions, ideas that hydration differences were the determining factor had already been advanced from considerations of protonation equilibria. These ideas have led to numerous calculations involving acidity functions, some of which are of interest for our understanding of the nature of strongly acid media. These theoretical aspects will therefore be discussed briefly in the present Section.

2.6.1. A CHEMICAL HYDRATION TREATMENT OF ACIDITY FUNCTIONS

The first theoretical approaches to the understanding of the concentration dependence of the acidity of aqueous strong acids were made by Bascombe and Bell (1957) and by Wyatt (1957) in terms of ideas about the chemical hydration of the proton. This has become known as the chemical hydration treatment and has been widely used to account for differences between various acidity functions.

The foundation stone of this theory is the observation that for a given value of the molality up to approximately $m = 10$ the Hammett acidity function H_0 has nearly the same value for a number of strong acids (Bascombe and Bell, 1957; Wyatt, 1957). This means that it is virtually independent of the anion and depends only upon the ratio of the number of hydrogen ions to the number of water molecules.

As a basis for further discussion of these ideas, some acidity functions from Section 2.4 have been plotted in Fig. 2.1, together with the logarithm of the hydrogen ion concentration. In calculating this, sulphuric acid was assumed to be fully dissociated into H^+ and HSO_4^- ions in the concentration range up to 70 per cent (i.e. the possibility of incomplete first dissociation and the second dissociation were neglected). Edward (1964) used the Raman spectral data of Young (see Chapter 1) in estimating $[H_3O^+]$, but the difference between the logarithms of the two estimates is minimal and may be ignored in view of the approximate nature of the treatment. It can be seen from Fig. 2.1 that the H_R acidity function decreases most steeply with acid concentration, followed by the H_R' acidity function, which was calculated from H_R using equation (2.22), p. 31.

Since it is now widely accepted that the proton is strongly hydrated in aqueous solution to H_3O^+ in the first instance, and then to larger aggregates, $H_9O_4^+$, by hydrogen bonding (with possible and probable further solvation of the aggregate), a protonation equilibrium may be represented by the equation

$$B.iH_2O + H^+.jH_2O \rightleftharpoons BH^+.kH_2O + (i+j-k)H_2O \qquad (2.40)$$

2. ACIDITY FUNCTIONS

in which all other species are assumed to be hydrated also. The number of water molecules released on protonation, h, is given by

$$h = i + j - k \tag{2.41}$$

and the equilibrium constant for this reaction may be written as

$$K = \frac{a_{BH^+} a_{H_2O}^h}{a_B a_{H^+}} = \frac{[BH^+] f_{BH^+} a_{H_2O}^h}{[B] f_B a_{H^+}} \tag{2.42}$$

all species being assumed hydrated. Taking logarithms one obtains

$$\log K = \log \frac{[BH^+]}{[B]} + \log \frac{f_{BH^+}}{f_B f_{H^+}} + h \log a_{H_2O} - \log [H^+] \tag{2.43}$$

Taking into account the general operational definition of any acidity function

$$H_i = \log K - \log \frac{[BH^+]}{[B]},$$

equation (2.43) may be rewritten in the form

$$H_i = \log \frac{f_{BH^+}}{f_B f_{H^+}} + h \log a_{H_2O} - \log [H^+] \tag{2.44}$$

It can be seen from this equation that in terms of the hydration theory the increasing protonating ability of the medium is due largely to the decreasing activity of water, because of the involvement of water primarily in solvating the protons (since j is probably the largest hydration number in equation 2.41). In the simplest case of the H_R' acidity function, the large carbonium ion with a highly delocalized charge may be assumed not to be hydrated itself and neither are the olefins. Therefore f_{BH^+}/f_B will in this case be close to unity (as experiments indeed show, Fig. 2.2), and H_R' may thus be regarded as the simplest acidity function (Yates, 1964).

The difference between the negative acidity function and the logarithm of the hydrogen ion concentration has been termed "excess acidity" by Perrin (1964). Hence the excess acidity may be written as:

$$-H_i - \log [H^+] = -\log \frac{f_{BH^+}}{f_B f_{H^+}} - h \log a_{H_2O} \tag{2.45}$$

In general the value of the activity coefficient term is not known. If one assumes as a first approximation that $f_{H^+} = f_{BH^+}$, a suggestion due to Bascombe and Bell (1957), and justified by the considerable charge delocalization in both hydrated cations, then "excess acidity" could be accounted for in terms of the activity coefficient of the neutral base and the total hydration change in the protonation reaction. Since $\log f_B$ for primary anilines changes by only about one unit up to 60 per cent sulphuric acid, and similarly for other bases

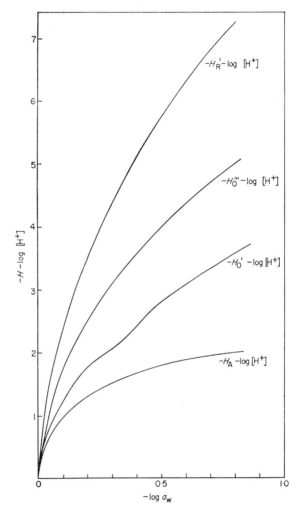

Fig. 2.3. Plots of "excess acidity" against the logarithm of water activity for some acidity functions.

(Edward, 1964), the major part of the difference between various acidity functions may be ascribed to the difference in the hydration numbers h.

In order to discuss this a little further, plots of the "excess acidity" for the four most divergent acidity functions H'_R, H'_0, H'''_0 and H_A against the logarithm of the activity of water have been given in Fig. 2.3. There is a dip in the curve for H_0', which is probably unimportant when the broad outline of phenomena is being considered. This apart, all the curves are of similar shape, and have been drawn to zero, because at infinite dilution the excess acidity becomes zero. The slopes of the lines are given by

$$d(-H_i - \log[\mathrm{H}^+])/d(-\log a_{\mathrm{H_2O}}) = h \qquad (2.46)$$

i.e. they are a measure of the hydration change in the protonation process at a particular acid concentration. It is apparent from the figure that there is no concentration region in which the lines are linear, and hence there is no constant hydration number h that can describe the hydration relationships. The slopes decrease with decreasing water activity, as they should, in view of the decreasing availability of water for proton hydration. The initial slopes are seen to be high, which implies not only primary and secondary solvation of the proton, but even tertiary solvation. It may be mentioned that, in an attempt to provide a chemical model to explain the acidity function of sulphuric acid/water mixtures, Högfeldt (1963) and Brock Robertson and Dunford (1964) postulate as the largest hydrate $\mathrm{H_{21}O_{10}^+}$.

The slopes are highest for the H_R' acidity function. The lower slopes for other acidity functions may then be ascribed mainly to increasing solvation of the conjugate acids of the indicators, because these species would form hydrogen bonds with water molecules more extensively than the uncharged bases, owing to their positive charge. As primary anilinium ions have three sites for hydrogen bonding (**2.1**), whereas tertiary anilinium ions have only one (**2.2**), the slopes at any acid concentration will be smaller for the "excess" H_0' function than for the "excess" H_0''' function, as observed. The smallest slopes are observed for the "excess" H_A function, which corresponds to strong solvation of the conjugate acids of primary amides used in defining the function. This can be accounted for in terms of a localized positive charge (as in anilinium ions) and an even greater number of sites for hydrogen bonding to water molecules than in the anilinium ions. Structure (**2.3**) satisfies both these conditions.

$$\mathrm{R-C} \underset{\underset{\mathrm{H_2O\cdots H}}{\overset{+}{\mathrm{N-H\cdots OH_2}}}}{\overset{\mathrm{O\cdots HOH}}{\diagup}} \mathrm{H\cdots OH_2}$$

(**2.3**)

The question of the protonation of amides will be discussed further in Chapter 3.

It is obvious from Fig. 2.3 that at acid concentrations above 45 per cent the differences between the various acidity functions will be linear functions of $\log a_{H_2O}$ over not too wide concentration ranges, as was in fact found by Taft (1960b) for $H_R - H_0$ between 44 and 60 per cent acid.

It follows from this discussion that for a base to obey a certain acidity function in its protonation behaviour, the balance of hydration numbers in equation (2.40) must be closely similar to that of the defining set of bases. Although the hydration numbers of the proton and the conjugate acid of the base are the most important, the hydration of the base also plays a part. In view of all these factors which vary from one type of base to another, it is purely coincidental if bases of chemically different types obey the same acidity function. The fact that the acidity function H_0 had assumed such importance over a number of years is due to its "average" position amongst other acidity functions, so that deviations of the behaviour of many bases from it did not appear too striking.

An alternative explanation of the differences between acidity functions is purely in terms of differences in activity coefficients in the defining equation

$$h_i = a_{H^+} \frac{f_B}{f_{BH^+}} \tag{2.47}$$

where h_i stands for any acidity function. This explanation has already been outlined in the preceding Section and is amenable to experimental testing. At a given acid concentration the activity of the proton is a constant, and differences between acidity functions stem from differences in activity coefficient ratios. Edward (1964) has shown that this line of approach can account almost quantitatively for differences between acidity functions. Its relationship with the hydration theory is reflected in the way in which variations with acid concentration of the activity coefficients of the conjugate acids of indicators are interpreted, namely in terms of hydration of conjugate acids (see formulae **2.1–2.3**).

The question of what happens to cation solvation at still higher acid concentrations where water molecules are progressively less available for hydration now arises. It is reasonable to suppose that at higher acid concentrations indicator cations form hydrogen bonds with undissociated acid molecules, whose concentration rises sharply above 85 per cent acid. Since sulphuric acid molecules are, however, much less basic than water molecules, this solvation is weaker, and therefore all acidity functions continue to rise at higher acid concentrations, and further into oleum solutions, in which sulphuric acid molecules are replaced by the still less basic disulphuric acid molecules.

2.6.2. THE ACIDITY FUNCTION H_0 IN THE VICINITY OF 100 PER CENT SULPHURIC ACID

The course of the H_0 acidity function in the vicinity of 100 per cent sulphuric acid, based on data in Tables 2.1 (p. 37) and 2.2 (p. 39) is shown in Fig. 2.4.

The shape of the curve is that of a typical acid/base titration curve in a self-dissociating solvent. On the aqueous side of the pure acid the main equilibrium determining the acidity of the medium is

$$H_2O + H_2SO_4 \rightleftharpoons H_3O^+ + HSO_4^- \tag{2.48}$$

In this concentration region water is fully (or almost fully) protonated by

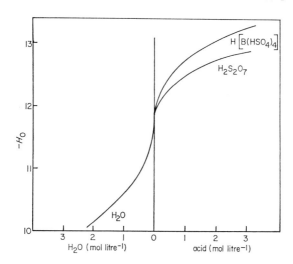

FIG. 2.4. The acidity function H_0 in the vicinity of 100 per cent sulphuric acid and in solutions of the strong acid $H[B(HSO_4)_4]$.

sulphuric acid. Cryoscopic work (see Section 3.2.5, p. 67) has shown, however, that in dilute oleum disulphuric acid is a weak acid dissociating according to

$$H_2S_2O_7 + H_2SO_4 \rightleftharpoons H_3SO_4^+ + HS_2O_7^- \tag{2.49}$$

In view of this the H_0 curve for a much stronger acid of the sulphuric acid solvent system, tetrahydrogenosulphatoboric acid, is also shown in Fig. 2.4 (based on a diagram published by Gillespie, 1963).

In aqueous solution the approximate pH difference between 1 M solutions of strong acid and strong alkali is 14 units, i.e. it corresponds to the pK_w of the solvent. An analogous estimate of the pK_{ap} of sulphuric acid from the H_0 values of 1 M solutions of water and $H[B(HSO_4)_4]$, as can be seen from Fig. 2.4, is approximately 2·2 units. This corresponds to an autoprotolysis constant of about $6·3 \times 10^{-3}$ mol² litre⁻². This estimate is much higher than the estimate of Hammett and Deyrup (1932), which was based only on the acidity function values for the slightly aqueous acid, and which gave a K_{ap} value of $2·4 \times 10^{-3}$ mol² litre⁻². If the H_0 jump between 1 M hydronium hydrogen sulphate and 100 per cent acid is doubled (assuming symmetry for the strong acid/strong base curve) a value of 1.9×10^{-3} mol² litre⁻² is obtained, which is a little lower than the estimate of Hammett and Deyrup (1932). In neither of these estimates was the ionic self-dehydration taken into account.

An analysis of the H_0 values in the vicinity of 100 per cent sulphuric acid showed that the following relationship holds between 90 and 99 per cent sulphuric acid (Brand, 1950):

$$H_0 + \log([H_2SO_4]/[HSO_4^-]) = \text{constant} \qquad (2.50)$$

where the concentration units may either be molarities or mole fractions. The validity of this equation implies approximate constancy of the activity coefficient ratio:

$$\frac{f_{BH^+} f_{HSO_4^-}}{f_B f_{H_2SO_4}} = \text{constant} \qquad (2.51)$$

Since the variation in the concentration of sulphuric acid molecules at these high concentrations is small, equation (2.50) may also be written as

$$H_0 = \text{const.} + \log[HSO_4^-] \qquad (2.52)$$

assuming HSO_4^- to be equal to the stoichiometric concentration of water. This equation was used by Brand *et al.* (1959) to obtain the HSO_4^- concentration in 100 per cent sulphuric acid from their measurements of the H_0 function in the vicinity of the 100 per cent acid. Their calculation gave the molality of these ions as 0.023_4, i.e. a K_{ap} value of 1.8×10^{-3} mol² litre⁻²—a little lower than the estimate of Hammett and Deyrup (1932), but in good agreement with the above estimate from the H_0 jump.

All these estimates are high compared with those from cryoscopy and conductance, to be discussed in the following Chapter. More recently Vinnik and Ryabova (1964) made an estimate from H_0, which came much closer to the estimates from cryoscopy and conductance, by using the Brand equation (2.50) between 95 and 99 per cent acid and assuming that the activity coefficients in > 99 per cent sulphuric acid may be taken to be calculable from the Debye-Hückel limiting law. Doubts have, however, recently been raised about the correctness of the analysis of cryoscopic data (Wyatt, 1969). The question will be discussed further in Section 3.2.5.

CHAPTER 3

Protonation of Very Weak Bases

3.1. Introduction

Most reactions that occur in strongly acid media are acid-catalysed, i.e. they involve a preprotonation of the reacting molecules, sometimes followed by more complex types of ionization. The interpretation of the kinetics of such reactions is facilitated by the knowledge of the extents of protonation of the reactants, and the formulation of the reaction mechanisms requires the knowledge of the protonation sites in cases where there is more than one alternative. This information is obtainable from studies of protonation of related compounds, which are stable under similar conditions. If a sufficiently wide range of basic compounds of a given class is studied, it is often possible to predict the basicity of the unstable members of the class quantitatively by use of linear free energy relationships. In some instances information about the reacting species themselves may be obtained by extrapolation of kinetic measurements to zero time, or by studying the compounds at temperatures too low for the reaction to occur. Protonation is also of importance when it serves to deactivate some compounds with respect to electrophilic substitution (particularly with respect to aromatic sulphonation in sulphuric acid media).

Apart from this direct relevance of basicity measurements to studies of the kinetics of reactions, studies of acid/base strength provide information on molecular structure which is of fundamental importance in chemistry in general, and therefore also highly relevant to questions of reaction mechanism. Many current ideas about the effect of structure on reactivity have developed from information on acid/base strength of compounds. The broader significance of studies of very weak bases in sulphuric acid/water mixtures and oleum solutions lies in the fact that these media offer an extremely wide and continuous range of acidity, joining on to the dilute aqueous range, in which a large number of normally neutral compounds are protonated. The measurements of the extents of protonation in these strongly acid media, in conjunction with the knowledge of the acidity functions that the various types of base obey, result in a large number of pK values, whose correlation often leads to new insights into the effects of structure on reactivity. These correlations may also prove useful in establishing the site of protonation in complex basic molecules.

Despite the ambiguities in the significance of pK values of very weak bases that may arise from the incorrect choice of the acidity function which they are believed to follow most closely, the pK values obtained by studies of the acid/base equilibria of these bases in strong aqueous acids are the only quantitative measures of their basicity having any degree of reliability. The results of numerous other methods that have been used to measure basicities of very weak bases, such as titrations in non-aqueous solvents, solubility and distribution measurements, studies of hydrogen-bonding spectral shifts, and of heats of mixing, often with non-protonic acids, have been reviewed recently in considerable detail by Arnett (1963). An examination of this vast material has shown that these other methods seldom, if ever, provide a quantitative measure of basicity, and that even orders of increasing or decreasing basicity deduced from different types of measurement often differ. In view of this, the development of methods for the study of acid/base equilibria in strong aqueous acids, and in particular in sulphuric acid/water mixtures and in pure sulphuric acid as solvent, has been a steady preoccupation of chemists since the turn of this century, and has been intensified in the past two decades. The advantages of sulphuric acid solutions for these determinations over other aqueous acids are in their stability, relatively low reactivity and ease of handling, apart from the already mentioned very wide range of acidity that they offer. In addition, for measurements in concentrated acid, it is frequently possible to compare directly several methods of determination.

One of the simplest and most widely used criteria of simple protonation in strongly acid media is that substances dissolved as simple bases are recoverable unchanged upon dilution with water. With sulphuric acid solutions this is usually done by pouring the solution onto ice, in order to minimize any secondary reactions which might occur in dilute acid at the higher temperatures resulting from the large heat of mixing evolved. Another criterion is the freezing-point depression of pure sulphuric acid, which for fully protonated bases is twice that for non-electrolyte solutes. Some basic substances do not satisfy these criteria, because they undergo more complex modes of ionization in concentrated acid (although in some instances of complex ionization dilution with water may regenerate the original compound). These more complex modes of ionization are also of very great interest in studies of chemical reactivity in strongly acid media. They will be treated separately in Chapter 4.

Some of the material in this Chapter is covered by the most recent review of the basicities of very weak bases by Arnett (1963). Also several reviews of the behaviour of solutes in concentrated sulphuric acid have been published by Gillespie and his collaborators over the past fifteen years (Gillespie and

Leisten, 1954a; Gillespie and Robinson, 1959 and 1965). In this rapidly developing field, however, a number of new results have become available in recent years, not only from studies in sulphuric acid/water mixtures and in pure sulphuric acid, but also from studies in superacid media, such as fluorosulphuric acid, which supplement information obtained for sulphuric acid, especially in regard to the structure of conjugate acids. Although superacid media are outside the scope of this monograph, some of the structural information obtained for these media will also be mentioned, in view of its direct relevance to the subject matter of this Chapter.

The first part of this Chapter will be devoted to a discussion of the methods used in determining basic strengths in strong aqueous acids in general, and in concentrated sulphuric acid in particular. The second part of the Chapter will present, firstly, the main current ideas on substituent effects and linear free-energy relationships, and secondly, the available information on the basic strengths of various types of bases and on the structure of their conjugate acids.

3.2. Methods of Determination of Basicities in Sulphuric Acid Media

Several methods of determination of pK values to be discussed in this Section are applicable to any strong acid/aqueous medium, whereas others have been developed for use in the region of pure sulphuric acid and are based on some of its specific properties. In the first category are mainly spectroscopic methods (ultraviolet spectrophotometry, nuclear magnetic resonance spectroscopy, and Raman spectroscopy) and the distribution (extraction) method. In the second category are the cryoscopic method and the conductometric methods. They will be discussed in this order.

3.2.1. THE SPECTROPHOTOMETRIC METHOD

Absorption in the ultraviolet spectral region of many aromatic compounds shows considerable changes when they become protonated in acid solution. Typical examples of this kind of behaviour are all Hammett indicators. The methods of study of acid/base equilibria of this type of compound have already been discussed in Chapter 2, in conjunction with the determination of acidity functions. The method and the problems encountered are exactly the same when the object of the measurements is the determination of an unknown pK value.

The basic quantity required is again the ionization ratio of the base, and the way from this to the pK_{BH^+} value is via the same equation used in defining the acidity function, except that this is now solved for the pK_{BH^+}:

$$pK_{BH^+} = H_0 + \log\frac{[BH^+]}{[B]} \quad (3.1)$$

assuming here that the base follows the H_0 acidity function.

The difficulties of calculating the ionization ratios from absorption spectra due to medium effects have already been discussed in Chapter 2. The absorptions of the free base and the conjugate acid must be determined at values of acidity functions at least two units removed from the value at half-protonation, to ensure that no more than 1 per cent of the other form is present.

In determining the pK_{BH^+} value of a new type of base, if this quantity is to have thermodynamic significance, one is faced with the problem of choosing the right acidity function, i.e. a function which is followed by the base under consideration. As a rule an acidity function defined by means of a closely structurally related type of indicator base would be chosen. The first requirement is that the slope of the plot of the logarithms of ionization ratios against the chosen acidity function should be minus unity. This is a necessary condition, but is not sufficient in general, although it seems to be so in the majority of cases. The second requirement is that the activity coefficient behaviour of the base studied and its conjugate acid should be closely similar to that of the indicator bases used in defining the acidity function. The activity coefficient behaviour of many bases is not known. Most studies have been carried out on indicator bases, as already described in Chapter 2. Therefore, if there is doubt concerning the choice of the right acidity function for a particular base, Arnett (1963) suggested that it would be advisable to report the percentage of sulphuric acid at half-protonation, as well as the value of the chosen acidity function. This would facilitate the conversion of the pK_{BH^+} value to a new scale, if a better acidity function for the particular type of base is subsequently defined.

The best known example of a base that satisfies the first requirement, i.e. gives a unit slope in the log$I$ vs. $-H_0$ plot, but does not satisfy the second requirement in sulphuric acid solution, is benzoic acid. Solubility measurements of Hammett and Chapman (1934) have shown that its solubility increases sharply at sulphuric acid concentrations which are insufficient for the protonation of the acid. Therefore the activity coefficient of the base shows here a sharp drop compared with aniline indicators, and consequently the H_0 value at half-protonation of this base is not equal to the thermodynamic pK_{BH^+} value. In perchloric acid solutions this anomaly in the activity coefficient behaviour is absent, and the thermodynamic pK_{BH^+} value for benzoic acid is obtainable without difficulty (Yates and Wai, 1965). It has been suggested (Arnett, 1963) that the activity coefficient behaviour of benzoic acid (and of nitrobenzene which behaves similarly) could be attributed to hydrogen bonding interaction between the base and the hydronium ion involving the two basic sites in the 1,3-position, as in (**3.1**):

$$C_6H_5-C\begin{subarray}{c}\diagup O\cdots H\diagdown \\ \diagdown O\cdots H\diagup\end{subarray}\overset{+}{O}-H$$
$$|$$
$$H$$

(**3.1**)

The same type of interaction would be expected in perchloric acid solutions, and hence this suggestion cannot account for the difference in the behaviour of benzoic acid in the solutions of the two acids. Hydrogen bonding with sulphuric acid molecules might, however, be the explanation, the following interaction (3.2) becoming possible in concentrated acid:

$$C_6H_5-C\begin{matrix}O\cdots H-O\\O\cdots H-O\\|\\H\end{matrix}S\begin{matrix}O\\O\end{matrix}$$

(3.2)

Hammett and Chapman (1934) did in fact find that a new solid phase is formed above 80 per cent acid, probably the 1:1 compound of benzoic acid with sulphuric acid. Perchloric acid cannot act as a bidentate hydrogen bond donor.

The standard method of determining the slopes of the $\log I$ vs. H_0 plots is the method of least squares. If the linear equation is represented by

$$\log I = mH_0 + c \tag{3.2}$$

then, apart from the slope m, the least squares fit yields the pK value directly, as the value of the acidity function at half-protonation, i.e. for $\log I = 0$,

$$pK_{BH^+} = H_0^{(1/2)} = -\frac{c}{m} \tag{3.3}$$

where $H_0^{(1/2)}$ means the value of the acidity function at half-protonation.

The spectrophotometric method using u.v. absorption is restricted almost exclusively to aromatic compounds, although even amongst this group there are some whose absorption is not changed sufficiently by protonation for this method to be applicable (e.g. p-nitrobenzoic acid and p-nitrobenzamide). The absorption of the carbonyl group in the lower ultraviolet can also be used in studying some aliphatic compounds containing this group, but difficulties are encountered owing to low absorption coefficients.

3.2.2. THE N.M.R. SPECTROSCOPIC METHOD

Protonation should lead to a downfield chemical shift of the nuclear magnetic resonance lines of nuclei in a basic molecule, because it results in a reduction of electron density in the molecule, especially in the vicinity of the basic centre. This was indeed found to be so by Grunwald *et al.* (1957a) for the protonation of methylamine. The change in the chemical shift with pH was found here to follow a sigmoid "titration curve", because the chemical shift of the rapidly exchanging base and conjugate acid is a weighted mean of the chemical shifts of the protons in the two separate species. The pH at the mid-point of the sigmoid curve corresponded approximately to the pK_a of the conjugate acid of methylamine. An early attempt by Grunwald *et al.* (1957b) to determine the basicity constants of methanol and ethanol by this method in sulphuric acid/water mixtures failed, because no sigmoid curves were obtained with these substances. The failure was ascribed to large medium effects on chemical shifts. The interest in the possibilities of this method for basicity determinations was revived again by Taft and Levins (1962), who studied the ^{19}F resonance of *p*-fluoroacetophenone and *p*-fluorobenzamide in sulphuric acid/water mixtures and deduced the pK_{BH^+} values from the point of maximum slope of the sigmoid curves. Good agreement with the u.v. spectrophotometric determination was found. Edward *et al.* (1962) studied the intramolecular chemical shift of the α- and the β-CH protons of propionic acid and propionamide in sulphuric acid/water mixtures and were able to deduce reasonable pK_{BH^+} values. Ethanol and diethylether were again found to give curves without an inflexion point under the same experimental conditions. No inflexion point could be obtained with tetrahydrofuran either (Arnett, 1963). The position therefore is that the n.m.r. method is applicable to some compounds, but not to others. Proton shifts are of widest interest, and have recently been used to estimate the basicities of several groups of compounds.

There are, in principle, two possibilities for the measurement of the changes of chemical shift caused by protonation: either the intramolecular shift between two groups may be used, one of which is closer to the basic centre than the other, or the intermolecular shift relative to a suitably chosen reference, internal or external. The advantage of the first method is that it does not require the introduction of any other substances into the solutions, but it suffers from the disadvantage that intramolecular shifts, being relative, are smaller than absolute downfield shifts, since resonances of all groups shift downfield when protonation takes place. The use of an external reference poses no problems, apart from the necessity to carry out diamagnetic susceptibility corrections, and even this may not be necessary (Deno and Wisotsky, 1963). The use of an internal reference does present certain problems. Firstly, the introduction of the internal reference may alter the acidity of the solutions, and, secondly, the reference itself may be affected by changes of acidity and by medium effects. The reference should preferably give only one sharp and strong line. Trimethylammonium and *t*-butylammonium ions have so far been used (Laughlin, 1967; Liler, 1969), because they are fully protonated in very dilute acid and remain so at still higher acidities. Also the tetramethylammonium ion was used (Armstrong and Moodie, 1968). All these references give a single sharp line, which does not seem to be subject to medium shifts, and they can be used at concentrations of 0·1 M or less. Even at these low concentrations, if the amine is introduced into already made up sulphuric acid/water mixtures, changes in acidity will result which are not negligible. Therefore, it has been suggested (Liler, 1969) that a 0·1 M solution of the chosen substituted ammonium sulphate should be used in preparing sulphuric acid/water mixtures, rather than pure water. The effect of such low concentrations of ammonium salts on the acidity of sulphuric acid/water mixtures is negligible compared with other sources of error inherent in the method.

The ionization ratios are obtained from the equation

$$I = (\Delta\nu - \Delta\nu_B)/(\Delta\nu_{BH^+} - \Delta\nu) \tag{3.4}$$

in which $\Delta\nu$, $\Delta\nu_B$ and $\Delta\nu_{BH^+}$ are the chemical shifts, relative to the reference, of the rapidly exchanging base and conjugate acid, and of the base and of the conjugate acid, respectively. The latter two quantities must be measured at such acidities of the medium at which the other form is present to the extent of 1 per cent at most, i.e. at acidity function values which are at least two units removed from the point of half-protonation. Chemical shifts are measurable with high accuracy. An accuracy of $\pm 0\cdot 5$ Hz can easily be achieved using a 60 MHz spectrometer, and better accuracies are possible with spectrometers operating at higher frequencies, which are becoming more generally available. Difficulties may arise in measuring the chemical shift of the conjugate acid if a medium drift with increasing acidity is observed, as was reported, for example, for n-butyramide in perchloric acid solutions (Armstrong and Moodie, 1968).

The estimation of ionization ratios from n.m.r. chemical shifts is thus both simple and accurate, and in this respect the n.m.r. method is superior to the u.v. spectrophotometric method. It has, however, a disadvantage in that the concentrations of the bases may have to be as high as $0\cdot 1$ M for comfortable detection of the lines (as compared with concentrations of 10^{-3} M to 10^{-4} M which are commonly used in spectrophotometric work). Therefore the pK values so obtained should preferably be described as apparent, rather than thermodynamic. With improvements in n.m.r. techniques, such as the development of spectrometers operating at higher frequencies, with larger sample tubes, and with the possibility of accumulating spectra, it should be possible to bring the concentrations of solutions down to the level used in spectrophotometric work.

3.2.3. THE RAMAN SPECTROMETRIC METHOD

In an effort to find a method suitable for the determination of basicities of common aliphatic compounds, Deno and Wisotsky (1963) have explored the possibility of using measurements of the intensities of Raman lines originating from the base and the conjugate acid. The high concentrations of the base that have to be used (5–20 per cent by weight) put a severe limitation upon the precision of the method. Also, in estimating areas of the bands, corrections for solvent background are up to a point arbitrary. The method has been used so far only to check estimates of basicity obtained by other methods (e.g. for acetone, by u.v. spectroscopy and n.m.r.). This it has done successfully, which demonstrates its inherent reliability.

3.2.4. THE DISTRIBUTION OR EXTRACTION METHOD

A distribution method for the determination of basicities, which is in principle applicable to any type of base which is sparingly soluble in water, has been suggested by Arnett and Wu (1960a) and further developed by Arnett et al. (1962). The distribution coefficients between aqueous acids and non-polar inert organic solvents are dependent on acid concentration, because protonation or formation of salts enhances the solubility of the base in the acid layer. Let the distribution constant of the base between water and the organic layer be given by

$$K_D = [B]_0/[B]_{aq} \qquad (3.5)$$

then the distribution ratio between the same organic solvent and an aqueous acid in which the base is partially protonated will be

$$D = \frac{[B]_0}{[B]_{aq} + [BH^+]} \qquad (3.6)$$

since the salt will be soluble in the polar aqueous layer only. Given the protonation constant

$$K_{BH^+} = h_0 \frac{[B]_{aq}}{[BH^+]} \qquad (3.7)$$

the distribution ratio becomes

$$D = K_D - Dh_0/K_{BH^+} \qquad (3.8)$$

and the ionization ratio of the base is given by

$$I = \frac{K_D - D}{D} \qquad (3.9)$$

Assuming that the base follows the H_0 acidity function, we obtain

$$pK_{BH^+} = H_0 - \log[D/(K_D - D)] \qquad (3.10)$$

and the slope of the logarithmic term on the right vs. H_0 should be close to unity. The method was first applied to aliphatic and aromatic ethers, which were found to follow H_0 closely. Experimentally the method was perfected by the use of gas/liquid chromatography for the analysis of the organic layer. The concentrations of the aqueous acid layer are obtained by difference.

An obvious limitation of this method is encountered for compounds which show solubility increases in acid solutions due to other effects apart from protonation. Such examples are benzoic acid and nitrobenzene, which show anomalous increases of solubility in aqueous sulphuric acid (discussed in Section 3.2.1). The distribution method thus appears to be suitable for certain types of compounds, but caution is needed in the interpretation of the results. A cautionary note has already been sounded by Deno and Wisotsky (1963), who regard the method as needing independent substantiation, because of its indirectness.

A closely related method of vapour pressure measurements, based on the distribution of volatile bases between the vapour phase and the acid solution, is applicable when Henry's law is obeyed by the unprotonated base and when the Henry's law constant does not change greatly with acid concentration (Jaques and Leisten, 1964). Under these conditions, since BH^+ is involatile, we may write

$$\frac{[B]}{[BH^+]} = \frac{p}{p_0 - p} \quad (3.11)$$

where p is the observed vapour pressure and p_0 would be the vapour pressure if all the base were unprotonated. This latter value has to be chosen so as to give a linear correlation between H_0 and $\log\{p/(p_0 - p)\}$. This obviously introduces some uncertainty into the method.

3.2.5. THE CRYOSCOPIC METHOD

The accuracy of both the cryoscopic and the conductometric methods of determination of basic strengths of weak bases in pure sulphuric acid depends on the knowledge of the concentrations of species arising from the self-dissociation of the solvent. There are at least three important self-ionization equilibria:

$$2H_2SO_4 \rightleftharpoons H_3SO_4^+ + HSO_4^- \quad (3.12)$$

$$2H_2SO_4 \rightleftharpoons H_3O^+ + HS_2O_7^- \quad (3.13)$$

$$H_2SO_4 + H_2S_2O_7 \rightleftharpoons H_3SO_4^+ + HS_2O_7^- \quad (3.14)$$

The first is autoprotolysis, the second was termed ionic self-dehydration by Gillespie (1950b) and corresponds to the dissociation $H_2SO_4 \rightleftharpoons H_2O + SO_3$, with further reaction of the two products with sulphuric acid, and the third is the dissociation of disulphuric acid as an acid in the sulphuric acid solvent. In addition there may be the equilibrium of incomplete protonation of water:

$$H_2O + H_2SO_4 \rightleftharpoons H_3O^+ + HSO_4^- \quad (3.15)$$

This equilibrium was postulated by Gillespie (1950a) on the basis of the cryoscopic behaviour of hydronium hydrogen sulphate as compared with that of metal hydrogen sulphates, and was at first thought of in terms of incomplete protonation. The more favoured view at present is complete protonation, accompanied by ion pair formation, possibly owing to the unusual hydrogen bonding ability of the H_3O^+ ion (Gillespie and Robinson, 1965).

The equilibrium constants of all these equilibria need to be known for a complete description of the solvent medium. It is impossible to obtain all these constants from cryoscopic data alone without making certain assumptions. The freezing-point depression curves for water (hydronium hydrogen sulphate), disulphuric acid and potassium hydrogen sulphate in pure sulphuric acid are shown in Fig. 3.1, together with the line for sulphuryl chloride in slightly aqueous sulphuric acid, in order to provide a basis for discussion.

In cryoscopic measurements to determine the extent of ionization of solutes,

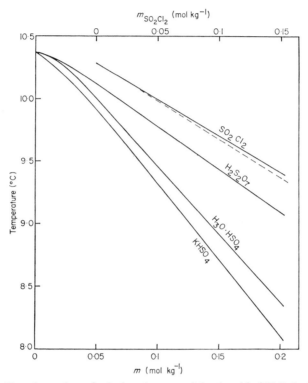

FIG. 3.1. Freezing points of solutions in pure sulphuric acid of $H_2S_2O_7$, $H_3O \cdot HSO_4$ and $KHSO_4$ (Bass and Gillespie, 1960) and of sulphuryl chloride in slightly aqueous acid (Gillespie et al., 1950a). The dotted line gives the ideal non-electrolyte slope.

in order to minimize the effect of self-dissociation, it was customary to work in slightly aqueous sulphuric acid (Hammett and Deyrup (1933), for example, used acid freezing at 9·8–10°C) and to calculate the van't Hoff's factor i from the equation

$$\frac{\Delta\theta}{m} = ik_f \qquad (3.16)$$

($\Delta\theta$—freezing-point depression, m—molality of solute, k_f—cryoscopic constant), assuming that the water present largely suppresses the self-dissociation of the solvent, until Brayford and Wyatt (1955) showed that the presence of the electrolyte hydronium hydrogen sulphate introduced complications in certain cases. The non-electrolyte solute sulphuryl chloride is apparently unaffected by the presence of this electrolyte, and the slope of the observed freezing-point depression line (Fig. 3.1) is close to the theoretical ideal slope of 6·12 K mol^{-1} kg. Simple bases which ionize to produce a cation and the solvent anion give slopes close to twice this figure.

Gillespie et al. (1950a) developed considerably the cryoscopic method in sulphuric acid, both experimentally and theoretically, pointing out that the van't Hoff's factor i is not simply a measure of the number of ionic (or neutral) species produced by one mole of solute, but a composite measure of both the number of these particles and the deviations of the solution from ideality. It was shown that the freezing-point depressions, $\Delta\theta$, are more correctly represented by the equation

$$\frac{\Delta\theta}{\Delta m_2} = vk_f(1 - 0.002\theta)\left\{1 + \frac{(2s - v)m_2}{2m_1}\right\} \tag{3.17}$$

in which v is the number of particles produced by one mole of solute, s is the number of solvent molecules used in the reaction with and in the solvation of the solute species, and m_1 and m_2 are the molalities of the solvent and solute, respectively. Gillespie (1950a and 1950b) further estimated the ionization constants of H_2O as a base and of $H_2S_2O_7$ as an acid, and the total molality of all self-dissociation species as 0·043. Hence the autoprotolysis constant was also obtained. These calculations were later refined, essentially by a method of trial and error, using more precise freezing-point depression data (Bass et al., 1960a), and the following values were arrived at as the best:

$$K_{ap}^{10} = [H_3SO_4^+][HSO_4^-] = 1.7 \times 10^{-4} \text{ mol}^2 \text{ kg}^{-2}$$

$$K_{id}^{10} = [H_3O^+][HS_2O_7^-] = 3.5 \times 10^{-5} \text{ mol}^2 \text{ kg}^{-2}$$

$$K_a^{10} = \frac{[H_3SO_4^+][HS_2O_7^-]}{[H_2S_2O_7]} = 1.4 \times 10^{-2} \text{ mol kg}^{-1}$$

These estimates have recently been questioned by Wyatt (1969), because they are based on a number of simplifications and corrections which, although apparently reasonable, do not necessarily lead to a quantitatively correct result. Cryoscopic data are insufficient to afford unique values of all these constants, but are rather consistent with a range of values. In particular, K_{ap} values of up to 70 per cent greater than that given above appear quite possible on the basis of Wyatt's calculations. Wyatt (1969) suggests that additional information for the evaluation of self-dissociation equilibrium constants may be obtained by precise measurements of the H_0 acidity function in solutions of hydronium hydrogen sulphate and metal hydrogen sulphates in sulphuric acid. (Some such measurements have already been reported by Gillespie and Robinson, 1965.) While this is possible in principle, it is questionable whether the accuracy required is experimentally attainable. Almost as a rule, nitro-compounds are used as indicators in this acidity region, and their activity coefficients are known to be particularly susceptible to ionic strength effects (Brayford and Wyatt, 1955). Wyatt (1969) favours the higher estimates of the concentrations of autoprotolysis species made by Hammett and Deyrup (1932) and by Brand *et al.* (1959) from acidity function measurements (Section 2.6.2, p. 56). An examination of the freezing-point depression curves in Fig. 3.1, however, shows that their variation with concentration of the basic solutes ($H_3O \cdot HSO_4$ and $KHSO_4$) becomes linear beyond about 0·1 molal, which suggests that the autoprotolysis is largely suppressed at these solute concentrations. This is consistent with the estimate of the autoprotolysis constant given above. The question will further be discussed in Section 3.2.6.

In view of the possible uncertainties in the autoprotolysis constant of sulphuric acid, the ν values obtainable for weak bases from cryoscopic data should be regarded as only approximate. The base ionization constant of weak bases in pure sulphuric acid is given by

$$K_b = \frac{[BH^+][HSO_4^-]}{[B]} = \frac{(\nu - 1)[HSO_4^-]}{(2 - \nu)} \qquad (3.18)$$

since the degree of dissociation of the solute $\alpha = \nu - 1$. $[HSO_4^-]$ is here the total concentration of HSO_4^- ions, including that remaining from the incompletely suppressed autoprotolysis of the solvent. These K_b values can be recalculated to pK_{BH^+} values referred to the H_0 acidity function scale as explained in Section 3.2.8.

Cryoscopic studies of solutions in dideuteriosulphuric acid have also led to estimates of the solvent self-ionization equilibrium constants (Flowers *et al.*, 1958). The following values were suggested:

$$K_{ap}^{14} = 2 \cdot 9 \times 10^{-5} \text{ mol}^2 \text{ kg}^{-2}$$
$$K_{id}^{14} = 5 \cdot 0 \times 10^{-5} \text{ mol}^2 \text{ kg}^{-2}$$
$$K_a^{14} = 2 \cdot 8 \times 10^{-3} \text{ mol kg}^{-1}$$

Since similar assumptions were made in obtaining these values as with the protio acid, they should be taken with equal reserve.

Several descriptions of the apparatus for cryoscopic studies in sulphuric acid are available in the literature (Hammett and Deyrup, 1933; Newman *et al.*, 1949; Gillespie *et al.*, 1950a). Commonly, the method of cooling curves, using a Beckmann thermometer or a platinum resistance thermometer, is employed. The solutions show a considerable tendency to supercooling, and to prevent this the wall of the freezing-point tube is touched by dry ice when temperatures close to the freezing point are reached (Gillespie *et al.*, 1950a). Whenever supercooling occurs, a supercooling correction should be applied, given by $\delta T = ks\theta$, where s is the amount of supercooling, θ is the observed freezing-point depression, and k is a constant that must be determined experimentally for the cryoscope used (Gillespie and Oubridge, 1956). Apart from this method of cooling curves, the more accurate equilibrium method of cryoscopy has also been applied to solutions in sulphuric acid by Bass and Gillespie (1960). It consists of equilibrating the solid solvent with the solution at a fixed temperature, and then analysing the solution to find the concentration of the solute. The easiest and the most accurate method of analysis is the measurement of the conductance of the solution. The concentration is then determined from a known conductance/concentration curve. A number of solutes were studied in this way by Bass and Gillespie (1960). A very recent review of the cryoscopic method by Gillespie and Robinson (1968) is available.

The usefulness of cryoscopic studies in sulphuric acid goes far beyond the determination of the degrees of protonation of weak bases, and is particularly great in studying complex modes of ionization of certain solutes. This application will be discussed in the next Chapter.

3.2.6. THE CONDUCTOMETRIC METHOD

The conductometric method for determining the strengths of very weak bases in sulphuric acid is based on the special mechanism of conduction operating in the pure acid as solvent. Weak bases are compared with strong bases, taking into account the fact that the unionized part also affects the conductance, as the non-electrolytes do. It is essential therefore to discuss first the mechanism of conduction in pure sulphuric acid and the conductometric behaviour of strong bases and of non-electrolytes. As correcting for the autoprotolysis of the solvent is as important for the conductometric method of determining basic strength as it is for the cryoscopic method, it will also be shown that an estimate can be made of the autoprotolysis constant of pure sulphuric acid from conductometric data on the basis of a few simple assumptions. The conductometric method itself will then be described.

3.2.6.1. *Mechanism of Conduction in Pure Sulphuric Acid*

As many of the properties of concentrated sulphuric acid discussed in Chapter 1 show, sulphuric acid is a highly associated liquid, with a high degree of internal order. Metal sulphates (M_2SO_4) and many organic weak bases ionize in this solvent according to the equations

$$M_2SO_4 + H_2SO_4 \rightleftharpoons 2M^+ + 2HSO_4^- \quad (3.19)$$

$$B + H_2SO_4 \rightleftharpoons BH^+ + HSO_4^- \quad (3.20)$$

producing the solvent anion, analogously to the behaviour of bases in water. Hammett and Lowenheim (1934) were the first to show by transport number measurements on solutions of barium sulphate in pure sulphuric acid that this solvent anion has an unusually high mobility, whereas the cation hardly contributes to the transport of electricity at all. Later, such measurements were carried out on solutions of a whole range of metal hydrogen sulphates by Gillespie and Wasif (1953a) and on solutions of some organic strong bases (acetic acid and benzoic acid) by Wasif (1955). All these measurements led to the same conclusion, that the anion is virtually the only conductor of electricity in these solutions. Transport numbers of cations vary between 0·030 for potassium and 0·003 for strontium. A similar value was found for the acetic acidium cation (0·022). The exceedingly low cation mobilities are readily ascribed to the high viscosity of the solvent, whereas the abnormally high mobility of the HSO_4^- ion is explained in terms of a proton-jump conduction mechanism, such as is generally accepted for the hydrogen and the hydroxide ions in water. Schematically the process may be represented as follows:

$$\begin{array}{c} O\diagdown\diagup OH \quad HO\diagdown\diagup O\cdots HO\diagdown\diagup O \\ S S S \\ O\diagup\diagdown O^-\cdots HO \quad \diagup\diagdown O\cdots HO\diagup\diagdown O \end{array}$$

The charge is thus transferred through the solution without migration of these ions, but the molecules taking part in the conduction have to rotate to some extent in order to take up the right position for passing on the proton. By analogy with aqueous solutions the solvent cation $H_3SO_4^+$, also present in oleum solutions, would be expected to conduct by the same mechanism. Evidence that this is so comes from the fact that the conductance minimum occurs very nearly at the composition of pure sulphuric acid. In fact, the slight displacement of the conductance minimum to the aqueous side of the 100 per cent acid suggests that this ion may be more highly conducting than the HSO_4^- ion. It was estimated from the position of the minimum conductance and conductance measurements on oleum solutions that the ratio of the mobilities of the two ions is 1·45 (Gillespie et al., 1960).

The facts established by transport number measurements are wholly consistent with electrical conductance measurements on solutions of metal hydrogen sulphates which show that the effect of cations on conductance is only a second order one (see Fig. 3.2). Solutions of all metal hydrogen sulphates show virtually the same conductivity up to about 0·1 M, and only at higher concentrations do the values diverge (Gillespie and Wasif, 1953c). The conductivity of hydronium hydrogen sulphate falls in line with the conductivities of other hydrogen sulphates, unlike its cryoscopic behaviour. Ion pairs $H_3O^+HSO_4^-$, which are believed to be responsible for the cryoscopic behaviour, are thought to be probably not sufficiently long-lived to affect conductance (Gillespie and Robinson, 1965). The viscosities and the densities of the same solutions were measured and discussed by Gillespie and Wasif

(1953b). It was shown that the divergence between the conductances of metal hydrogen sulphates at higher concentrations can largely be eliminated in plots of $\kappa \times \eta^{1/2}$. This is an indication that viscosity affects the mobility of the anion, most likely through its effect on molecular rotation, which is probably the slow stage of the conduction process.

Molar conductances of metal hydrogen sulphates, calculated by substracting the contribution of the solvent autoprotolysis ions by Flowers *et al.* (1960a),

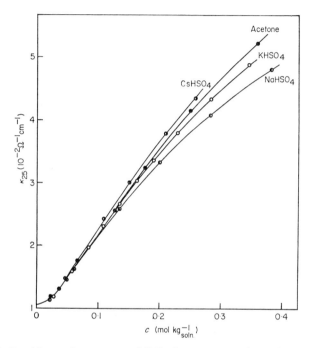

FIG. 3.2. Specific conductances at 25°C of some strong bases in pure sulphuric acid (Bass *et al.*, 1960a).

were found not to follow the conductance equation of Robinson and Stokes (1955), p. 151, but to decrease rapidly with increasing electrolyte concentrations. It was thought difficult to account for this in terms of the proton-jump mechanism of conduction, and this led Wyatt (1961) to suggest that there is an additional contribution to conductance in the pure acid by "asymmetric dissociation", i.e. by a bias in the dissociation process in the direction of the electric field. An expression for this kind of conduction was derived:

$$\kappa_D = 3.733 \times 10^{-11} k_R [H_3SO_4^+][HSO_4^-](\phi a^2/T) \quad (3.21)$$

where k_R is the rate constant for recombination of the autoprotolysis ions, ϕ the effective internal field at an external field of 1 V cm^{-1}, and a the distance of approach of solvated solvent self-ions needed for their neutralization to occur. It is important to note that this contribution to conductance is proportional to the K_{ap} value. It was estimated that for pure sulphuric acid this may amount to 40 per cent of the total conductance, and with this the mobility of the HSO_4^- ion remains constant up to about 0·05 mol litre^{-1}.

The ionic contribution to conductance by solvent autoprotolysis ions is given by

$$\kappa_i = 10^{-3}(\lambda_+ c_+ + \lambda_- c_-) \tag{3.22}$$

where c_+ and c_- stand for $[H_3SO_4^+]$ and $[HSO_4^-]$ in order to simplify the equation, and λ_+ and λ_- are the respective ionic conductances. Since $c_+ = c_- = K_{ap}^{1/2}$ in the pure solvent,

$$\kappa_i = 10^{-3} K_{ap}^{1/2}(\lambda_+ + \lambda_-) \tag{3.23}$$

K_{ap} being here in mol^2 litre^{-2}. The ionic conductance thus depends on $K_{ap}^{1/2}$.

This difference in the dependence of the two conductance contributions on K_{ap} affords a means in principle of distinguishing between the two mechanisms, or of detecting the "asymmetric dissociation" contribution, if a means of varying the autoprotolysis constant (i.e. the ionic product) of the solvent could be found. This may be achieved by introducing an inert diluent into the self-dissociated solvent, since the ionic product K_{ap} is given by

$$K_{ap} = K'_{ap} c^2 \tag{3.24}$$

where c is the molarity of the solvent molecules, and K'_{ap} is the equilibrium constant for autoprotolysis (equation 3.12, p. 67):

$$K'_{ap} = \frac{c_+ c_-}{c^2} \tag{3.25}$$

This possibility will further be discussed in connection with the conductance of non-electrolyte solutions. At this point it may be noted that an underestimate of K_{ap} necessarily leads to an overestimate of the mobilities of the solvent self-ions, if purely ionic conduction is assumed. The higher estimates of K_{ap} that Wyatt (1969) favours more recently eliminate the large variation of equivalent conductances of the HSO_4^- ions in dilute solution, because they do not require in the first instance such high Λ values in the region of the pure acid, as were calculated by Flowers et al. (1960a). The need for an "asymmetric dissociation" contribution to conductance is thereby also eliminated (Wyatt, 1969).

Even accepting the low autoprotolysis constant of sulphuric acid, as estimated by Gillespie (1950b) and Gillespie et al. (1960), the case for an "asymmetric dissociation" contribution is not seen to be clear if "excess" equivalent ionic conductances of the H_3O^+ ion in water (in which the "asymmetric dissociation" contribution should be negligible, because of its small ionic product—(Wyatt, 1961)) and the equivalent ionic conductances of the HSO_4^- ions in pure sulphuric acid are compared on the basis of an approximate mole ratio of the conducting ion to the solvent. This ratio should be the determining factor in proton-jump conduction. Such a comparison is shown in Fig. 3.3. "Excess" equivalent conductances of the H_3O^+ ion were calculated from the equivalent conductances of aqueous hydrochloric acid and the cation transport numbers, using data from Robinson and Stokes (1955)—pp. 361 and 159, respectively—and were corrected for a normal conduction of 90 Ω^{-1} cm^2 (assumed independent of concentration)—ibid., p. 116. The equivalent ion conductances of the HSO_4^- ions in sulphuric acid are those calculated by

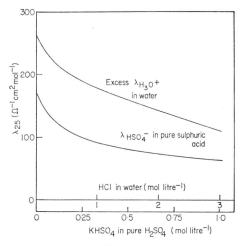

FIG. 3.3. A comparison of proton-jump conductances of H_3O^+ ions in water and HSO_4^- ions in sulphuric acid on an approximate mole ratio basis.

Flowers et al. (1960a), using their estimate of the autoprotolysis constant. Both curves are plotted up to an approximate mole ratio of conducting ion to solvent of 0·057 (i.e. 3:52 for HCl in water and 1:17·6 for $KHSO_4$ in sulphuric acid). It can be seen that the general trend of both curves is the same. The percentage drop in ion mobility up to the highest concentration plotted is 58 per cent for the H_3O^+ ion in water and 62 per cent for the HSO_4^- ion in sulphuric acid, i.e. both are of the same order of magnitude. The similarity of proton-jump conduction behaviour in the two solvents is such that there is little ground to suppose that there are any special effects in sulphuric acid solution.

3.2.6.2. Conductometric Behaviour of Strong Bases

As already pointed out, the conductance of strong electrolytes, metal hydrogen sulphates or fully protonated bases, shown in Fig. 3.2, is wholly consistent with the finding that HSO_4^- ions are virtually the only conductors of electricity in these solutions. A common characteristic of all the conductance curves is an inflexion point at about 0·16 M. If this inflexion point has any significance at all, this must be that it separates the region in which conductance is wholly or largely due to the solvent autoprotolysis ions (with a possible contribution from "asymmetric dissociation") from the region in which autoprotolysis is largely suppressed and the conductance is due predominantly to the HSO_4^- ions arising from the strong base. It can be seen from Fig. 3.2 that potassium hydrogen sulphate shows an "average" behaviour amongst strong bases.

3.2.6.3. *Conductometric Estimate of the Autoprotolysis Constant*

In view of the uncertainty in the values of solvent self-dissociation constants of pure sulphuric acid, which was pointed out by Wyatt (1969), and the doubts concerning the presence of the "asymmetric dissociation" contribution raised in Section 3.2.6.1, it is not out of place to try to obtain an estimate of the autoprotolysis constant from conductometric data alone, assuming ionic conduction. Conductance data at 10 and 25°C were used previously by Gillespie and Wasif (1953c) and by Gillespie *et al.* (1960) to estimate the autoprotolysis constant at 25°C on the basis of the then accepted cryoscopic value at 10°C, which, according to Wyatt (1969), appears to be doubtful. There is good evidence that the mobilities of all other ions formed in the self-dissociation of sulphuric acid, apart from autoprotolysis ions, are very small (Gillespie and Wasif, 1953d). This makes conductance particularly suited for the estimation of the autoprotolysis constant.

If the conductance of the pure acid and dilute solutions in it is primarily due to the solvent self-ions, i.e. if it is given by equation (3.23), p. 74, the ionic conductances and the K_{ap} value cannot both be determined from conductance alone without any assumptions. The simplest assumption that can be made, in the absence of a detailed theory of proton-jump conduction, is that the ion conductances should vary with concentration in dilute solution in a smooth and uniform manner, as they do in more concentrated solutions.

Taking potassium hydrogen sulphate as a typical strong electrolyte, the conductance of its solutions should fall to zero at infinite dilution in the absence of solvent autoprotolysis. When such an extrapolation from the vicinity of the inflexion point to zero conductance at zero concentration is attempted, it turns out that there is little ambiguity in the way the line can be drawn, if the resultant curve is to yield uniformly and smoothly decreasing values of the ionic conductance of the HSO_4^- ions. Such an extrapolation is shown in Fig. 3.4(a) and the insert (b) shows the corresponding equivalent ionic conductances of the HSO_4^- ions, calculated from the hypothetical and the real parts of the curve, the whole of the conductance being ascribed to this ion. The value at infinite dilution, λ_0, corresponds to the slope of the tangent

3. PROTONATION OF VERY WEAK BASES 77

to the hypothetical part of the curve at zero concentration. The values of λ_- obtained in this way are the same within 2 per cent as the values quoted by Flowers et al. (1960a). Whether the slope of the λ_- vs. $c^{1/2}$ line should correspond to the Debye-Hückel-Onsager slope or not, is a matter for dispute,

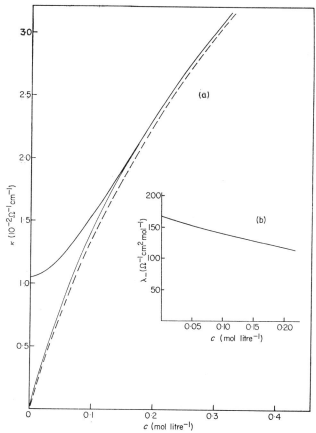

FIG. 3.4. (a) Specific conductance of $KHSO_4$ in pure sulphuric acid (dotted line: extrapolation assuming no self-dissociation of the solvent). (b) Equivalent ionic conductances of HSO_4^- ions corresponding to the dotted line.

because for an ion that conducts wholly by proton-jumps at least some of the effects present in migratory conduction, e.g. the electrophoretic effect, must be absent.

One can now represent the conductance of pure sulphuric acid as equivalent to the conductance of a solution of potassium hydrogen sulphate in the hypothetical undissociated solvent:

$$c_+ \lambda'_+ + c_- \lambda'_- = c_s \lambda_- \quad (3.26)$$

where c_+ and c_- are the molarities of the autoprotolysis ions in the pure acid and λ'_+ and λ'_- their equivalent conductances, c_s is the concentration of the hypothetical solution of potassium hydrogen sulphate having the same conductance as the pure acid and λ_- is the equivalent conductance of the HSO_4^- ion in such a solution. Since $c_+ = c_-$ in the pure acid, the symbol c' may be used for both, and since $\lambda_+ = 1 \cdot 45 \lambda_-$ (Gillespie et al., 1960) we may rewrite equation (3.26) in the form

$$c' \times 2 \cdot 45 \lambda'_- = c_s \lambda_- \qquad (3.27)$$

The change in λ_- with concentration is relatively small and therefore as a first approximation λ_- may be taken to be equal in the pure acid and in the hypothetical solution of potassium hydrogen sulphate. Hence

$$c' = \frac{c_s}{2 \cdot 45} \qquad (3.28)$$

From the graph in Fig. 3.4(a), $c_s = 0 \cdot 07 \pm 0 \cdot 007$ mol litre^{-1} and this gives $c' = 0 \cdot 0285$ mol litre^{-1}. At this concentration of HSO_4^- ions $\lambda'_- = 160$ Ω^{-1} cm^2 as compared with $\lambda_- = 149$ Ω^{-1} cm^2 in the hypothetical $0 \cdot 07$ M solution of KHSO$_4$ (both from Fig. 3.4(b)). Inserting these values into equation (3.27) we obtain $c' = 0 \cdot 0265$ mol litre^{-1}. This gives a K_{ap} value of $7 \cdot 0 \times 10^{-4}$ mol^2 litre^{-2} or $2 \cdot 1 \times 10^{-4}$ mol^2 kg^{-2}, which may be compared with the estimate of Gillespie et al. (1960), who obtained $K_{ap} = 2 \cdot 4 \times 10^{-4}$ mol^2 kg^{-2}, using cryoscopic data at 10°C and conductometric

TABLE 3.1

Summary of estimates of the ionic product of sulphuric acid

Method	$K_{ap} \times 10^4$ mol^2 kg^{-2}			Reference
	10°C	25°C	40°C	
Cryoscopic	1·7			Gillespie (1950b)
Cryoscopic	1·56			Gillespie and Oubridge (1956)
Cryoscopic	1·7			Bass et al. (1960a)
Cryoscopic and conductometric	(1·7)	2·9		Gillespie and Wasif (1953d)
Cryoscopic and conductometric	(1·7)	2·4	3·2	Gillespie et al. (1960)
Cryoscopic	2·85	4·5		Wyatt (1969)
Conductometric		2·3		Present text
Thermochemical		3·4		Kirkbride and Wyatt (1958)
Acidity function		7·3		Hammett and Deyrup (1932)
Acidity function		5·4		Brand et al. (1959)
Acidity function		2·18		Vinnik and Ryabova (1964)
Acidity function		5·5		Present text, Chapter 2

data at 10 and 25°C. It may easily be calculated from the value of K_{ap} obtained that at the inflexion point (0·16 M) the concentration of HSO_4^- ions arising from autoprotolysis is only 2·5 per cent of the total HSO_4^- concentration and hence that the total contribution to conductance from autoprotolysis ions is about 6 per cent. At 0·3 M this is only about 2 per cent. The autoprotolysis corrected curve is also shown in Fig. 3.4(a) as a dotted line. A second approximation to c' may be obtained from the corrected curve and comes to $0·028 \pm 0·003$ mol litre^{-1}. This gives $K_{ap} = (7·8 \pm 1·7) \times 10^{-4}$ mol^2 litre^{-2} or $K_{ap} = (2·3 \pm 0·5) \times 10^{-4}$ mol^2 kg^{-2}, which is the same within the limits of error as the estimate of $2·4 \times 10^{-4}$ mol^2 kg^{-2} made by Gillespie et al. (1960).

The conclusion that follows from this analysis is that conductometric behaviour of strong bases in sulphuric acid is consistent with the autoprotolysis constant estimated by Gillespie et al. (1960) assuming ionic conduction only. It will be shown in the next section that the same is true of non-electrolyte behaviour.

A summary of all estimates of the ionic product of sulphuric acid is given in Table 3.1.

3.2.6.4. Conductometric Behaviour of Non-electrolytes

A number of substances which, according to cryoscopic measurements, do not undergo any type of ionization in 100 per cent sulphuric acid have been found to depress its conductance. Conductances of a few such solutes are shown in Fig. 3.5. It was first suggested by Gillespie and Solomons (1957) that if the conductance of these solutions is due to autoprotolysis ions, it should fall off with the molarity of the solvent c according to the equation

$$\kappa = 10^{-3}(\lambda_+ + \lambda_-)(K'_{ap})^{1/2} c \qquad (3.29)$$

where λ_+ and λ_- are the equivalent ionic conductances of the autoprotolysis ions and K'_{ap} is the equilibrium constant for autoprotolysis (equation 3.25, p. 74). If ionic conductances may be assumed to be largely unaffected by the presence of the non-electrolyte in fairly dilute solution, then the slope $d\kappa/dc$ should be the same for all non-electrolytes and should be calculable from the known values of λ_+, λ_- and K_{ap}. It was found that conductances of all non-electrolytes indeed fall close together when plotted against the molarity of the "free" acid—allowing that solvation of solutes removes some solvent molecules from effective participation in the conduction process. If a solvation number of one is assumed for sulphuryl chloride and a solvation number of one per nitro group for polynitro compounds, the lines for sulphuryl chloride, trinitrobenzene and trinitrotoluene fall very close together. For 1-chloro-2,4-dinitrobenzene and 2:5-dichloronitrobenzene it is necessary in addition to assume hydrogen bonding to the negative chlorine substituents.

Curves for four of these solutes are plotted vs. the molarity of "free" sulphuric acid in Fig. 3.6.

It is now possible, using the K_{ap} value estimated in the preceding section and the λ_- values from Fig. 3.4(b), to calculate the theoretical slope of the lines of κ vs. the molarity of the "free" solvent. Using λ values for the pure acid (i.e. $\lambda_+ = 232\ \Omega^{-1}\ cm^2$ and $\lambda_- = 160\ \Omega^{-1}\ cm^2$) and $K_{ap} = 7 \cdot 8 \times 10^{-4}\ mol^2$

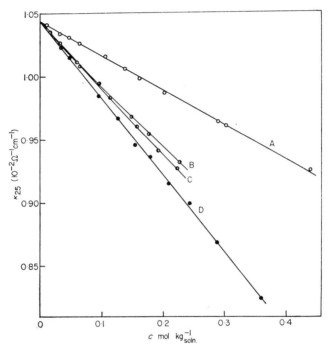

FIG. 3.5. Specific conductances of some non-electrolytes in pure sulphuric acid: A, sulphuryl chloride; B, 2:5-dichloronitrobenzene; C, 1-chloro-2:4-dinitrobenzene; D, trinitrobenzene. (A and D, Gillespie and Solomons, 1957; B and C, Liler, 1962.)

litre^{-2}, the calculated slope is $5 \cdot 9 \times 10^{-4}$. This theoretical slope is also indicated in Fig. 3.6 by a dashed line. It can be seen that it is very close to the experimental slope for all non-electrolytes. We may conclude therefore that the behaviour of non-electrolytes is quite consistent with the quantities derived in the preceding section and that non-electrolytes do not affect significantly the mobility of the solvent self-ions in dilute solution. This may be regarded as a confirmation of the purely ionic mechanism of conduction in sulphuric acid.

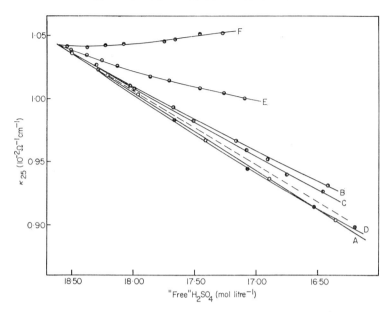

FIG. 3.6. Specific conductances of some non-electrolytes in pure sulphuric acid plotted against the molarity of "free" sulphuric acid. A, sulphuryl chloride; B, 2:5-dichloronitrobenzene; C, 1-chloro-2:4-dinitrobenzene; D, trinitrobenzene. Dashed line—theoretical (see text). Also weak bases: E, m-chloronitrobenzene; F, o-chloronitrobenzene (Liler, 1962).

3.2.6.5. *Conductometric Determination of the Basicities of Very Weak Bases*

A simple method for the determination of basic strengths of weak bases in sulphuric acid has been suggested by Gillespie and Solomons (1957) based on the fact that the HSO_4^- ion is effectively the only conducting ion in solutions of bases in sulphuric acid. Solutions of weak bases show lower conductances than solutions of strong bases at a given molarity of the base, or else they show the same conductances at higher concentrations. When the conductances are the same for the solution of a strong base at the concentration c_s and for the solution of a weak base at the concentration c_w, then the relationship

$$c_{BH^+} = \alpha c_w = c_s \tag{3.30}$$

must apply, where α is the degree of dissociation of the weak base. At equal conductance both solutions have the same HSO_4^- concentration, and the degree of suppression of autoprotolysis is also the same (only autoprotolysis of the solvent need be considered in view of the approximate nature of the treatment). The base ionization constant of the weak base is given by

$$K_b = \frac{[BH^+][HSO_4^-]}{[B][H_2SO_4]} = \frac{\alpha[HSO_4^-]}{(1-\alpha)[H_2SO_4]} \tag{3.31}$$

where [HSO_4^-] is the total concentration of the HSO_4^- ions. The concentrations of the HSO_4^- ions are obtained from the expression for the autoprotolysis constant, $K_{ap} = [H_3SO_4^+][HSO_4^-]$, and the equation

$$[HSO_4^-] = c_s + c_- \tag{3.32}$$

where c_- is the contribution from the unsuppressed solvent autoprotolysis. Since $c_+ = c_-$, we have

$$(c_s + c_-)c_- = K_{ap} \tag{3.33}$$

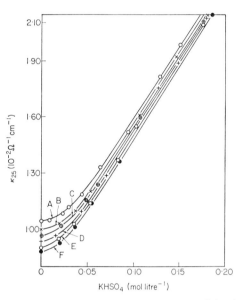

FIG. 3.7. Specific conductances of $KHSO_4$ in pure sulphuric acid (A) and in H_2SO_4/SO_2Cl_2 mixtures. The concentrations of SO_2Cl_2 (mol kg_{soln}^{-1}) are: B, 0·1462; C, 0·2682; D, 0·3849; E, 0·5054, and F, 0·5980 (Liler, 1962).

whence c_- may be calculated from known values of c_s and K_{ap}. This simple method is quite adequate for bases which are extensively ionized (50 per cent or more), but for weaker bases the dilution of the solvent and the consequent reduction in the contribution by autoprotolysis ions must be taken into account, because the limiting behaviour of very weak bases is that of non-electrolytes which depress the conductance of the pure solvent. Very weak bases may depress the conductance of the pure solvent, but not to the same extent as non-electrolytes do. In order to take into account this dilution effect, it was suggested by Liler (1962) that the conductances of very weak bases should be compared not with those of a strong base in pure sulphuric acid (this becomes impossible when the solute depresses conductance), but with those of a strong base in sulphuric acid diluted with an inert diluent to the same solvent molarity as that in the solution of the very weak base. Conductances of potassium hydrogen sulphate in mixtures of sulphuric acid with sulphuryl chloride were measured for this purpose, the first being taken as an "average" strong electrolyte and the latter as a typical non-electrolyte, which is least likely to disrupt the structure of the solvent, in view of the similarity of its molecular structure with that of sulphuric acid. The conductances of these solutions are shown in Fig. 3.7.

The procedure for determining the degrees of dissociation of very weak bases remains the same in this modified method as in the original one, except that the solvent molarity in the solution of the weak base must be calculated first, and a sulphuric acid/sulphuryl chloride solvent mixture of equal molarity with respect to sulphuric acid chosen for comparison (or an interpolation made on a large-scale plot of the data of Fig. 3.7). Figure 3.6 also shows conductances of two solutes (*o*-chloronitrobenzene and *m*-chloronitrobenzene) to which only the modified method is applicable (Liler, 1962).

3.2.6.6. Conductometric Measurements in Pure Sulphuric Acid

A few points regarding the measurement of conductances in pure sulphuric acid must be made. Owing to the high specific conductance of the acid itself and the solutions of all solutes in it, the conductivity cells used must have relatively large cell constants (values of 20–40 cm^{-1} are suitable). Cells of a basic Jones and Bollinger design were adapted for work with sulphuric acid solutions by Gillespie *et al.* (1957). Contact with the open atmosphere must be avoided in manipulating these solutions in order to minimize the absorption of moisture. For the same reason work with larger volumes of solvent is preferable. Platinization of the platinum foil electrodes is desirable, but as platinum black may catalyse the decomposition of some organic solutes, work with bright platinum electrodes may be necessary and is adequate for all but the most accurate measurements. The temperature coefficient of the conductance of sulphuric acid is rather high (about 4 per cent per degree at 25°C) and therefore good temperature control is essential in conductance measurements.

3.2.7. THE CONDUCTOMETRIC TITRATION METHOD

Apart from disulphuric acid, some other substances behave as acids in pure sulphuric acid, e.g. perchloric acid, fluorosulphuric acid, chlorosulphuric acid, and complex acids such as tetrahydrogenosulphato arsenic acid and tetrahydrogenosulphato boric acid. They all dissociate to produce the solvent cation, i.e.

$$HClO_4 + H_2SO_4 \rightleftharpoons H_3SO_4^+ + ClO_4^- \qquad (3.34)$$
$$HB(HSO_4)_4 + H_2SO_4 \rightleftharpoons H_3SO_4^+ + B(HSO_4)_4^- \qquad (3.35)$$

Their dissociation constants have been determined by comparing the conductances of their solutions with those of disulphuric acid, whose K_a value is known from cryoscopy, using a method analogous to the conductometric method for the determination of the strengths of very weak bases (Barr et al., 1961). The following order of decreasing strengths was found

$$HSO_3F > HAs(HSO_4)_4 > HSO_3Cl > HClO_4$$

Perchloric acid causes a negligible change in the conductance of pure sulphuric acid and thus falls into the region where the simple conductometric method is not applicable. A method for taking into account the dilution effect in solutions of acids has not yet been developed. Trichloroacetic acid and trifluoroacetic acid depress the conductance of the pure acid and thus behave either as non-electrolytes or as very weak bases. Tetrahydrogenosulphato boric acid is rather strong (Barr et al., 1961).

The addition of the solution of a base to the solution of an acid in pure sulphuric acid leads to a decrease in conductance, because the highly conducting $H_3SO_4^+$ ion is removed by neutralization with the HSO_4^- ion in the reverse of the autoprotolysis reaction until a minimum is reached beyond which the conductance increases again as more base is added. It has been shown by Flowers et al. (1960b) that there is a relation between the position of the conductance minimum and the strengths of the acid and the base involved. The position of the minimum conductance may be expressed in terms of the ratio $r = n_b/n_a^i$, where n_b is the added number of moles of the base, and n_a^i is the initial number of moles of the acid. For the reaction of a base with disulphuric acid, whose initial concentration is taken as 0·1 m, r at the conductance minimum is given by

$$r_{min} = 0.56\left(1 + \frac{0.018}{K_b}\right) \tag{3.36}$$

whence the K_b of the weak base can be obtained. Conversely, using the solution of a strong base, the strength of the acid may be determined from the position of the minimum. Positions of the minima have been calculated for various strengths of the acid as follows:

K_a	∞	1	10^{-1}	10^{-2}	10^{-3}
r_{min}	1·01	0·98	0·89	0·44	0·08

In this way the dissociation constant of tetrahydrogenosulphato boric acid was found to be 0·3 mol kg^{-1} (Flowers et al., 1960b), subsequently confirmed by conductometric measurements which gave 0·2 mol kg^{-1} (Barr et al., 1961). This makes it the strongest acid of the sulphuric acid solvent system so far studied.

If both the acid and the base are weak the position of minimum conductance is given by

$$r_{\min} = \left(1 + \frac{0.018}{K_b}\right)\left[\left(\frac{K_a}{K_a + 0.013}\right) + \frac{0.0007}{m_a^i}\right] \quad (3.37)$$

where m_a^i is the initial molality of the acid. If K_a or K_b is known, the other can be calculated from this equation. Basicity constants for a number of weak bases obtained using this method are in good agreement with those obtained from cryoscopy and conductance. The method can also be used to study more complex types of ionization in pure sulphuric acid. This will be discussed in Chapter 4.

3.2.8. CORRELATION OF BASICITY DETERMINATIONS IN PURE SULPHURIC ACID WITH H_0

In order to convert K_b values found by cryoscopic, conductometric and conductometric titration methods to pK_{BH^+} values based on the H_0 acidity function, it is necessary to know the ionization ratio of the weak base at a given concentration of the solution and the H_0 of the solution. This is a function of the mole ratio of the HSO_4^- ion to undissociated sulphuric acid, as discussed in Section 2.6.2, p. 58. From the definition of K_b (equation 3.31, p. 81) and the definition of the acidity function, $H_0 = pK_{BH^+} + \log[B]/[BH^+]$, it follows directly that

$$pK_{BH^+} = \log K_b + H_0 - \log[HSO_4^-]/[H_2SO_4] \quad (3.38)$$

For bases which are extensively protonated in pure sulphuric acid, the recalculation of K_b values obtained by cryoscopy or conductance to pK_{BH^+} values may be carried out neglecting the change in the molarity of the solvent using the equation

$$H_0 = -9.20 + \log(m_{HSO_4^-}/m_{H_2SO_4}) \quad (3.39)$$

which holds in the region 95–99 per cent acid (Vinnik and Ryabova, 1964). For bases which are only slightly protonated, the H_0 value of a particular solution is best obtained from a plot of H_0 vs. $\log(m_{HSO_4^-}/m_{H_2SO_4})$ for acid concentration > 99 per cent. The ratio of molarities for pure sulphuric acid is given by the square root of K'_{ap} (equation 3.25, p. 74). In solutions of weak bases it must be calculated taking into account the autoprotolysis of the solvent, i.e. the concentration of HSO_4^- ions must be obtained from the equation

$$[HSO_4^-] = \alpha c_w + \frac{K_{ap}}{[HSO_4^-]} \quad (3.40)$$

where c_w is the concentration of the weak base, α is its degree of protonation, and K_{ap} refers to the molarity of "free" sulphuric acid in the particular solution of the very weak base. The K_{ap} value given by Flowers et al. (1960a) may be used, since this is consistent with cryoscopic and conductometric data (Section 3.2.6). This method of correlating K_b values obtained in the pure acid with pK_{BH^+} values referred to the acidity function H_0 has been developed by Liler (1962) with particular reference to conductometrically determined basicity constants.

3.3. Basicities of Very Weak Bases

3.3.1. INTRODUCTION

3.3.1.1. *Classification*

Basic behaviour is observed most commonly in compounds which have lone electron pairs to offer for bonding with the proton (*n*-bases), but it is also found in compounds which contain unsaturation in the form of double bonds or conjugated double bond systems ($\bar{u}$-bases).

The first group, the *n*-bases, comprises compounds containing elements of Groups V and VI in the Periodic Table. These elements contain paired electrons in their atomic valence shells, which remain uninvolved or are only partly involved in bonding in compounds. They are thus available for interaction with protons. Some such compounds also contain double bonds and could in principle also act as $\bar{u}$-bases. Compounds containing elements of Group VII also offer paired electrons in the valence shells of these atoms, but, owing to the high electronegativity of these elements, their simple protonation is not observed even in strongly acid media. They do, however, act as hydrogen bond acceptors (see, for example, Arnett, 1963). Although this kind of behaviour is described as basic in a broader sense, the term basicity will here be used only with reference to proton addition resulting in the formation of a covalent bond.

The second group of bases, the $\bar{u}$-bases, are mostly unsaturated hydrocarbon compounds, and therefore also termed carbon bases. Their protonation leads to a partial or complete destruction of the $\bar{u}$-electron system with the formation of σ-bonds, the resulting cations being called carbonium ions. Therefore the term $\bar{u}$-basicity strictly applies to weaker interactions of these compounds with hydrogen bond donors, in which the proton is only weakly embedded in the $\bar{u}$-electron cloud without destroying it. Simple protonation of $\bar{u}$-bases by aqueous acids is in many instances not readily amenable to study, because of secondary reactions that follow. When protonation leads to the formation of sufficiently stable carbonium ions, the equilibria can be studied as for other basic species. A detailed discussion of carbonium ions is not intended in the present monograph, because of the very recent publication of an extensive monograph on the subject by Bethell and Gold (1967) and of a series of volumes on carbonium ions edited by Olah and von Schleyer (1968). A brief survey of the most important carbonium ion equilibria in aqueous acids is, however, necessary, not only for the sake of completeness but as a basis for the understanding of some reaction mechanisms to be discussed in Chapter 5.

The major part of this section will be devoted to n-bases. There are a large number of such bases and a classification would therefore be desirable. It would be logical to classify the bases according to the atom being protonated, the basic site. This classification would unfortunately run into difficulty, because there are instances where more than one basic site occurs in compounds of closely similar structure, depending on experimental conditions and minor variations of structure, or where the site of protonation is controversial. In all such instances it is still true to say that a given atom (say N or O) is the primary source of the electrons which make the protonation possible, even when the proton is not added directly onto the electron-pair donor, but onto an atom conjugated with it. Therefore a simple classification according to the Group in the Periodic Table of the main electron donor and according to the complexity of the basic group will be adopted here.

The behaviour of amino nitrogen bases will be discussed first. This will be followed by a discussion of bases containing nitrogen in more complex groups. This choice is made because the behaviour of these bases in acid solution links up most directly with the behaviour of weak bases in dilute aqueous solution, which are all, with few exceptions, nitrogen bases. Oxygen bases and bases containing other elements in Group VI will be discussed next. Some of the oxygen bases on protonation give cations in which the positive charge partly resides on the carbon. Hence these cations represent a transition towards carbonium ions. Simple protonation of unsaturated hydrocarbons to give carbonium ions will finally be discussed, followed by a few examples of very weak inorganic bases.

Carbonium ions also arise in complex ionizations of carbinols and other compounds in acid solution. Chapter 4 deals with this method of generating them. However, the term "complex ionization", originally used in that context, is interpreted here more broadly to cover more complex modes of ionization of a number of other compounds.

3.3.1.2. *Factors Determining Acid/Base Strength*

Most simple bases in dilute aqueous solutions are compounds containing amino-nitrogen. Structural effects on the basicities of these compounds have been extensively studied (see, for example, the review by Brown *et al.*, 1955). The amount of material collected is so large that predictions of changes of basic strength with structural change can be made with a considerable degree of reliability (Clark and Perrin, 1964). These predictions are mainly based on linear free-energy relationships, the most important of which is the Hammett equation. Some excellent reviews are available on this subject (Jaffé, 1953; Pal'm, 1961; Wells, 1963; Shorter, 1969) and a monograph has recently been published (Wells, 1968). Linear free-energy relationships were originally developed for substituted benzene derivatives (Hammett, 1940), but were later extended by Taft to aliphatic compounds also (reviewed by Taft, 1956) and by other authors to other types of compounds and reactions (e.g. by Kabachnik, 1956, to the dissociation of substituted phosphinic acids).

In the wide acidity range available in sulphuric acid/water mixtures and dilute oleum many other groups display basic properties. It would facilitate the discussion of acid/base strengths of this wide range of bases if some group characteristics were recalled at this stage.

Qualitative ideas about the inductive ($\pm I$) and the mesomeric ($\pm R$) effects of groups (the term "polar" effect is often used to mean the first only, but also to embrace both) have been given a quantitative basis in the Hammett equation, which takes the form

$$\log(K/K_0) = \rho\sigma \qquad (3.41)$$

for equilibria, and

$$\log(k/k_0) = \rho\sigma \qquad (3.42)$$

for reaction rates, where K and K_0 are the equilibrium constants for the substituted and the unsubstituted benzene derivative involved in a given reaction, and k and k_0 the respective rate constants. The parameter σ is a characteristic of the substituent, defined in terms of its effect on the dissociation constants of benzoic acid

$$\sigma \equiv \log(K/K_0) \qquad (3.43)$$

where K and K_0 are the dissociation constants of the substituted and the unsubstituted acid, respectively, and the constant ρ is a measure of the sensitivity of the reaction to the effect of the substituents. This was set equal to unity for the standard benzoic acid series.

On the basis of their σ values substituents can be classified into electron-releasing groups, which decrease the strength of benzoic acid and hence have negative σ values, and electron-attracting groups, which increase the strength of benzoic acid and hence have positive σ values. Both electron release and electron attraction can occur by inductive and mesomeric mechanisms, and σ values are dependent on whether the substituent is in the *meta* or the *para* position, the effects in both positions being free of complications due to steric interactions. Numerous correlations of data have demonstrated that the constants for *meta* substituents (σ_m) show little variation from one reaction series to another. Constants for *para* substituents (σ_p) have been found to vary more or less from one reaction to another, because they reflect also the mutual interaction between the *para* substituent and the reaction centre through the benzene ring. This has led to attempts to define true constants, which would hold when certain structural limitations apply. Thus σ^+ constants have been defined for situations where substituents conjugate through the aromatic system with an electron-acceptor group as the reaction site. Conversely, σ^- constants have been defined to apply to situations where substituents conjugate through the aromatic system with a mesomerically electron-donating reaction centre.

TABLE 3.2

Substituent constants for some common substituents (from the compilation by Pal'm, 1961)†

Substituent	σ_m	σ_p	σ_p^+	σ_p^-
CH_3	−0·069	−0·17	−0·311	
$C(CH_3)_3$	−0·120	−0·197	−0·256	
CH_2Cl	+0·14	+0·184	+0·01	
$CHCl_2$	+0·185	+0·212		
CCl_3	+0·407	+0·454		
CH_3CO	+0·306	+0·502		+0·874
F	+0·337	+0·062	−0·073	
Cl	+0·373	+0·227	+0·114	
Br	+0·391	+0·232	+0·150	
I	+0·352	+0·276	+0·135	
OH	+0·121	−0·37	−0·92	
OCH_3	+0·115	−0·268	−0·778	
OC_2H_5	+0·150	−0·250		
C_6H_5	+0·06	−0·01	−0·179	
NH_2	−0·16	−0·66	−1·3	
$NHCH_3$	−0·302	−0·84		
NH_3^+	+0·634			
CN	+0·678	+0·66	+0·659	+1·00
CF_3	+0·43	+0·54	+0·612	
NO_2	+0·710	+0·778	+0·790	+1·270
COOH	+0·355	+0·265	+0·421	+0·728
$COOCH_3$	+0·315	+0·436	+0·489	+0·636
$CONH_2$	+0·280			+0·627

† Preference has been given to the more recent or confirmed values.

A selection of substituent constants is given in Table 3.2. For the *meta* position the constants defined for either acceptor or donor reaction centres (σ_m^+ and σ_m^-) are little different from Hammett's σ_m values, defined for the carboxyl group as the reaction centre, which is itself a relatively weak acceptor group. For the *para* position the "through resonance" or conjugation between the substituent and the reaction centre is dependent on the nature of both, and therefore σ_p^+ and σ_p^- values are somewhat variable, and have recently been described as representing a "sliding scale" of values (Shorter, 1969).

The constants for *meta* substituents on the other hand measure an interaction of the substituent with the benzene ring, which is transmitted to the reaction centre in the *meta* position purely inductively. This does not mean that the substituent itself interacts with the benzene ring purely inductively. The benzene ring can act either as an electron-donor or as an electron-acceptor in mesomeric interactions with substituents. The reaction centre in the *meta* position may be regarded as a probe, reflecting changes in the electron density in the benzene ring at that position produced by the substituent, without

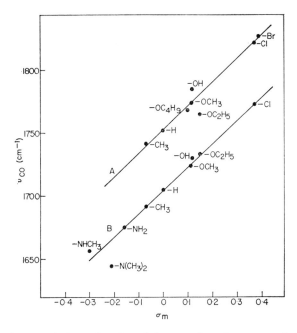

FIG. 3.8. The correlation of carbonyl frequencies of compounds RCOX with σ_m constants of groups X: A, acetyl compounds; B, benzoyl compounds (Liler, 1967).

interacting directly with it. The side chain containing the reaction centre interacts likewise with the benzene ring, but this interaction remains constant within a reaction series. Evidence that this is so, at least as a first approximation, is the good constancy of σ_m constants found in a variety of reactions.

It has been shown more recently (Liler, 1967) that these same σ_m constants are determining for the effect of substituents upon the carbonyl frequency in compounds RCOX, because a good linear correlation exists between σ_m constants of X and the carbonyl frequency (Fig. 3.8). The carbonyl frequency is an isolated, mass-insensitive vibration, which is sensitive to changes of electron density on the carbonyl carbon in the first instance. These changes are further reflected in the changes of electron density on the carbonyl oxygen, and hence in the ionization potentials of carbonyl compounds (Walsh, 1946). The correlation of Fig. 3.8 means that the interactions of the substituents possessing a lone-electron pair with the benzene ring and with the carbonyl group are qualitatively the same, as shown by the formulae (3.3) and (3.4).

(3.3)   (3.4)

Alternatively, one might say that the phenyl ring and the carbonyl group are electron acceptors, or "electron sinks" of the same kind, i.e. that the relative importance of the inductive and mesomeric interactions is the same for both.

Since the mesomeric interaction in (3.3) and (3.4) is relatively less important than conjugation through the benzene ring with a *para* substituent which is itself an "electron sink", as in (3.5), an interaction measured by the σ_p^+

(3.5)

constants which are all more negative than the σ_m constants, it was suggested by Liler (1967) that such an interaction should preferably be described as partial coordination, rather than conjugation, in order to distinguish between the two situations which are not equivalent. The benzene ring carrying a *para* substituent, which is itself an electron sink, may be described as a composite electron sink, consisting of two conjugated simple electron sinks. The need for a separate term for the latter type of interaction (3.5) was felt for some time. The term "through resonance" was used, for example, by Clark and Perrin (1964), and "cross-conjugation" by Shorter (1969).

Carbonyl frequencies have been correlated linearly with vibrational frequencies of the nitroso group (Bellamy and Williams, 1957), which means that the same inductive and mesomeric effects apply to this group as to the carbonyl group, and are therefore also determined by the σ_m constants of substituents X in compounds XNO. There are also indications that substituent effects upon the $>C=CH_2$ group are of the same character. All these groups are planar, involving an sp^2 hybridized atom. More limited correlations with carbonyl frequencies are found for the asymmetric stretching frequency of the nitro group (Bellamy and Williams, 1957). The symmetric and asymmetric frequencies of the sulphonyl group are not correlated with the carbonyl frequencies, because the tetrahedral structure of the sulphonyl group does not allow mesomeric effects to operate upon the SO bonds to the same extent as in the planar carbonyl group (Bellamy and Williams, 1957). According to Exner (1963) the mean of the symmetric and the asymmetric stretching frequency of this group correlates with Taft's constants for purely inductive interactions, σ_I (see p. 94).

Thus, it can be seen that the significance of σ_m constants extends outside the sphere of aromatic chemistry, and that these constants are characteristics of groups which measure their ability for inductive and mesomeric interaction with a number of electron sinks. For all such electron acceptors the relative importance of the inductive and the mesomeric interaction with a given substituent is the same, although the absolute magnitude of the substituent effect will vary from one acceptor group to another. Hence, if one represents the σ_m constants as a sum of these two effects

$$\sigma_m = \sigma_I + \sigma_R$$

where σ_I stands for the inductive and σ_R for the resonance component of the substituent effect, there will be a sensitivity factor associated with σ_m in describing quantitatively the magnitude of the effect of the substituent upon the electron density in the acceptor groups. (For example, ν_{NO} in nitroso compounds is more sensitive to substituent effects than ν_{CO} in carbonyl compounds (Bellamy and Williams, 1957). This sensitivity factor may also be described as a measure of the "depth" of the electron sink. Therefore, this factor will also determine the effect of various electron sinks upon the electron density on the substituent X. Attempts to estimate the relative importance of the inductive and the mesomeric effects for various substituents have been made by Taft (1960a), and some of his figures are reproduced in Table 3.3.

TABLE 3.3

The separation of inductive and resonance components of σ_m constants (Taft, 1960a)

Substituent	σ_m	σ_I	σ_R
Cl	+0·37	+0·47	−0·10
OCH$_3$	+0·12	+0·29	−0·17
H	0	0	0
CH$_3$	−0·07	−0·05	−0·02
NH$_2$	−0·16	+0·10	−0·26

The relevance of this discussion to questions of acid/base strength is two-fold. The interaction between donor and acceptor groups is reflected in the ability of both to donate electron pairs in protonation. The basicity of an

3. PROTONATION OF VERY WEAK BASES

acceptor group, such as carbonyl, will be determined by the donor substituent. On the other hand, the basicity of donor groups, which all contain lone electron pairs, will be more or less affected by the strength of adjacent groups as electron acceptors. Alternatively, it may be said that the depth of the electron sinks will determine the availability of the lone electron pairs on the substituents X for protonation. This is particularly important for donor groups such as —NH_2 (or —NR_2) and —OH (or —OR). Therefore this discussion will be concluded by quoting an example of the effect of donor–acceptor interactions within molecules upon acid/base strength, which was discussed by Brown *et al.* (1955), because it is one of the rare cases in which the inductive and the resonance components of the effect can be separated in a relatively straightforward manner.

In the following series of amino bases, starting from an aliphatic tertiary

	(3.6)	(3.7)	(3.8)
pK_{BH^+}	10·58	7·79	5·06
ΔpK_{BH^+}		2·79	2·73

amine (**3.6**), the introduction of a phenyl ring adjacent to the N atom leads to a reduction in basicity. This is only half as great when the electron withdrawal is by inductive effect only as in quinuclidine (**3.7**) in which the mesomeric interaction is virtually eliminated by the geometry of the compound, as it is when both the inductive and the mesomeric withdrawal are operative, as in dimethyl aniline (**3.8**). The statistical and solvation effects are the same in all three. This led Webster (1952) to suggest that the effect of the phenyl sink on the basicity of an amino group was due half to inductive and half to resonance effects.

Turning now to the phenyl electron sink, the presence of an amino group substituent increases the electron density within the ring, as reflected in the negative σ_m value for the —NH_2 group (–0·16). The effect in the *para* position is considerably enhanced (see Table 3.2).

Even in the absence of possible conjugation between a substituent in the *para* position and the reaction centre, the effect of the substituent from the *para* position has been shown to be greater by a factor of approximately 1·14 than the effect from the *meta* position (Exner, 1966). For a series of substituted benzoic acids in 50 per cent methanol, for inductive effect only, it was found that

$$\log \frac{K_p}{K^\circ} = 1 \cdot 14 \log \frac{K_m}{K^\circ} \tag{3.44}$$

where $K°$ is the dissociation constant of the unsubstituted benzoic acid. This result calls for a reconsideration of all earlier attempts to separate the substituent effects of *meta* and *para* substituents into inductive and mesomeric components, which were based on the assumption that the inductive effects from the *meta* and the *para* positions are the same.

There have been three attempts to arrive at substituent constants for purely inductive interaction. Following a suggestion of Ingold, Taft proposed in the early nineteen-fifties a method based on the rates of acidic and basic hydrolysis of esters of the type RCOOR', in which R' is constant and R is the variable substituent (Taft, 1956). The polar (inductive) substituent constant for the group R is defined by

$$\sigma_I = [\log(k/k_0)_B - \log(k/k_0)_A]/2\cdot48 \tag{3.45}$$

where k's are the rate constants for basic (B) and acidic (A) hydrolysis, and the subscript zero refers to $R = CH_3$, the methyl group being taken as standard. The factor 2·48 puts σ_I values on about the same scale as Hammett's σ values. This definition is based on the assumption that polar effects of substituents are much greater in the alkaline than in the acidic hydrolysis, the steric and resonance effects being the same in both. The constants so defined have been successful in correlating a considerable amount of data. In the aliphatic series these constants measure the inductive effect of groups only. Constants for inductive effects of substituents (σ') have also been defined by Roberts and Moreland (1953) for 4-X-bicyclo-2.2.2-octane-1-derivatives in which inductive effects only can operate. There is a parallelism between these values and the corresponding σ_I values for groups $X—CH_2—$, which also can interact only inductively ($\sigma' = 0\cdot45\sigma_I$).

There are objections to Taft's assumptions concerning the equality of steric effects in acidic and basic hydrolysis, owing to opposite charges of transition states and consequent differences in solvation (Shorter, 1969). The definition of each constant also requires four rate measurements. Recently Charton (1964) has therefore proposed a redefinition of inductive substituent constants based on the dissociation constants of substituted acetic acids in water, a series in which steric effects are also small. The constants are now defined by the equation

$$\sigma_{I,x} = b(pK_{a,x}) + d \tag{3.46}$$

in which $pK_{a,x}$ refers to the X-substituted acid. This definition has the advantage that only one pK measurement is sufficient to obtain a new σ_I value. Also, this definition is attractive because it is directly comparable to that of Hammett substituent constants. The standard substituent on this scale is hydrogen ($\sigma_I = 0$), and b is taken as $-0\cdot25$, in order to make the new σ_I values commensurate with Taft's σ_I scale. Charton's redefinition of the σ_I values thus anchors the scale to the hydrogen substituent in the series of substituted acetic acids, but leaves them on a basis comparable to Hammett's σ values. σ_I values for some common substituents are given in Table 3.4.

TABLE 3.4

Primary σ_I substituent constants (Charton, 1964)

Substituent	σ_I	Substituent	σ_I
Et	−0·05	I	0·39
Me	−0·05	t-Bu	−0·07
H	0·00	OH	0·25
PhCH$_2$	0·04	MeO	0·25
Ph	0·10	NHAc	0·28
F	0·52	CN	0·58
Cl	0·47	EtCONH	0·25
Br	0·45	H$_2$NCONH	0·21

This very brief account of a rather confused subject is meant only to provide essential minimum background for the discussion of some relationships of acid/base strength and of substituent effects on the basicity of very weak bases in general. Fundamentally the same ideas are used in discussions of reaction mechanisms and therefore this theoretical basis is of importance for the understanding of Chapter 5 as well.

3.3.2. BASES INVOLVING N^{III} AND P^{III} AS THE BASIC CENTRE

Nitrogen in the oxidation state +3 is the major source of basicity of organic compounds in aqueous solution, because its lone electron pair is readily available for protonation under these conditions, even when it is involved in resonance interactions with electron sinks, such as the benzene ring. Its basicity is, however, reduced below the level detectable in aqueous solution in a range of compounds, in which either a number of electron-withdrawing substituents are present, or the amino nitrogen is adjacent to more powerful (deeper) electron sinks or is involved in strongly resonance stabilized symmetrical structures. Such bases will be discussed in the present section. Unless otherwise stated all pK_{BH^+} values refer to 25°C.

3.3.2.1. *Amines*

There are very few instances of amino groups which remain incompletely protonated in 100 per cent sulphuric acid. Guanidine, which is one of the most strongly basic compounds in aqueous solution (pK_a^{20} = 13·65), owing to resonance stabilization of the symmetric cation (**3.9**), is protonated further only with difficulty for the same reason. Williams and Hardy (1953) have shown

$$H_2\overset{+}{N}=C\overset{NH_2}{\underset{NH_2}{\Big\langle}} \longleftrightarrow H_2N-C\overset{\overset{+}{NH_2}}{\underset{NH_2}{\Big\langle}} \longleftrightarrow \text{etc.}$$

(3.9)

cryoscopically that guanidine can acquire a second proton to the extent of only 27 per cent in a 0·08 M solution of guanidinium perchlorate in 99·9 per cent sulphuric acid.

Aromatic amines (substituted anilines) are all fully protonated in pure sulphuric acid, but an example of an aromatic amino group which is not has been found by Newman and Deno (1951a) in tri-p-aminophenyl carbinol and the N,N-dimethyl derivative. While two amino groups in the carbonium ions are largely protonated in 1 N hydrochloric acid (Adams and Rosenstein, 1914), the last amino group is still not protonated in concentrated sulphuric acid. This large difference in basicity is ascribed to the very stable quinonoid structure of the carbonium ion

(3.10)

in which the last amino group already bears a positive charge by donating its electrons to the carbonium centre.

A range of negatively substituted anilines become protonated in sulphuric acid/water mixtures and have been used as indicators in defining the H_0 acidity function (see Chapter 2). Their pK_{BH^+} values are therefore known with a considerable degree of accuracy. The values in Table 3.5 are based on the review by Vinnik (1966). A certain number from the corresponding range of N,N-dimethyl anilines were also studied in establishing the H_0''' acidity function (Arnett and Mach, 1964). The values are also given in Table 3.5. The pK_{BH^+} values of a few more tertiary amines, used in addition to tertiary anilines in defining H_0''', are given in Table 3.6. All these values are true pK_{BH^+} constants, assuming that H_0 and H_0''' are thermodynamically exact acidity functions.

The vast majority of substituted aniline bases in Table 3.5 contain at least one *ortho*-substituent, which introduces complications, especially in the N,N-dialkyl derivatives, owing to steric inhibition of resonance. It is therefore difficult to analyse in detail substituent effects on the basicity of these bases. A statistical factor of 3 is involved in comparing the basicities of primary and tertiary amines, which makes tertiary amines more basic by 0·48 units. The much larger increases in basicity of 3–4 units, observed in going from primary

TABLE 3.5

pK_{BH^+} values of some substituted aniline indicators at 25°C
(determined in sulphuric acid/water mixtures)

Amine	—NH$_2$†	—N(CH$_3$)$_2$‡	—N(C$_2$H$_5$)$_2$§
p-Nitroaniline	+1·0		
Diphenylamine	+0·77		
o-Nitroaniline	−0·33		
4-Chloro-2-nitroaniline	−1·02		
5-Chloro-2-nitroaniline	−1·55		
2,5-Dichloro-4-nitroaniline	−1·90		
2-Chloro-6-nitroaniline	−2·43		
4-Nitrodiphenylamine	−2·50		
2,6-Dichloro-4-nitroaniline	−3·2§		
2,4-Dichloro-6-nitroaniline	−3·28		
2,4-Dinitroaniline	−4·45	−1·0†	+0·21
2,6-Dinitroaniline	−5·64		
2,6-Dinitro-4-methylaniline	—	−1·66	
4-Chloro-2,6-dinitroaniline	−6·25	−3·12	
2,4-Dinitro-1-naphthylamine	−6·5§	−2·59	
6-Bromo-2,4-dinitroaniline	−7·09		
3-Methyl-2,4,6-trinitroaniline	−8·34		
3-Bromo-2,4,6-trinitroaniline	−9·77§		
2,4,6-Trinitroaniline	−9·98	−6·55	−5·71

† Data from the review by Vinnik (1966).
‡ Data of Arnett and Mach (1964).
§ Values obtained in perchloric acid/water mixtures (Vinnik, 1966).

to tertiary amines, are predominantly due to steric inhibition of resonance of the —NR$_2$ group by the *ortho*-substituents, which increases the availability of the lone electron pair for protonation. A positive inductive effect of the alkyl substituents acting in the same direction is also present.

TABLE 3.6

pK_{BH^+} values of some tertiary amine indicators at 25°C (Arnett and Mach, 1964)

Amine	pK_{BH^+}
N-(2,4-Dinitrophenyl)-piperidine	−0·38
N-Methyl-4-nitrodiphenylamine	−3·42
N-Methyl-4′-bromo-4-nitrodiphenylamine	−4·21
N-Methyl-2,4-dinitrodiphenylamine	−6·19
N-Methyl-4-bromo-2′,4′-dinitrodiphenylamine	−6·93
N-Methyl-x′,4′-dibromo-2,4-dinitrodiphenylamine	−8·17
N-Methyl-2,4,2′,4′-tetranitrodiphenylamine	−10·56

Substituted primary diphenylamines were also studied in sulphuric acid/water mixtures containing 20 volume per cent of ethanol (Dolman and Stewart, 1967), in which they are more soluble. The acidity function they define was mentioned in Section 2.4.1.4, p. 40. The unsubstituted base has a pK_{BH^+} of 0·78 (Perrin, 1965), and bases carrying one or more negative substituents are very weakly basic (e.g. 4,4'-dinitrodiphenylamine has pK_{BH^+} of −6·21). Their basicities correlate well with σ and σ^- substituent constants with $\rho = 3·36$. The basicities of monosubstituted diphenylamines correlate also very well with those of equally substituted anilines with unit slope (Dolman and Stewart, 1967). The plot of the basicities of 4-nitro-3'(or 4')-substituted diphenylamines against the basicities of monosubstituted diphenylamines is also linear with a somewhat lower slope (0·863).

Polybasic amines are also fully protonated in 100 per cent sulphuric acid, regardless of the relative positions of amino groups. Thus o-phenylenediamine gives an *i* factor of three, indicating complete diprotonation (Gillespie and Leisten, 1954a), and all four nitrogen atoms in hexamethylenetetramine were found to be protonated (Gillespie and Wasif, 1953c). Only strong resonance stabilization can prevent complete protonation of amino groups in 100 per cent sulphuric acid, as in the examples mentioned above.

3.3.2.2. *Pyrroles and Indoles*

The basicity of pyrrole and its methyl derivatives has been studied recently by u.v. spectrophotometry, for some in aqueous buffer solutions (Abraham *et al.*, 1959) and for most in sulphuric acid/water mixtures (Whipple *et al.*, 1963, and Chiang and Whipple, 1963). Pyrrole itself trimerizes in dilute acid (Potts and Smith, 1957) but in more concentrated acid, in which it is more extensively protonated, dilute solutions are sufficiently stable for a spectrophotometric study (Chiang and Whipple, 1963). Nuclear magnetic resonance examination of the solutions of substituted pyrroles in concentrated hydrochloric and sulphuric acid shows that proton addition occurs on the α-carbon

(3.11) (3.12)

atom, so that the cation structure is (**3.11**) and not (**3.12**), although the possibility that rapid *N*-protonation may occur in dilute acid solutions is not ruled out. The donor of the electron pair in (**3.11**) is the nitrogen atom, and the cations are ammonium ions rather than carbonium ions. Some β-protonation also occurs, since all the hydrogen atoms are readily exchanged with deuterium in acid solution (Abraham et al., 1959). Relative protonation rates were discussed by Chiang and Whipple (1963).

Pyrroles do not obey the H_0 acidity function, but obey rather more closely the H_I function defined for indoles. The pK_{BH^+} values of some methyl substituted pyrroles are given in Table 3.7. In general methyl substitution increases basicity, but there is no clear additivity of substituent effects. α-Methyl groups direct the proton to the opposite α-position and β-methyl groups to the adjacent α-site (Chiang and Whipple, 1963).

TABLE 3.7

pK_{BH^+} values of methyl-substituted pyrroles (Chiang and Whipple, 1963)

Compound, pyrrole	Molarity of H_2SO_4 at half-protonation	pK_{BH^+}
2,3,4-Trimethyl-	†	3·94
2,3,4,5-Tetramethyl-	†	3·77
2,4-Dimethyl-	0·0081	2·55
2,3,5-Trimethyl-	0·009	2·00
3,4-Dimethyl-	0·164	0·66
1,2,5-Trimethyl-	0·54	−0·24
2-Methyl-	0·70	−0·21
2,5-Dimethyl-	1·10	−0·71
3-Methyl-	1·50	−1·00
1-Methyl-	4·14	−2·90
Pyrrole	5·34	−3·80

† Obtained in aqueous buffer solution by Abraham et al. (1959).

With some pyrroles the proportion of β-protonation is considerable and may be determined from the n.m.r. spectrum of the fully protonated pyrrole in sulphuric acid. It was demonstrated for 1-phenyl-2,5-dimethylpyrrole that both α- and β-protonation follow the same acidity function, the ratio $\beta:\alpha$ being 0·19, independent of acid concentration (Whipple et al., 1963). The slopes $d(\log I)/d(c_{H_2SO_4})$ for N-phenylpyrroles approach those for the H'_R indicators (Chiang et al., 1967). N-phenyl substitution reduces the basicity of pyrrole considerably.

On the basis of u.v. and n.m.r. spectra it was concluded that protonation of indoles occurs predominantly at the 3-position, so that the cation has structure (3.13) (Hinman and Whipple, 1962). Deuterium exchange occurs generally at positions 1 and 3, and in some cases at position 2.

Studies of the variation of protonation ratios with acid concentration (Hinman and Lang, 1964) led to the conclusion that within this series of closely

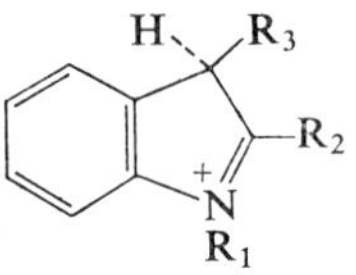

(3.13)

similar bases (24 indoles were studied) the $\log I$ vs. acid concentration slopes vary. All bases fell into three groups: indole and others unsubstituted in the hetero ring, the 1,3-dialkyl substituted derivatives, and the largest class, all those with alkyl groups in other positions. In terms of this latter class the H_I function was defined and the pK_{BH^+} values of these bases obtained (Table 3.8). For indoles which do not obey the H_I acidity function pK_{BH^+} values based on that scale were also determined (e.g. for indole a value of -3.5 was obtained). These values have no thermodynamic significance and will not be reported here.

TABLE 3.8

pK_{BH^+} values of some indoles at 25°C (Hinman and Lang, 1964)

Indole	Molarity of H_2SO_4 at half-protonation	pK_{BH^+}
1,2-Dimethyl-	0.09	+0.30
2,5-Dimethyl-	0.10	+0.26
2-Methyl-	0.76	−0.28
2-Ethyl-	0.92	−0.41
1,2,3-Trimethyl-	1.22	−0.66
2,5-Dimethyl-3-n-propyl-	2.31	−1.40
2,3-Dimethyl	2.46	−1.49
1-Methyl-	3.70	−2.32
1,2-Dimethyl-5-nitro-	4.60	−2.94
2-Methyl-5-nitro-	5.54	−3.58
3-t-Butyl-	5.91	−3.84
3-Ethyl-	6.53	−4.25
3-n-Propyl-	6.66	−4.34
3-Methyl-	6.95	−4.55
Indole-3-acetic acid	9.27	−6.13
Tryptamine	9.51	−6.31

3.3.2.3. *Azobenzenes*

The basicities of azobenzenes were studied by Jaffé and his collaborators (1958–59) in 20 volume per cent ethanol and 80 per cent sulphuric acid/water mixtures, and more recently by Reeves (1966). The spectrophotometric measurements of Jaffé and Gardner (1958) on monosubstituted azobenzenes were extended to disubstituted azobenzenes (3.14) by Yeh and Jaffé (1959a),

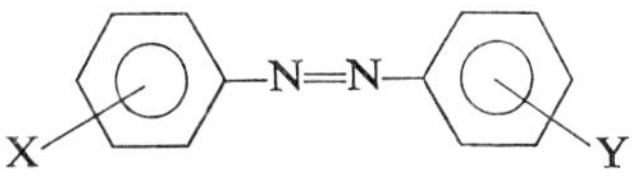

$$X-\langle\bigcirc\rangle-N=N-\langle\bigcirc\rangle-Y$$

(3.14)

with the object of defining an acidity function in these media (see Section 2.4.1.4, p. 40). The pK_{BH^+} values of azobenzene and monosubstituted azobenzenes (Table 3.9) are correlated with the σ^+ substituent constants, with $\rho = 2 \cdot 2$,

TABLE 3.9

pK_{BH^+} values of substituted azobenzenes in 20 per cent ethanol–80 per cent sulphuric acid/water mixtures at 25°C
(Jaffé and Gardner, 1958; Yeh and Jaffé, 1959a, b)

	−pK_{BH^+}				
X	Y = m-NO$_2$ (1)	Y = H (2)	Y = p-OCH$_3$ (3)	Y = OH (4)	Y = p-SO$_3^-$ (5)†
4-N(CH$_3$)$_2$		(−3·3)‡			−3·49
4-OEt	2·48	1·28			
4-OMe	2·54	1·36	0·73	0·56	1·49
4-OH					1·45
4-Me	3·83	2·35	1·03	0·84	
3-Me	4·32	2·70			
H	4·63	2·90	1·36	1·02	3·13
4-Cl					3·55
4-Br	5·04	3·47	1·74	1·42	
3-Br	5·52	3·83	2·06	1·67	
3-COCH$_3$	5·69				
4-COCH$_3$	5·96	3·98	2·23		
4-CN	6·50	4·52			
3-NO$_2$	6·57	4·63	2·54		
4-NO$_2$		4·70			
4-NMe$_3^+$					4·95

† According to Reeves (1966).
‡ From Perrin (1965).

which is a little smaller than that for anilines (2·75). This was interpreted as indicating that the proton is attached equally to both nitrogen atoms of the azo group in a kind of bridge structure (3.15). This structure requires that the conjugate acid should have a *cis* conformation. The arguments leading to

$$\text{Ph-N=N-Ph} \quad \text{(with bridging H)}$$

(3.15)

this conclusion have been questioned by Webster (see Yeh and Jaffé, 1959b), who prefers the idea of a tautomeric equilibrium with the proton on the one or the other of the nitrogen atoms.

Extension of this work to disubstituted azobenzenes (Yeh and Jaffé, 1959a) yielded several independent reaction series, in which one of the benzene rings was substituted by a constant substituent (Table 3.9). An analysis of substituent effects in this Table (Yeh and Jaffé, 1959b) showed that the pK_{BH+} values of the first two series are better correlated with σ^+ constants, whereas those of series (3) and (4) correlate better with σ constants. This was believed to be consistent with the bridged cation structure (3.15) (Yeh and Jaffé, 1959b). These observations can, however, be equally explained by assuming that the mesomeric effect of the strongly electron-releasing p-MeO and p-OH groups is dominant and directs the proton to the more distant nitrogen by resonance stabilization of the cation (3.16). The effect of substituents in the adjacent ring

$$\text{RO-C}_6\text{H}_4\text{-N=NH}^+\text{-C}_6\text{H}_5 \longleftrightarrow \text{RO}^+=\text{C}_6\text{H}_4=\text{N-NH-C}_6\text{H}_5$$

(3.16)

would then be governed by the σ constants. Conversely, if the substituent is totally unable to stabilize the cation in this way (as NO_2 and H are), the substituents in the other ring then become dominant if they can and require a correlation with σ^+ constants. The obvious consequence is the existence of tautomeric equilibria in certain intermediate cases.

A p-amino substituent enhances the basicity of azobenzene very considerably ($pK_{BH+} = 2\cdot82$—Bascombe and Bell, 1959). The conjugate acid of p-dimethylaminoazobenzene represents a tautomeric mixture of the ammonium (3.17) and the quinonoid-azonium (3.18) forms. Several estimates of

$$\text{Ph-N=N-C}_6\text{H}_4\text{-}\overset{+}{\text{N}}\text{H(CH}_3)_2 \rightleftharpoons \text{Ph-NH-N=C}_6\text{H}_4=\overset{+}{\text{N}}\text{(CH}_3)_2$$

(3.17)   (3.18)

the tautomeric equilibrium constant $K_T = [\mathbf{3.17}]/[\mathbf{3.18}]$, made by Yeh and Jaffé (1959c), lie in the region of 2–3.

More recently the effect of ionic solubilizing groups (such as SO_3^- and

N(CH$_3$)$_2$) on the indicator behaviour of a number of substituted azobenzenes and 2-(*para*-substituted phenylazo)-1-naphthols was studied by Reeves (1966) in aqueous sulphuric acid. The series examined include anionic, dianionic and zwitterionic bases, some of which exist as tautomeric equilibria. The basicities of azobenzenes with the p-SO$_3^-$ group as a constant substituent (Table 3.9) are correlated with σ^+ substituent constants of the *para* substituents in the other ring. This is consistent with the inability of the SO$_3^-$ group to conjugate with the azo group through the benzene ring, owing to its tetrahedral geometry.

3.3.2.4. *Imines and Nitriles*

The basicities of aldimines and ketimines (Schiff bases) do not appear to have been measured, but n.m.r. spectra of their conjugate acids have been recorded recently by Olah and Kreienbühl (1967) in superacid systems (HSO$_3$F, HSO$_3$F—SbF$_5$, or D$_2$SO$_4$—SbF$_5$, with SO$_2$ as diluent). The results indicate that there is no free rotation around the C—N bond, and therefore support the immonium structure of the cation (**3.19**) in preference to the aminocarbonium ion (**3.20**). The contribution from the latter may only be minor. The chemical

$$\begin{array}{c} R_1 \\ R_2 \end{array}\!\!C\!=\!N\!\begin{array}{c} R_3 \end{array} \xrightarrow{H^+} \begin{array}{c} R_1 \\ R_2 \end{array}\!\!C\!=\!\overset{+}{N}\!\begin{array}{c} R_3 \\ H \end{array} \qquad (3.19)$$

$$\begin{array}{c} R_1 \\ R_2 \end{array}\!\!\overset{+}{C}\!-\!N\!\begin{array}{c} R_3 \\ H \end{array} \qquad (3.20)$$

shift and coupling data indicate that the immonium form is predominant, even when R$_1$ or R$_2$ or both are phenyl groups, which would help to delocalize the positive charge on the carbon atom. The cations exist in the *cis* and *trans* forms, the *trans* form being characterized by larger coupling constants between the added proton and the *trans* R group (for R$_1$ = H, J = 17·4 Hz).

For benzaldehyde semicarbazone a pK_{BH^+} value of $-1\cdot1$ was reported, based on spectrophotometric measurements in perchloric acid (Wolfenden and Jenks, 1961). Its basicity is rather insensitive to substituent effects, which suggests that protonation occurs on the terminal amide group.

Nitriles are very weak bases. A cryoscopic investigation of Hantzsch (1909) had shown that they are incompletely protonated in pure sulphuric acid. Liler and Kosanovic (1959) attempted to determine conductometrically the basicity of acetonitrile and benzonitrile in 100 per cent sulphuric acid, but found that conductances at 25°C varied with time owing to hydrolysis. The variation was, however, sufficiently slow to allow an extrapolation of conductances to zero time to be made. Using the conductometric method (see Section 3.2.6) K_b values were obtained for acetonitrile (0·157) and for benzonitrile (0·069), which, recalculated to the acidity function scale of Table 2.1 (p. 37), correspond to pK_{BH^+} values of $-10\cdot02$ and $-10\cdot34$ respectively. These values were later confirmed by Deno and Wisotsky (1963) by Raman spectra and by Deno *et al.* (1966b) by n.m.r. spectra. They found chloroacetonitrile to be half-protonated in 30 per cent oleum, which corresponds to a pK_{BH^+} value of $-13\cdot9$.

In 100 per cent sulphuric acid nitriles undergo hydrolysis to produce amides. The rate of the process was studied by both conductance (Liler and Kosanović, 1959) and n.m.r. (Deno *et al.*, 1966b).

The structures of protonated nitriles were recently studied by n.m.r. spectroscopy in superacid media at low temperature (Hogeveen, 1967; Olah and Kiovsky, 1968a). The spectra show that the ions have a linear configuration, because only one isomer was observed in all cases. This means that the structure of the cations is that of carbammonium ions R—C≡$\overset{+}{N}$H, and not that of iminocarbonium ions R—$\overset{+}{C}$=N—H, which should give rise to two isomers. The latter form makes only a minor contribution to the resonance structure of the cations.

3.3.2.5. *Amides and Imides*

Amides have been regarded as weak bases for a long time, but until recently most information on their basicities was based on titrations in glacial acetic acid, with uncertainties involved in the recalculation of the results to pK values in water. Since the cryoscopic investigations in 100 per cent sulphuric acid by Oddo and Scandola (1909c), it was known that they are fully monoprotonated in this medium. The amide group protonation of glycineamide is, however, incomplete ($i = 2\cdot7$), whereas phthalimide is fully monoprotonated (O'Brien and Niemann, 1957). The adducts of amides with hydrogen halides have been well known for a long time. The crystalline acetamide hydrogen sulphate (m.p. $43\cdot9°$C) was reported more recently (Tutundžić *et al.*, 1954).

Attempts to determine the pK_{BH^+} values of amides in high dielectric constant media of strong aqueous acids have been made relatively recently, beginning with the work of Goldfarb *et al.* (1955), who made use of the changes in carbonyl absorption with acidity (mainly for aliphatic compounds) but, owing to low absorption, these results could not be very accurate. The basicities of substituted benzamides were studied by u.v. spectroscopy (Edward *et al.*, 1960) and were subsequently recalculated to the H_A acidity function scale (Yates and Stevens, 1965). Benzamide has a pK_{BH^+} value on this scale of $-1\cdot74$, and pK_{BH^+} values of substituted benzamides correlate with the σ substituent constants, with $\rho = 0\cdot92$ (Yates and Stevens, 1965). This was interpreted by Edward *et al.* (1960) as indicating that amides are *N*-protonated, because the dissociation of the conjugate acids of amides (3.21) is closely similar to that

$$\text{R—C}\underset{NH_3^+}{\overset{O}{\diagup}} \qquad \text{R—C}\underset{OH}{\overset{O}{\diagup}} \qquad \text{R—}\overset{+}{\text{C}}\underset{NH_2}{\overset{OH}{\diagup}}$$

(3.21)　　　　　(3.22)　　　　　(3.23)

of carboxylic acids (3.22). The alternative possibility is *O*-protonation (3.23). The basicities of substituted aromatic carbonyl compounds (see Sections 3.3.4.3 to 3.3.4.7) are, however, all correlated by σ^+ constants. Moreover, the pK_{BH^+} value of benzamide deviates strongly from the pK_{BH^+} *vs.* carbonyl frequency correlation followed by all other aromatic carbonyl bases (see Section 3.3.4.7 and Fig. 3.12).

The question of *O-* vs. *N*-protonation has been for many years a subject of controversy, a detailed account of which may be found in reviews by Katritzky and Jones (1961) and Arnett (1963), who decided firmly in favour of *O*-protonation. On the whole, preference was given to spectroscopic information, which is itself ambiguous. Inadequate consideration was given to acid/base strengths, partly because accurate pK_{BH^+} values became available only in recent years. Therefore acid/base strengths of amides will be considered more fully here.

A renewed attempt was made by Edward and Wang (1962) to determine the pK_{BH^+} value of an aliphatic amide, propionamide, by u.v. spectrophotometry, and a value of -0.9 was obtained. More rapid progress in obtaining further data was achieved by use of the n.m.r. method. The chemical shifts of groups in the spectra of aliphatic amides undergo a downfield shift when amides are protonated, giving good sigmoid curves of chemical shift *vs.* H_A, from which the pK_{BH^+} values can be deduced (see Section 3.2.2). Propionamide was studied by Edward *et al.* (1962), butyramide by Armstrong and Moodie (1968), dimethylacetamide by Haake *et al.* (1967), and a range of substituted

TABLE 3.10

The pK_{BH^+} values of substituted acetamides, *N*-methylacetamides and *N,N*-dimethylacetamides at 33·5°C (Liler, 1969)

Amide	—NH$_2$	—NHCH$_3$	—N(CH$_3$)$_2$
Acetamide	−0·93	−0·72	−0·36
Propionamide	−0·99	−0·85	−0·62
n-Butyramide	−1·03		
Chloroacetamide	−2·80		
Cyanoacetamide	−3·69		
Phenylacetamide	−1·68		
Dichloroacetamide	−4·21	−3·84	−3·73
iso-Butyramide	−1·26		
Trimethylacetamide	−1·43		
Trichloroacetamide		(−4·9)†	

† Only an estimate because the H_A acidity function has not been defined beyond −4·79.

acetamides and some of their N-methyl derivatives by Liler (1969). In the latter study chemical shifts of C—CH protons and N—CH$_3$ protons were both measured and concordant values of pK_{BH^+} obtained from both. The results are reported in Table 3.10.

Although the interpretation of small pK differences is usually uncertain, the differences arising from N-methyl substitution in acetamide and propionamide are not only in direction, but also in order of magnitude, close to differences calculated on a purely statistical basis, assuming N-protonation. The inductive effects of N-methyl groups thus appear to be small.

The pK_{BH^+} values of substituted acetamides are linearly correlated with those

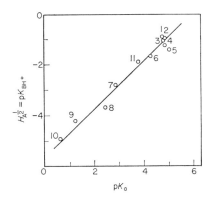

FIG. 3.9. The correlation of pK_{BH^+} values of substituted acetamides with pK_a values of the corresponding carboxylic acids: 1, Acetamide; 2, Propionamide; 3, n-Butyramide; 4, iso-Butyramide; 5, Trimethylacetamide; 6, Phenylacetamide; 7, Chloroacetamide; 8, Cyanoacetamide; 9, Dichloroacetamide; 10, Trichloroacetamide (estimate for the N-methyl derivative); 11, Formamide.

of the corresponding carboxylic acids, with a slope of 0·94 (Fig. 3.9). A point for formamide is also included in the Figure (Liler, unpublished). The slope of the line is close to $\rho = 0·92$ found for substituted benzamides (Yates and Stevens, 1965). The deviations of points for methyl substituted acetamides towards higher acidities of the conjugate acid suggest that steric effects are greater in the dissociation of amide cations than in the dissociation of carboxylic acids. This is consistent with strong solvation of a cation with a localized charge. The point for cyanoacetamide deviates probably because of hydrogen bonding of the cyano group to the highly acidic solvent.

Since it has been pointed out in Section 3.3.1.2 that electron withdrawal by the carbonyl group is part inductive and part mesomeric, the relative importance of the two effects being the same as for the phenyl group, a comparison of pK values from Table 3.10 with those of amides with steric inhibition

3. PROTONATION OF VERY WEAK BASES

of resonance is of considerable interest. Such amides have been synthesized by Pracejus (1959) and their pK values determined (Pracejus et al., 1965) with the following results:

	p$K_{BH^+}^{22}$
2,2-Dimethyl-quinuclidone-6	5·33
exo-2,2,6-Trimethyl-quinuclidone-7	5·60
endo-2,2,6-Trimethyl-quinuclidone-7	5·60

Comparing these values with that for quinuclidine (pK_{BH^+} = 10·58), it can be seen that a reduction in basicity of 5–5·3 units is accountable in terms of the inductive effect alone. An unhindered amide with most closely comparable inductive and solvation effects is N,N-dimethylacetamide, for which Table 3.10 gives pK_{BH^+} = −0·36. This means that the additional electron withdrawal by the mesomeric mechanism accounts for 5·69 to 5·99 pK units. The situation is analogous to that of quinuclidine, benzoquinuclidine and dimethylaniline mentioned in Section 3.3.1.2 and may be represented in the same way:

pK_{BH^+}	10·58	5·33	−0·36
ΔpK_{BH^+}		5·25	5·69

This is a confirmation of the conclusion that the phenyl group and the carbonyl group are electron sinks of the same kind, which was reached on the basis of the correlation of carbonyl frequencies with σ_m constants (Section 3.3.1.2). The pK differences are about twice as great for the carbonyl group as an electron sink as for the phenyl group. The carbonyl group is therefore twice as effective an electron sink as the phenyl group.

In the controversy over the site of protonation of amides there are very few facts that have not been interpreted in terms of both N- and O-protonation. Thus the above pK differences between quinuclidones and ordinary amides are believed by Pracejus et al. (1965) to be due to the fact that quinuclidones are N-protonated, whereas amides are O-protonated. One fact which appears to be inexplicable in terms of O-protonation is the deviation of the pK_{BH^+} values of amides from the relationship between pK_{BH^+} value and carbonyl frequency followed by other carbonyl bases (see Section 3.3.4.7).

There are some other major pK differences that are consistent with N-protonation. For example, the introduction of a phenyl group instead of ethyl enhances the acidity of an alcohol, just as it enhances the acidity of an ammonium ion. A similar effect would be expected for the carbonyl group and is in fact found, as the following pK differences show:

	pK_a	ΔpK_a		pK_{BH^+}	ΔpK_{BH^+}
Ethanol	16·6		Ethylamine	10·7	
Phenol	10·0	6·6	Aniline	4·6	6·1
Acetic acid	4·7	11·9	Acetamide	−0·9	11·6

In each case the effect of the carbonyl electron sink is almost twice as great as the effect of the phenyl electron sink.

Urea, urethane and N-methyl urethane are also simply monoprotonated in 100 per cent sulphuric acid (Holstead et al., 1953), whereas tetramethylurea is probably diprotonated to the extent of 10 per cent in a 0·05 M solution (Gillespie and Robinson, 1965). The pK_{BH^+} value of urea is usually quoted as 0·1 and that of thiourea as −1 (Clark and Perrin, 1964). A recent attempt to determine the pK_{BH^+} of N-ethylurea by n.m.r. was unsuccessful because of the very small intramolecular chemical shift ($\Delta\nu_{CH_2}$-$\Delta\nu_{CH_3}$) caused by protonation (Armstrong and Moodie, 1968). The basicities of urethane (pK_{BH^+} = −3·03), N-methylurethane (pK_{BH^+} = −2·92) and N,N-dimethylurethane (pK_{BH^+} = −3·16) were determined by this method (Armstrong and Moodie, 1968). Available n.m.r. evidence shows all these compounds to be O-protonated (Olah and Calin, 1968). The symmetrical urea molecule is strongly resonance stabilized. This symmetry is not destroyed by O-protonation, which is therefore favoured. O-protonation of carbamic acid and

$$\begin{array}{c} R \\ R \end{array} \!\!\!\! N \!\!=\!\! \overset{+}{C} \!\!\! \begin{array}{c} OH \\ OR' \end{array}$$

(3.24)

urethanes leads to cations (3.24) in which the charge is spread over four centres. This considerable spreading of charge is an important internal stabilizing factor. An exception is the ethyl N,N-diisopropyl compound which is N-protonated (Armstrong et al., 1968b), probably owing to steric hindrance to planarity of structure (3.24).

Structural information obtained from n.m.r. spectra on the site of protonation of amides is in many ways ambiguous. In dilute acid where amides become protonated, the N—CH$_3$ doublet in the spectrum of N-methylacetamide (due to NH—NCH$_3$ coupling, J = 5 Hz) collapses to a singlet (Takeda and Stejskal, 1960), and also the resonances of the cis and trans methyl groups (3 per cent cis present in the unprotonated amide) both merge on protonation (Liler, 1969). Likewise the two resonances of the cis and the trans methyl groups in the spectrum of N,N-dimethylacetamide (0·157 ppm apart) collapse in dilute acid into a singlet (Berger et al., 1959), but in the spectrum of N,N-dimethyl-formamide the same is observed only at higher temperature (43°C)—Fraenkel and Franconi (1960). All N-methyl- and N,N-dimethyl-amides listed in Table 3.10 show singlet amide group resonances in dilute acid (Liler, 1969). All these facts are explicable in terms of NH exchange and free rotation about the C—N bond, and thus represent clear evidence for N-protonation in dilute acid. However, studies of the n.m.r. spectra of amide cations in 100 per cent sulphuric acid (Fraenkel and Niemann, 1958) and in fluorosulphuric acid at low temperature (Gillespie and Birchall, 1963; Birchall and Gillespie, 1963) have shown that the two N-methyl groups in the cations of N,N-dimethyl-amides become again non-equivalent in these media (although only 0·117 ppm or less apart), that in the cations of N-methylamides in these media the N-methyl group is coupled to one NH proton only ($J = 5$ Hz), and that the captured proton is non-equivalent with the other NH protons in the spectra of unsubstituted amides and N-methylamides. All these facts are apparently consistent with O-protonated cation structures. However, no clear identification has been made of the *cis* and the *trans* N-methyl groups in the cations of N-methylformamide, which should be present if O-protonation occurs— 8 per cent *cis* is present in the unprotonated amide (La Planche and Rogers, 1964; Birchall and Gillespie, 1963). Also, in the spectrum of formamide in pure fluorosulphuric acid at 25°C coupling of NH protons to the ^{14}N nucleus is apparent with $J = 60 \pm 4$ Hz (Gillespie and Birchall, 1963), which is usually associated with tetrahedral symmetry around that nucleus, and thus again suggests N-protonation. It is therefore conceivable that the spectra in concentrated acids signify that the N-protonated cations interact with the solvent by hydrogen bonding to produce solvates of fixed structure, the hydrogen-bonded proton being non-equivalent with the other NH protons. If these spectra are due to O-protonated species, then a change from N-protonation in dilute acid to O-protonation in concentrated acid occurs, which may be associated with decreasing availability of water for cation hydration.

The n.m.r. spectra of urea and N-methyl substituted ureas in superacid media were studied by Olah and White (1968). Ureas were found to be diprotonated in these media giving cations of structure (**3.25**):

$$\begin{array}{c} R \\ | \\ R-\overset{+}{N} \\ \diagdown \\ C=OH \\ \diagup \\ R-\overset{+}{N} \\ | \diagdown H \\ R \end{array} \qquad R = CH_3 \text{ or } H$$

(**3.25**)

The protonation constants of lactams were obtained by Huisgen *et al.* (1957) from non-aqueous titrations and by Virtanen and Södervall (1967) from u.v. spectra, and those of thiolactams were also determined from u.v. spectra

(Edward and Stollar, 1963). Only values based on u.v. spectra are given in Table 3.11. Pyrrolidone and N-methylpyrrolidone obey the H_A acidity function reasonably well (slopes 1·06 and 1·12). The slopes of $\log I$ vs. $-H_0$ plots for thiolactams are somewhat greater than unity, in contrast to amides, which give slopes considerably less than unity. This may be connected with the fact that thiolactams are S-protonated, thus giving cations with a delocalized charge.

TABLE 3.11

pK_{BH^+} values of some lactams and thiolactams at 25°C

Lactam	pK_{BH^+}†	Thiolactam	pK_{BH^+}‡
2-Pyrrolidone	−0·94	Thiopyrrolidone	−2·0
N-Methyl-2-pyrrolidone	−0·92	Thiopiperidone	−1·4
		Thiocaprolactam	−1·6

† Data of Virtanen and Södervall (1967).
‡ Data of Edward and Stollar (1963).

Evidence from n.m.r. spectra for S-protonation of thioamides appears to be unambiguous (Birchall and Gillespie, 1963).

The change in basicity of thiolactams with ring size parallels the change in basicity of the corresponding lactams, as found by Huisgen et al. (1957), according to the equation

$$pK_{BH^+} \text{ (thiolactam)} = 0{\cdot}62 pK_{BH^+} \text{ (lactam)} - 1{\cdot}8$$

(Edward and Stollar, 1963). The basicities of lactams were, however, determined in glacial acetic acid and are uncertain.

3.3.2.6. *Sulphonamides*

The basicities of some aliphatic sulphonamides were studied recently by Laughlin (1967) using the n.m.r. method. Aromatic sulphonamides are too sparingly soluble for the application of this method. The results are given in Table 3.12. The pK_{BH^+} values were calculated on the assumption that the protonation of these compounds follows the H_0 acidity function. The data for N,N-dimethyl sulphonamide fit, in fact, the H_0''' acidity function much better ($pK_{BH^+} = -7{\cdot}2$ on that scale).

There is complete agreement concerning the site of protonation of sulphonamides. They are protonated on the nitrogen according to Birchall and Gillespie (1963) and Laughlin (1967). The question of whether the considerable reduction in the basicity of the amino group by the adjacent tetracoordinate sulphur

3. PROTONATION OF VERY WEAK BASES

TABLE 3.12

pK_{BH^+} values of some sulphonamides (Laughlin, 1967)

Sulphonamide	% H_2SO_4 at half-protonation	pK_{BH^+}
N-methylmethane-	71·4	−6·0
N-ethylmethane-	71·4	−6·0
N,N-dimethylmethane-	67·6	−5·5

is purely inductive or not has been discussed by Laughlin (1967), who concludes that the N—S bond probably has some partial double bond character. The downfield chemical shift of the S—CH$_3$ protons caused by protonation is in fact larger than that of the N—CH$_3$ protons, suggesting that nitrogen already carries a partial positive charge in the amide itself. This observation is consistent with a model involving appreciable double bonding (**3.26**) but does not

$$CH_3-\overset{\overset{O}{\uparrow}}{\underset{\underset{O}{\downarrow}}{S}}-N\!\!<\quad\longleftrightarrow\quad CH_3-\overset{\overset{O^-}{\uparrow}}{\underset{\underset{O}{\downarrow}}{S}}=\overset{+}{N}\!\!<$$

(**3.26**)

rule out a purely inductive effect (Laughlin, 1967). It was pointed out in Section 3.3.1.2 that the sulphonyl group is not an electron sink of the same kind as the carbonyl group, but there may still be some partial coordination between N and S.

The basicities of some aromatic sulphonamides were determined spectrophotometrically (Virtanen and Maikkula, 1968). Benzenesulphonamide was found to behave as a Hammett base (p$K_{BH^+} = -6·64$), but its N-methyl and N-ethyl derivatives appear to follow the H_A acidity function more closely. They are all more basic than benzenesulphonamide itself, but the interpretation of relatively small pK differences is uncertain.

3.3.2.7. N-Nitrosamines and Oximes

An attempt to measure the base strength of N-nitrosamines in aqueous sulphuric acid by u.v. spectrophotometry was made by Layne et al. (1963). No simple protonation behaviour was found, but rather a complex series of changes. It was concluded that N-nitrosamine molecules exist in acid solution in four spectroscopically distinguishable forms, with the proportion of each determined by the acid concentration. Some of these may be complexes with sulphuric acid or with bisulphate ions.

Oximes are very weak bases and decompose in acid solution. In an attempt to determine their basicities by u.v. spectrophotometry in aqueous hydrochloric acid, extrapolation of spectra to zero time was necessary (Ellefsen and Gordon, 1967). Nonetheless, the absorption curves show good isosbestic points. Diacetylmonoxime and dimethylglyoxime were studied in this way and pK_{BH^+} values of $-3\cdot16$ and $-0\cdot94$ were obtained, respectively. The structure of the protonated oxime group was assumed to be

$$\mathrm{\underset{}{>}C{=}\overset{+}{N}\underset{H}{\overset{OH}{\diagup}}},$$

which was confirmed by n.m.r. spectra in superacid media (Olah and Kiovsky, 1968a).

3.3.2.8. *Phosphines and Arsines*

Information on the basicities of phosphines, mostly based on titrations in partially aqueous and non-aqueous media, has been reviewed by Arnett (1963). Trimethylphosphine is strongly basic ($pK_{BH^+} = 8\cdot80$) (Perrin, 1965). According to potentiometric titrations in 50 per cent ethanol (Davies and Addis, 1937) phenyldimethylphosphine has a pK_{BH^+} of $4\cdot32$, i.e. it is a little less basic than dimethylaniline. Mesityldimethyl phosphine and mesityldimethyl arsine have pK_{BH^+} values of $6\cdot51$ and $2\cdot11$, respectively. Accordingly, the basicity of all these compounds is well pronounced.

The only investigation of the behaviour of phosphines in sulphuric acid media is a cryoscopic study related by Gillespie and Robinson (1959). Triphenylphosphine is fully protonated in pure sulphuric acid, but slowly undergoes some further reaction, probably sulphonation and/or oxidation.

3.3.3. BASES INVOLVING MORE COMPLEX GROUPS CONTAINING N, P AND As

Compounds to be discussed in this part contain nitrogen, phosphorus and arsenic in the oxidation state $+5$. Here belong some organic oxides and sulphides, azoxycompounds and nitrocompounds. All pK_{BH^+} values are for 25°C unless otherwise stated.

3.3.3.1. *Amine Oxides*

The oxides of tertiary aliphatic amines are relatively strong bases, whose basicity can be measured by potentiometric titration. They are weaker bases than the parent amines by 5–6 pK units (Nylén, 1941). Information on phenyl amine oxides appears not to be available.

Pyridine oxides have been studied more recently (Jaffé and Doak, 1955; Johnson *et al.*, 1965; Klofutar *et al.*, 1968). A pK difference between pyridine and pyridine oxides of about 4–5 units makes pyridine oxides very weak bases, which become protonated in very dilute aqueous acid or in more strongly acidic solutions, depending on the nature of the substituents present. For the stronger bases the potentiometric method in aqueous solution is suitable, but for the weaker bases the spectrophotometric method in aqueous sulphuric acid has been used by all the authors mentioned. The earlier measurements of Jaffé and Doak (1955) were essentially confirmed by Johnson et al. (1965), who also showed that pyridine oxides follow the H_0 acidity function. A plot of

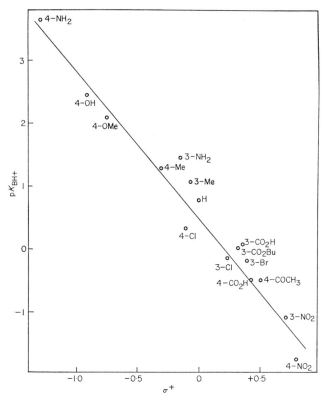

FIG. 3.10. The plot of pK_{BH^+} values of substituted pyridine-1-oxides vs. σ^+ constants.

the pK_{BH^+} values of pyridine oxides vs. σ constants gave a ρ value of 2·09, not very different from that for the dissociation of phenols ($\rho = 2\cdot11$) (Jaffé and Doak, 1955). It was also apparent from the Hammett plot that 4-substituents possessing a lone electron pair (—NH_2, —OH, —OCH_3) require enhanced σ values, and therefore all available pK_{BH^+} data for 2-, 3- and 4-substituted pyridine oxides are replotted vs. σ^+ constants in Fig. 3.10. The improved correlation shows that structures such as (**3.27**) and (**3.28**) play an important part in the resonance stabilization of protonated pyridine oxides. The basicity

$$\overset{+}{H_2N}=\langle\rangle\overset{\cdot\cdot}{N}-OH \qquad \overset{+}{HO}=\langle\rangle\overset{\cdot\cdot}{N}-OH$$

(3.27)   (3.28)

and tautomerism of these latter compounds were studied by Gardner and Katritzky (1957). Their results are also included in Fig. 3.10. The ρ constant now obtained is 2.33. Johnson et al. (1965) believe that σ^- constants should be used for the 4-acetyl and 4-nitro derivatives, whereas Nelson et al. (1967) suggest a new set of σ constants for use in correlating the reactivities of pyridine-1-oxides.

3.3.3.2. Phosphine Oxides and Arsine Oxides

Phosphines are less basic than amines, and arsines even less so than phosphines (see Section 3.3.2.8). The order of basicity of oxides differs considerably, phosphine oxides being the weakest bases. Information on their basicities became available only very recently. Haake et al. (1967) studied the basicities of dimethylphenylphosphine oxide and diphenylmethylphosphine oxide using the n.m.r. chemical shift method, whereas Klofutar et al. (1968) studied triphenylphosphine oxide using the solvent extraction method and found that it behaves as a Hammett base ($pK_{BH^+} = -2\cdot10$). The first two are of similar basicity, but give slopes of considerably less than unity in the plots of $\log I$ vs. $-H_0$. Triphenylarsine oxide also behaves as a Hammett base with $pK_{BH^+} = 0\cdot99$ (Klofutar et al., 1968). Trimethylarsine oxide is much more basic ($pK_{BH^+} = 3\cdot75$), and trimethylstibine oxide even more so, having $pK_{BH^+} = 5\cdot36$ (Nylén, 1941). There is thus a basicity minimum with phosphine oxides, due to the back donation of the oxygen p electrons to the vacant d orbitals of phosphorus. The back donation is less effective with arsine and stibine oxides, probably because of the lower electronegativity of arsenic and antimony.

Dimethylphosphinic acid and dimethylphosphonate have pK_{BH^+} values of $-4\cdot07$ and $-5\cdot22$ respectively (Haake et al., 1967).

As with amines and amine oxides, phosphine oxides are also some 5–6 pK units less basic than the corresponding phosphines. Some of the corresponding thiocompounds do not follow the H_0 acidity function, and hence their basicity could not be determined. Both dimethylphenylphosphine sulphide and diphenylmethylphosphine sulphide are half-protonated at $H_0 = -4\cdot5$ (Haake et al., 1967).

3.3.3.3. Azoxybenzenes

The basicities of a series of variously substituted azoxybenzenes were studied spectrophotometrically in 20 per cent ethanolic sulphuric acid by Hahn and

Jaffé (1962). The basicity of azoxybenzene itself is 3·55 units less than that of azobenzene, as compared with 4·5 units for pyridine and pyridine-1-oxide. The similarity of this pK difference suggests that azoxybenzenes are probably O-protonated, as Hahn and Jaffé (1962) also conclude from an analysis of substituent effects. The azoxybenzenes studied form several reaction series, with all substituents in *para* positions, as shown in Table 3.13.

TABLE 3.13

pK_{BH^+} values of azoxybenzenes at 25°C (Hahn and Jaffé, 1962)

X	Y = OMe	Y = Me	Y = H	Y = Cl	Y = Br
MeO	−5·23		−6·10		
EtO			−6·04		
Me		−5·47	−6·04		−6·90
H	−6·15	−6·16	−6·45	−6·96	−6·94
Cl				−7·69	
Br		−6·95	−7·01		−7·77
NO₂			−9·83		

3.3.3.4. Nitrocompounds

The basicity of nitrocompounds has been known since Hantzsch (1909) showed that p-nitrotoluene was extensively protonated in pure sulphuric acid. Cherbuliez (1923) and then Masson (1931) obtained a 1:1 adduct of nitrobenzene with sulphuric acid, and similar adducts have more recently been obtained for nitrotoluenes and nitromethane (Liler, 1959).

Nitrocompounds become protonated in the region of pure sulphuric acid and their protonation equilibria were used in extending the acidity function H_0 into oleum solutions in the spectrophotometric studies of Brand (1950) and Brand *et al.* (1952). The cryoscopic method (Gillespie and Robinson, 1957b) and the conductometric method (Gillespie and Solomons, 1957; Liler, 1962) were also used to estimate their basicities. Some negatively substituted nitrocompounds and polynitrocompounds are only slightly protonated or unprotonated even in 100 per cent sulphuric acid, and it was therefore necessary in the conductometric estimation to take into account the dilution effect discussed in Section 3.2.6. The results of all three methods agree closely, as Table 3.14 shows. The pK_b values obtained from this Table correlate very well with σ^+ constants, giving $\rho = 2·77$ (Liler, 1962). This high ρ value probably reflects the high polarizability of the nitro group. There is no clear *ortho* effect in the protonation of nitrobenzenes, which suggests that there is no extensive solvation of the protonated nitro group, probably owing to the delocalization of the charge into the benzene ring.

TABLE 3.14

Basic ionization constants (K_b) of substituted nitrobenzenes (Liler, 1962)†

	Method			
	Conductometric		Cryoscopic	Spectrophotometric
Substituent	(25°C)	(10°C)	(10°C)	(18°C)
p-t-Bu			1×10^{-2}	1×10^{-2}
p-Me	$1 \cdot 05 \times 10^{-2}$		$9 \cdot 2 \times 10^{-3}$	$9 \cdot 1 \times 10^{-3}$
o-Me	$7 \cdot 3 \times 10^{-3}$		$6 \cdot 1 \times 10^{-3}$	
m-Me	$2 \cdot 8 \times 10^{-3}$		$2 \cdot 0 \times 10^{-3}$	$2 \cdot 5 \times 10^{-3}$
H	$1 \cdot 3 \times 10^{-3}$	$1 \cdot 0 \times 10^{-3}$	$1 \cdot 1 \times 10^{-3}$	$9 \cdot 6 \times 10^{-4}$
p-Cl	$5 \cdot 7 \times 10^{-4}$	$4 \cdot 3 \times 10^{-4}$	3×10^{-4}	$4 \cdot 2 \times 10^{-4}$
o-Cl	$2 \cdot 4 \times 10^{-4}$	$1 \cdot 8 \times 10^{-4}$		
m-Cl	$1 \cdot 3 \times 10^{-4}$	$9 \cdot 5 \times 10^{-5}$		$1 \cdot 1 \times 10^{-4}$
4-Me-3-NO$_2$	$5 \cdot 3 \times 10^{-5}$			$2 \cdot 9 \times 10^{-5}$
o-NO$_2$ } m-NO$_2$ }	$2 \cdot 8 \times 10^{-5}$	$1 \cdot 6 \times 10^{-5}$		
p-NO$_2$	$2 \cdot 9 \times 10^{-5}$	$1 \cdot 7 \times 10^{-5}$		
3-NO$_2$-4-Cl	Non-electrolyte			3×10^{-6}

† K_b ($=[BH^+][HSO_4^-]/[B][H_2SO_4]$) is given here, because this is directly obtained from conductometric and cryoscopic work. Conversion to pK_{BH^+} (see Section 3.2.8) gives for nitrobenzene a value of $-12 \cdot 07$ (based on H_0 data in Table 2.1, p. 37).

Yates and Thompson (1967) showed that there was a linear relationship between the asymmetric stretching frequency of the nitro group and the basicities of substituted nitrobenzenes or the σ^+ constants of the substituents. This is accounted for by the fact that the protonation of the nitro group is itself an asymmetric process, in which one of the two nitrogen-oxygen bonds is decreased in bond order, while the other is simultaneously increased:

$$-\overset{+}{N}\underset{O^{\frac{1}{2}-}}{\overset{O^{\frac{1}{2}-}}{\lessgtr}} \xrightarrow{H^+} -\overset{+}{N}\underset{O}{\overset{OH}{\lessgtr}}$$

The relationship is given by the equation

$$\nu_{as} = -10 \cdot 53 pK_{BH^+} + 1395 \text{ cm}^{-1}$$

The symmetric stretching frequency does not correlate with either σ or σ^+ constants of the substituents.

Nitromethane is unstable in pure sulphuric acid. The conductances of its solutions increase slowly with time. Gillespie and Solomons (1957) report a

3. PROTONATION OF VERY WEAK BASES   117

K_b value of $2 \cdot 6 \times 10^{-3}$ mol kg^{-1}, based on the initial conductance values. When the solution of nitromethane in sulphuric acid is poured into water after some hours, a strongly reducing solution is obtained (probably containing hydroxylamine) and some carbon monoxide appears to be evolved. Aliphatic nitrocompounds are known to react with mineral acids (Hass and Riley, 1943). Deno et al. (1966a) found by the n.m.r. method that nitromethane is half-protonated in 23 per cent oleum, and nitroethane likewise. 2-Nitropropane partially decomposes on mixing. The knowledge of the behaviour of aliphatic nitrocompounds in sulphuric acid and oleum is thus still very incomplete.

In superacid media (HSO$_3$F-SbF$_5$) nitroalkanes and variously substituted nitrobenzenes are fully protonated. The rotation of the protonated nitro group was studied by n.m.r. spectroscopy (Olah and Kiovsky, 1968b).

3.3.3.5. Miscellaneous Compounds Containing P^V

Phosphoric acid is fully protonated in 100 per cent sulphuric acid according to the equation

$$H_3PO_4 + H_2SO_4 = H_4PO_4^+ + HSO_4^- \quad (3.47)$$

Potassium pyrophosphate abstracts some water from the pure acid according to

$$K_4P_2O_7 + 7H_2SO_4 = 4K^+ + 2P(OH)_4^+ + HS_2O_7^- + HSO_4^- \quad (3.48)$$

(Gillespie and Robinson, 1965). The basicity constant of phosphoric acid is not known.

Dimethyl phosphinic acid (**3.29**) and dimethyldithiophosphinic acid (**3.30**) have been found by the n.m.r. method to be half-protonated at H_0 values of $-4\cdot3$ and $-5\cdot72$ respectively (Haake et al., 1967), and methyl diphenylphosphinate (**3.31**) at $-4\cdot8$ (Haake and Hurst, 1966). Only compound (**3.30**) approaches the behaviour of a Hammett base. Compounds (**3.29**) and (**3.31**) give slopes of log I vs. $-H_0$ of 0·35 and 0·46 respectively. This could be due to the solvation of the cations by hydrogen bonding.

$$\begin{array}{ccc}
\text{O} & \text{S} & \text{O} \\
\| & \| & \| \\
\text{CH}_3\text{—P—OH} & \text{CH}_3\text{—P—SH} & \text{C}_6\text{H}_5\text{—P—OCH}_3 \\
| & | & | \\
\text{CH}_3 & \text{CH}_3 & \text{C}_6\text{H}_5 \\
(\textbf{3.29}) & (\textbf{3.30}) & (\textbf{3.31})
\end{array}$$

Some work on phosphonitrilic chlorides, (NPCl$_2$)$_n$, by D. R. Smith, reported by Paddock (1964), shows them to be very weak bases (pK_{BH^+} values of $-8\cdot2$ and $-8\cdot6$ are reported for the trimer and the tetramer respectively). The number of protons taken up in pure sulphuric acid is far short of the number of N atoms present in the ring (e.g. for $n = 3$, $i = 1\cdot2$).

3.3.4. BASES INVOLVING O AND S AS THE BASIC CENTRE

Unlike bases containing nitrogen, hardly any compounds containing oxygen are sufficiently basic to show basicity in aqueous solution and to give salts stable to hydrolysis. Some stable salts have been known about since the turn of the century, when Collie and Tickle (1899) found that α,α-dimethyl-γ-pyrone gives salt-like compounds with a number of acids. It was soon shown by cryoscopic studies in 100 per cent sulphuric acid that a number of oxygen-containing compounds behave as simple bases (Hantzsch, 1908; Oddo and Scandola, 1908). Crystalline compounds with sulphuric acid have since been reported for 1,4-dioxane (Paternò and Spalino, 1907), ethers (Tchelintzev and Kozlov, 1914), phenols, benzophenone and a number of carboxylic acids (Kendall and Carpenter, 1914), some dicarboxylic acids (Tutundžić et al., 1954) and others. All these compounds are hygroscopic and easily hydrolysed. The 1:1 compound of 1,4-dioxane with sulphuric acid (m.p. 101°C) has recently been subjected to X-ray analysis (Hassel and Roemming, 1960). It was found to contain endless chains of alternating dioxane and sulphuric acid molecules, which probably explains its high melting point. Each sulphuric acid molecule is linked by hydrogen bonding to two neighbouring dioxane molecules.

In 1935, when Flexser, Hammett and Dingwall (1935) began spectrophotometric investigations of the basicities of some oxygen bases (acetophenone, anthraquinone, benzoic and phenylacetic acids), quantitative information on the strength of oxygen bases was virtually non-existent. A lot of indirect information on the basicities of oxygen compounds is now available in the literature, and very many compounds have been studied spectrophotometrically and by other methods in sulphuric acid/water mixtures. Apart from the review by Arnett (1963), recent reviews dealing specifically with all aspects of oxygen basicity are available (Gerrard and Macklen, 1959; Mavel, 1961). More information has since become available and a lot of data have been correlated and systematized. This both justifies and facilitates a detailed survey of the basicities of oxygen bases in the present monograph.

There are no oxygen bases that show clear basic behaviour in aqueous solution. The strongest oxygen base known, 2,6-dimethyl-4-pyrone, is as weakly basic as the amides (pK_{BH^+} = −0.28, see Section 3.3.4.8). It was shown to be fully monoprotonated in pure sulphuric acid (Oddo and Scandola, 1910). 4,6-Dimethyl-2-pyrone has also been shown cryoscopically to be fully protonated in 100 per cent sulphuric acid (Wiley and Moyer, 1954), but its basicity does not seem to have been determined. The unravelling of the structure of their cations has played a very important part in the development of the theory of resonance (see, for example, a recent historical account by Nenitzescu, 1968). Their structures are resonance hydrids (**3.32**) and (**3.33**) in which the positive charge is shared between the two oxygen atoms.

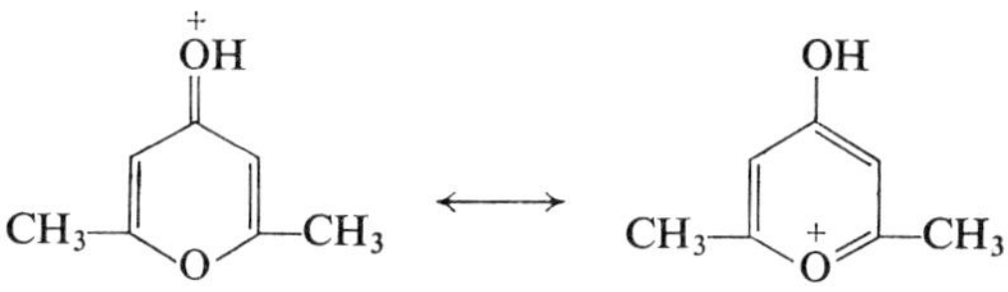

3. PROTONATION OF VERY WEAK BASES

(3.32)

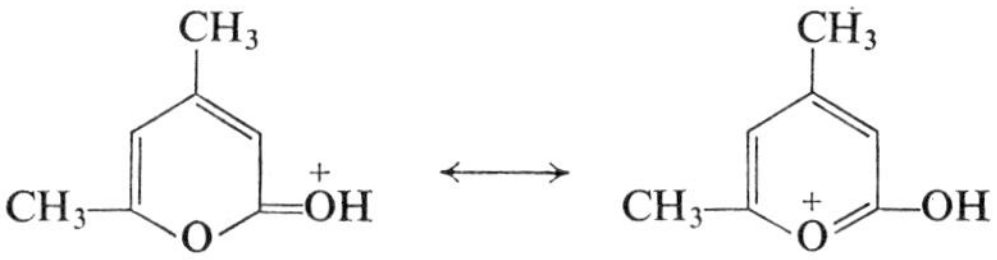

(3.33)

All oxygen bases may thus be said to become protonated only in moderately or strongly acid solutions. Many have been studied in sulphuric acid/water mixtures, and some remain incompletely protonated even in 100 per cent sulphuric acid. Their protonation will here be discussed in groups of closely structurally related compounds, including their sulphur analogues. Finally, the basicities of some compounds containing more complex groups involving elements of Group VI will be reviewed. All pK_{BH^+} values are for 25°C, except when otherwise stated.

3.3.4.1. *Alcohols and Phenols*

As was mentioned earlier (Section 3.2), the determination of the basicity of aliphatic compounds is often difficult. Alcohols fall into this category, with the additional complication that they react with sulphuric acid to form sulphates or carbonium ions in concentrated acid, and both reactions may be accompanied by further changes. In dilute acid alcohols are, however, sufficiently stable to allow their protonation equilibria to be studied. Only in recent years has agreement been reached between the several methods employed. Arnett and Anderson (1963) used solvent extraction coupled with gas/liquid chromatography to determine the basicities of several primary, secondary and tertiary alcohols. It is still uncertain whether alcohols behave as Hammett bases, and therefore their basicities are given in Table 3.15 as the percentage of sulphuric acid at half-protonation, as well as $H_0^{(1/2)}$ ($=pK_{BH^+}$ if Hammett base behaviour is found). These values were essentially confirmed by a Raman investigation of methanol (Deno and Wisotsky, 1963) and by solubility studies on some phenyl-substituted alcohols by Deno and Turner (1966)—results also included in Table 3.15. The application of the n.m.r. method to ethanol by Edward *et al.* (1962) showed a continuous monotonic change of the internal chemical shift between the methyl and the methylene protons from 20–90 per cent sulphuric acid. Although difficult to interpret, this appeared to indicate a pK_{BH^+} value of -4.8. Deno and Wisotsky (1963)

TABLE 3.15

Basicities of aliphatic alcohols

Alcohol	Weight % H_2SO_4 at half-protonation	$H_0^{(1/2)}$	Method	References†
Methanol	41	−2·5	Raman	a
n-Butanol	39·5	−2·3		
iso-Butanol	36·5	−2·2	Distribution	b
t-Butanol	41·5	−2·6		
t-Amylalcohol	38·5	−2·3		
8-Phenyloctanol	41	−2·3	Solubility	c

† a. Deno and Wisotsky (1963); b. Arnett and Anderson (1963); c. Deno and Turner (1966).

claimed that these results could be reinterpreted to indicate a pK_{BH^+} between −2 and −3 (changes of chemical shift at higher acidities being ascribed to changing solvation).

Nuclear magnetic resonance spectra of the cations of primary and secondary alcohols in superacid solution at low temperature (in FSO_3H-SbF_5-SO_2 at −60°C) have been reported by Olah and Namanworth (1966).

Phenols undergo sulphonation in concentrated sulphuric acid, but in more dilute sulphuric acid/water mixtures the fast reaction is simple protonation, and sulphonation may be minimized by working with chilled solutions. The pK_{BH^+} value of phenol determined spectrophotometrically under these conditions is −7·04 (Arnett and Wu, 1960b). However, this figure is doubtful because in aqueous perchloric acid the spectra do not change in any simple way, and protonation appears to occur only at $H_0 < -9$ (Yates and Wai, 1965).

The protonation site of phenol was concluded to be the oxygen atom from a comparison of u.v. spectral changes caused by the protonation of phenol and of aniline (Arnett and Wu, 1960b). The alternative is protonation on the *para* carbon atom. Good n.m.r. evidence has been obtained for this in superacid media at low temperature. This evidence will be discussed in the following section in connection with aromatic ethers, in which the same duality of protonation sites is possible.

The basicities of thioalcohols and thiophenols have apparently not been studied, but protonated thiols have been observed by n.m.r. spectroscopy in superacid media at low temperature (Olah et al., 1967a).

3.3.4.2. *Ethers, Aromatic Ethers and Their Sulphur Analogues*

What has been said regarding the determination of the basicities of alcohols applies equally to ethers, except that ethers do not react readily with sulphuric acid at higher concentrations. Cryoscopic studies thus show them to be simply protonated (Hantzsch, 1908; Oddo and Scandola, 1910). The basicities of a series of aliphatic ethers have been determined by Arnett and Wu (1960a) and by Arnett et al. (1962) by the distribution method, and a phenyl-substituted ether was studied by Deno and Turner (1966) by the solubility method. The results are given in Table 3.16. Vapour pressure measurements confirm the value for diethyl ether (Jaques and Leisten, 1964).

TABLE 3.16

Basicities of some ethers

Ether	Weight % H_2SO_4 at half-protonation	$H_0^{(1/2)}$	Method	References†
Dimethyl-	54·4	−3·83	Distribution	a
Methyl-ethyl-	54·2	−3·82	Distribution	a
Diethyl-	52·3	−3·59	Distribution	a
Diisopropyl-	58·6	−4·30	Distribution	a
Methyl-n-butyl-	51·3	−3·50	Distribution	a
Methyl-t-butyl-	45·2	−2·89	Distribution	a
Methyl-4-phenylbutyl-	49	−3·2	Solubility	b
Tetrahydrofuran		−2·08	Distribution	a
Tetrahydropyran		−2·79	Distribution	a

† a. Arnett and Wu (1960a) and Arnett et al. (1962); b. Deno and Turner (1966).

An attempt to determine the basicity of ethyl ether from a study of the variation of the internal chemical shift in the ethyl group with acid concentration did not result in a clear sigmoid curve, but indicated a superposition of a solvent effect upon the effect of protonation (Edward et al., 1962). A pK_{BH^+} value of −6·2 was deduced from the spectra, which was apparently confirmed by an indicator study in acetic acid solution (Edward, 1963). There is thus still no general agreement on the basicity of ethers.

The basicities of some cyclic ethers were also studied by the distribution method (Arnett and Wu, 1962). They all fall in the range between −2 and −3. The value for 1,4-dioxane (−3·16) was confirmed by a Raman study which gave pK_{BH^+} = −3·4 (Deno and Wisotsky, 1963). An n.m.r. study of the intramolecular shift of the α- and β-protons in tetrahydrofuran with the change in acid concentration again showed no inflexion point (Arnett, 1963).

Nuclear magnetic resonance studies of protonated aliphatic ethers in superacid media at low temperatures were reported by Olah and O'Brien

(1967). Ethers cleave rapidly in these media if one of the alkyl groups is tertiary.

According to Arnett (1963) the lower aliphatic sulphides (and mercaptans) are 3–5 pK units less basic than the corresponding ethers (and alcohols). Tetrahydrothiophene is more basic than diethylsulphide. The structures of protonated sulphides have been studied in superacid media at low temperature by n.m.r. spectroscopy (Hogeveen, 1966; Olah et al., 1967a).

The same difficulties as in the study of phenols due to sulphonation are encountered in the study of the protonation of phenolic ethers (Arnett and Wu, 1960b). Optical densities at higher acid concentrations had to be extrapolated to zero time, and neither the optical density of the free base nor of the conjugate acid could be observed directly. Therefore the pK_{BH^+} values obtained (all in the range −5·4 to −7·4, depending on the alkyl group) must be taken with reserve. The u.v. spectra of phenol and anisole in aqueous perchloric acid, where there can be no complication due to sulphonation, were found not to change with acidity in any simple way, and protonation seems to occur only at $H_0 < -9$ (Yates and Wai, 1965).

The structure of the cations obtained by the protonation of phenols and aromatic ethers has been the subject of a recent controversy. It is well established that in HSO_3F-SbF_5 mixtures anisole is protonated on the *para* carbon atom, giving a resonance stabilized hybrid (**3.34**):

(3.34)

This is the *p*-methoxybenzenonium ion, in which the positive charge is partly carried by the oxygen and partly by the benzene ring (Birchall et al., 1964). More recently n.m.r. evidence has been adduced for the existence of an *O*-protonated cation in addition to the *C*-protonated species in HF-BF_3 mixtures, the ratio of their concentrations being strongly dependent on temperature (Bronwer et al., 1966). The proportion of the *O*-protonated form increases with decreasing temperature, but the *C*-protonated ion predominates even at −80°C. In concentrated sulphuric acid and in fluorosulphuric acid at room temperature the n.m.r. spectrum is the same, and anisole is not recoverable from these solutions unchanged, owing to rapid formation of sulphonation products (Ramsey, 1966). A possible changeover from *O*-protonation in 77 per cent sulphuric acid to *C*-protonation in more concentrated acid has been suggested (Kresge and Hakka, 1966), but this is ruled out by the above result. The interpretation of u.v. spectra in aqueous strong acids remains ambiguous (Ramsey, 1966).

The n.m.r. evidence for the site of protonation of phloroglucinol (1,3,5-benzenetriol) and its ethers appears to be unambiguous. The spectrum of phloroglucinol in 70 per cent perchloric acid consists of two lines of equal area, indicating C-protonation (Kresge et al., 1962), which results in the formation of the benzenonium ion (3.35):

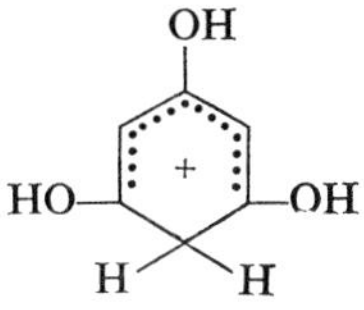

(3.35)

The question of whether these bases follow the H'_R, rather than the H_0 acidity function, was therefore considered. The slopes of $\log I$ vs. $-H_0$ plots were found to decrease with the number of free hydroxyl groups in the molecule, suggesting cation stabilization by hydrogen bonding of these groups with the solvent (Schubert and Quacchia, 1962). Only the 1-methoxy-3,5-dihydroxybenzene obeys the H_0 acidity function satisfactorily, and none obey the H'_R acidity function. The pK_{BH^+} values estimated on the H_0 scale are within the range $-3 \cdot 6$ to $-3 \cdot 8$ for these compounds.

There is no information on the basicities of aromatic sulphides. A cryoscopic study in 100 per cent sulphuric acid (Szmant and Brost, 1951) showed that, with the exception of 4,4'-dinitrodiphenyl sulphide, phenyl and benzyl sulphides become oxidized in 100 per cent sulphuric acid and possibly sulphonated. 4,4'-Dinitrodiphenyl sulphide is recoverable from solution, shows an i factor of 2·3, and is therefore presumably simply protonated. These results were essentially confirmed by Gillespie and Passerini (1956). A brilliant red colour is characteristic of aromatic sulphides in sulphuric acid solution. This is probably due to resonance stabilization of the sulphonium cation (3.36).

(3.36)

The oxygen analogue of phenyl sulphide gives a colourless solution in cold 100 per cent sulphuric acid. Here no resonance stabilization of the cation is possible (Szmant and Brost, 1951). According to Gillespie and Passerini (1956) 4-aminodiphenyl sulphide gives initially an i factor of 4, which suggests complete protonation of both the sulphur and the amino group, and rapid sulphonation of the unsubstituted ring. This increases with time to $i = 5$, owing to sulphonation according to

$$PhSC_6H_4NH_2 + 4H_2SO_4 = HO_3SC_6H_4{}^+SH.C_6H_4NH_3{}^+ + H_3O^+ + 3HSO_4{}^- \qquad (3.49)$$

Some oxidation probably also occurs.

The behaviour of 2:4-dinitrobenzene-sulphenyl chloride in 100 per cent sulphuric acid was recently studied (Robinson and Zaidi, 1968) and, contrary to earlier opinion (Karasch et al., 1953), found to be consistent with incomplete simple protonation. Thus highly negatively substituted sulphur shows weakly basic properties in pure sulphuric acid.

3.3.4.3. Aldehydes and Ketones

Aliphatic aldehydes are readily hydrated in dilute acid solution and also undergo aldol condensation. Simple protonation studies are therefore not possible. Benzaldehyde and a number of substituted benzaldehydes have been studied spectrophotometrically in sulphuric acid/water mixtures (Stewart and Yates, 1959). They obey the H_0 acidity function reasonably well. Benzaldehyde has a pK_{BH^+} value of $-7\cdot60$, and the effects of substituents are correlated by the σ^+ constants with $\rho = 1\cdot85$. The pK_{BH^+} values of 1- and 2-naphthaldehyde are by $0\cdot65$ and $0\cdot31$ units, respectively, less negative, and 9-phenanthraldehyde by $0\cdot60$ units, whereas 1- and 9-anthraldehydes are considerably more basic (Culbertson and Pettit, 1963). The basicities of these compounds agree with predictions based on molecular orbital theory.

The difficulties of studying the basicities of aliphatic ketones are not quite so great as with other aliphatic compounds, because the weak carbonyl u.v. absorption can be used. A number of aliphatic and alicyclic ketones were studied by this method (Campbell and Edward, 1960). For acetone an ionization ratio of unity was found in 80 per cent sulphuric acid, corresponding to a pK_{BH^+} of $-7\cdot7$. This estimate was confirmed by the Raman and the n.m.r. methods (Deno and Wisotsky, 1963), as well as by distribution measurements (Salomaa and Keisala, 1966). Unstrained cyclic ketones are a little more basic, but cyclobutanone is more basic by $2\cdot3$ units.

Acetophenone was one of the first compounds to be studied by the u.v. method (Flexser et al., 1935). A series of substituted acetophenones has been studied more recently (Stewart and Yates, 1958). These compounds behave as Hammett bases. The parent compound has a pK_{BH^+} value of $-6\cdot72$, and substituent effects are best correlated by the σ^+ constants with $\rho = 2\cdot17$. The only deviating substituents are p-OH and p-OMe, which are believed to form strong hydrogen bonds with the solvent. Aminosubstituted acetophenones have also been studied (Lutskii and Dorofeev, 1963), as well as some dihydroxy derivatives (Arnand, 1967).

The behaviour of benzophenones is in several respects different from that of benzaldehydes and acetophenones. First of all, benzophenones deviate from the behaviour of Hammett bases and define an acidity function which decreases less steeply with acid concentration than the H_0 acidity function (Bonner and Phillips, 1966). On this scale the parent compound has a pK_{BH^+} of -5.70. Secondly, the sensitivity of the protonation reaction to substituent effects is considerably less. The pK_{BH^+} values of a number of monosubstituted and polysubstituted compounds follow a satisfactory correlation with $\Sigma\sigma^+$, but the smaller substituent influence reflected in a low ρ value (1·09) suggests that the resonance effect is not fully operative. This is consistent with the non-planarity of the benzene rings (owing to their mutual repulsion), which leads to a greatly reduced overlap of the $\bar{u}$-orbitals of the rings and the carbonyl carbon (Bonner and Phillips, 1966). A higher reaction constant (3·5) is predicted, assuming unhindered resonance (Liler, 1966a). The fact that benzophenones follow an acidity function which decreases less steeply with acid concentration than does H_0 could also be ascribed to less effective resonance in the cation, because this would result in a more localized charge and hence stronger solvation of the cation than occurs in protonated acetophenones and benzaldehydes. The effect on the basicity of joining the *ortho* positions of the two rings by bridges of various sizes was also determined and discussed in terms of steric and electronic effects (Stewart et al., 1963).

β-Benzoylnaphthalene and *p*-benzoyldiphenyl were amongst the original Hammett indicators with half-protonation at 73·5 and 74·5 per cent acid respectively, corresponding to pK_{BH^+} values of -6.45 and -6.6 (Hammett and Deyrup, 1932). The slopes of their $\log I$ vs. acid concentration plots are closely parallel with those of primary anilines.

In benzyl phenyl ketones, substituted in the benzyl ring, the methylene group cuts out the resonance interaction of the substituent with the reaction centre, and the pK_{BH^+} values are correlated by the σ constants (Fischer et al., 1961). The pK_{BH^+} value of the parent compound is -7.3, and the ρ value is only 0·72, i.e. the sensitivity of the reaction to substituent effects is reduced compared with benzophenones by the intervening methylene group.

A number of ketones containing double bonds conjugated with the carbonyl group have also been studied. Dibenzalacetone was, in fact, one of the first ketones whose change in light absorption with acid concentration (halochromism) was ascribed to the formation of carbonium salts (Hantzsch, 1922). Benzalacetophenones (chalcones) were studied more recently (Noyce and Jørgenson, 1962). The *cis* compound undergoes an acid-catalysed isomerization (see Chapter 5). Owing to low solubility, the determination of basicity was carried out in 5 per cent dioxane/aqueous sulphuric acid in which an H_0 scale was established by Noyce and Jørgenson (1961 and 1962). The parent compound has a pK_{BH^+} value of -5.10, and substituent effects are correlated satisfactorily with σ^+ constants with $\rho = 1\cdot60$. *Cis*-chalcones isomerize too rapidly for direct study, except *cis*-4-nitrochalcone ($\mathrm{p}K_{\mathrm{BH^+}} = -6\cdot75$). The basicities of other *cis*-chalcones were obtained indirectly by the hydrogen bonding method. On the average, *cis*-chalcones are less basic than the *trans* compounds by about $0\cdot7$ pK units. For β-phenylchalcone a $\mathrm{p}K_{\mathrm{BH^+}}$ value of $-5\cdot65$ was found, and thus the additional phenyl ring does not provide any added resonance stabilization of the cation but acts mainly inductively.

The furan, thiophen and selenophen analogues of chalcone are more basic, the enhancement of basicity resulting from the substitution of the phenyl ring by 2-furyl, 2-tienyl, and 2-selenienyl, decreasing in that order (Tsukerman *et al.*, 1965).

Some diketones were studied cryoscopically in 100 per cent sulphuric acid (Wiles and Baughan, 1953). While benzil and dibenzoylmethane behave as monoacid bases, the insertion of further methylene groups between the two carbonyl groups leads to an increase in the i factor. Dibenzoylpropane and higher members of the series behave as diacid bases. This behaviour is readily interpreted in terms of electrostatic theory. However, the enolizable β-diketones have been shown by Eistert *et al.* (1954) to form enol-oxonium cations, which are resonance stabilized (**3.37**):

$$\underset{\text{(3.37)}}{\mathrm{R-\underset{\underset{OH}{|}}{C}=CH-\underset{\underset{^+OH}{\|}}{C}-R \leftrightarrow R-\underset{\underset{^+OH}{\|}}{C}-CH=\underset{\underset{OH}{|}}{C}-R}}$$

If R is a phenyl group, the charge is spread further into the ring, and the basicity of these compounds is strongly enhanced. The $\mathrm{p}K_{\mathrm{BH^+}}$ value of dimedon, for example, is $-0\cdot82$, whereas that of acetylacetone is $-4\cdot4$. Crystalline addition compounds with acids are readily obtainable (e.g. dibenzoylmethane hydrogen sulphate, m.p. 100°C).

Unsaturated conjugated aryl ketones give highly coloured solutions in sulphuric acid and very high i factors. Conflicting views about their mode of ionization, reviewed by Gillespie and Leisten (1954a), were all shown to be wrong, and the i-factors were accounted for accurately by simple carbonyl protonation, accompanied by rapid sulphonation of the aromatic ring (Gillespie and Leisten, 1954b). The intense colours of the highly conjugated ketones in sulphuric acid are a result of charge delocalization along the chain of conjugated double bonds. Each sulphonation increases the i factor by approximately 2, because of the protonation of water set free in the reaction

$$\mathrm{X-H + 2H_2SO_4 = X-SO_3H + H_3O^+ + HSO_4^-} \tag{3.50}$$

The basicities of a large number of polycyclic ketones and quinones, determined spectrophotometrically, were reported by Handa (1955). The $\mathrm{p}K_{\mathrm{BH^+}}$ values for most range between -5 and -8. Some quinones, bianthrone, diphenoquinone and *p*-benzoquinone are said to be destroyed by concentrated sulphuric acid. Anthraquinone was one of the original Hammett indicators ($pK_{BH^+} = -9$) and was shown by Hammett and Deyrup (1933) to be only monoprotonated in sulphuric acid. On the basis of their absorption spectra $\Delta^{10,10'}$-bianthrone, helianthrone and *meso*-naphthobianthrone may be protonated on one or on both carbonyl oxygen atoms in concentrated sulphuric acid (Herbert et al., 1952). The same seems to be true of dibenzanthrone and some of its methoxy derivatives (Durie and Shannon, 1958). Methoxyanthraquinones are monoprotonated in sulphuric acid, except when methoxy groups are *ortho* to both carbonyl groups (Wiles and Baughan, 1953), in which case they become diprotonated, probably because of the stabilization of the cations by hydrogen bonding, as in (**3.38**). Alkoxyl groups are usually not

(3.38)

protonated by concentrated sulphuric acid, if they are attached to molecules containing other electron-attracting basic centres (Wiles, 1956).

A conductometric investigation of the solutions of 1,1'-dianthrimide in pure sulphuric acid has shown that it accepts three protons, one by the imino group, and two by two of the carbonyl groups (Arnesen and Langmyhr, 1964).

Acetylferrocene and diferrocenyl ketone are considerably more basic than other aromatic ketones, with pK_{BH^+} values $-2 \cdot 80$ and $-2 \cdot 55$ respectively (Arnett and Bushick, 1962). This demonstrates the ability of the ferrocenyl nucleus to stabilize carbonium ions.

While the basicities of aliphatic aldehydes could not be determined in aqueous acid, in the highly acidic anhydrous acid media they are fully protonated, and the structures of their cations have been studied by n.m.r. spectroscopy in superacid media (Olah et al., 1967c). Both *cis* and *trans* isomers are observed for acetaldehyde:

80 per cent *cis*   20 per cent *trans*

With higher members of the series only the *cis* form appears. The cations of aliphatic ketones have likewise been studied (Olah *et al.*, 1967d). The large deshielding of the OH proton suggests that the positive charge remains largely on the oxygen, i.e. that the cations are carboxonium ions and not carbonium ions. The chemical shift of the OH proton of some *para*-substituted acetophenones shows a correlation with their pK_{BH^+} values (Birchall and Gillespie, 1965). In asymmetrically substituted ketones two isomers of the cation are observed (Olah *et al.*, 1967d) and likewise in asymmetrically substituted benzophenones (Sekuur and Kranenburg, 1966), although in the latter case this is believed to be due to the fact that both phenyl rings cannot be coplanar with the carbonyl group simultaneously.

3.3.4.4. *Carboxylic Acids*

It has been known since the work of Hantzsch (1908) and Oddo and Casalino (1917) that many carboxylic acids are simply protonated in 100 per cent sulphuric acid. Some aryl carboxylic acids were later found to ionize in a more complex manner (Newman and Deno, 1951b). Benzoic acid, *p*-nitrobenzoic acid and phenylacetic acid have been amongst the first compounds whose basicities were determined by u.v. spectrophotometry in sulphuric acid/water mixtures (Flexser *et al.*, 1935). The determination of the basicities of aliphatic acids was more difficult owing to low u.v. absorption, but some data are now available.

The protonation of acetic acid was studied spectrophotometrically by Goldfarb *et al.* (1955) and that of propionic acid by Edward and Wang (1962), and their pK_{BH^+} values estimated as $-6 \cdot 75$ and $-7 \cdot 43$ respectively. The introduction of electron-withdrawing substituents reduces the basicity and thus monochloro-, dichloro- and trichloroacetic acid are only partly protonated even in 100 per cent sulphuric acid. Their basicities were recently determined conductometrically (Liler, 1965b). Trichloroacetic acid actually depresses the conductivity of the 100 per cent acid, but compared with the typical non-electrolyte sulphuryl chloride appears to be slightly ionized as a base (p$K_{BH^+} = -13 \cdot 1$).

The amino group of amino-acids is protonated and the ammonium positive centre, if close to the carboxyl group, leads to its incomplete protonation in 100 per cent sulphuric acid (O'Brien and Niemann, 1951). So, for example, the following *i* factors have been found for amino acids:

$$H_3\overset{+}{N}(CH_2)_n COOH: n = 1, i = 2 \cdot 2; n = 2, i = 2 \cdot 7; \text{ and } n = 3, i = 2 \cdot 9$$

(Williams and Hardy, 1953).

Likewise, double protonation of dicarboxylic acids is approached only when the two carboxyl groups are sufficiently far apart. Even the first protonation of oxalic acid is incomplete, viz., $i = 1 \cdot 3$, increasing with time, owing to

3. PROTONATION OF VERY WEAK BASES   129

decomposition which yields water in solution (Wiles, 1953). Its K_b^{25} was estimated conductometrically (from an extrapolation of the increasing conductance to zero time) and found to be $1 \cdot 02 \times 10^{-3}$ (Liler, 1963). Malonic acid is just monoprotonated (Wiles, 1953), but succinic acid is extensively diprotonated, whereas glutaric and adipic acids are fully diprotonated (Robinson and Quadri, 1967b). 2-Pyrone-5-carboxylic acid gives an i factor of only 1·83, suggesting that even the pyrone is incompletely protonated, and the carboxyl group not at all (Wiley and Moyer, 1954).

For fumaric acid a K_b of 0·013 mol kg^{-1} was reported (Robinson and Quadri, 1967b), but the results of a spectrophotometric study of the protonation of fumaric and maleic acid are in disagreement with this (Pospišil et al., 1968).

Aromatic acids have also been studied extensively. The pK_{BH^+} value for benzoic acid obtained spectrophotometrically in sulphuric acid/water mixtures (−8·0) (Stewart and Yates, 1960) does not represent a thermodynamic pK value because of the peculiarity in the activity coefficient behaviour of benzoic acid at acidities smaller than required for protonation, as discussed in Section 2.5, and on p. 62. In the protonation region itself benzoic acid and substituted benzoic acids parallel the behaviour of Hammett bases, which means that the relative pK_{BH^+} values of substituted benzoic acids are thermodynamically significant. In perchloric acid/water mixtures, however, benzoic acid shows no peculiarity, and the pK_{BH^+} value obtained in these media (−8·85) is thermodynamically significant (Yates and Wai, 1965). The basicities of the *meta* and *para* substituted benzoic acids are best correlated by the σ^+ constants (Stewart and Yates, 1960), and this is taken as evidence for their carbonyl protonation, because resonance interaction of the substituent with the carbonium ion centre is possible as in (3.39). The reaction constant $\rho = 1 \cdot 30$, when

(3.39)

all pK_{BH^+} values are corrected to the H_0 scale of Table 2.1, p. 37.

According to a spectrophotometric study, *o*-, *m*- and *p*-phthalic acids have pK_{BH^+} values of −6·50, −7·75 and −8·24, respectively (Stewart and Granger, 1961). This suggests that *p*-phthalic acid should be fully monoprotonated in pure sulphuric acid. An i factor of 2·2 has in fact been reported (O'Brien and Niemann, 1957), confirming this conclusion and suggesting a small degree of second protonation.

The pK_{BH^+} values of a number of carboxylic acids are plotted against their pK_a values in water in Fig. 3.11 (Liler, 1965b). The graph includes phenylacetic acid having a pK_{BH^+} value of −8·4 (Flexser et al., 1935, corrected to the new

H_0 scale) and oxalic acid, the pK_{BH^+} value of which was taken as approximately equal to that of nitrobenzene on the basis of conductometric data (Liler, 1963). Both the pK_a and the pK_{BH^+} values of oxalic acid have been statistically corrected, and the latter is expressed on the new H_0 scale. It can be seen that there is a good linear correlation, the slope of the line being 1·40. This is closely similar to the above quoted slope of the analogous plot for substituted benzoic acids ($\rho = 1\cdot30$). This is to be expected, because the group transmitting the substituent effects is the same in both series. The fact that the value of the slope is greater than unity is not so readily accounted for. Possibly the partial positive charge on the carbon in the conjugate acids makes it more susceptible to substituent effects than is the neutral carboxyl group.

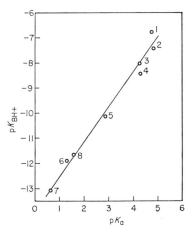

FIG. 3.11. The plot of pK_{BH^+} values of carboxylic acids vs. their pK_a values in water. Acids: 1, acetic; 2, propionic; 3, benzoic; 4, phenylacetic; 5, monochloroacetic; 6, dichloroacetic; 7, trichloroacetic; 8, oxalic (statistically corrected).

The basicities of some *ortho*-substituted benzoic acids have also been studied (Stewart and Granger, 1961). Alkyl groups decrease the basicity, the more so the greater their bulk. Oxygen-containing *ortho*-substituents (except OH), including NO_2 and COOH groups, have a base-strengthening effect, presumably because of hydrogen bonding in the conjugate acids, as for example in *o*-phthalic acid (**3.40**). Because this structure is resonance stabilized,

(3.40)

3. PROTONATION OF VERY WEAK BASES 131

there is a base-strengthening of 1·74 pK units compared with p-phthalic acid. A second spectral change is observed for o-phthalic acid at 104 per cent sulphuric acid, owing to anhydride formation. Di-*ortho*-substituted benzoic acids often show complex ionization (Treffers and Hammett, 1937). This will be discussed in the next Chapter.

Higher fatty acids give soap-like solutions in cold concentrated sulphuric acid (McCulloch, 1946). Micelle formation has recently been studied (Steigman and Shane, 1965).

The protonation of substituted o-benzoylbenzoic acids leads to complex changes in the u.v. spectra, and therefore pK_{BH^+} values of only moderate accuracy were obtained for these compounds (Noyce and Kittle, 1965a). Substituent effects lead to the conclusion that protonation occurs on the ketonic oxygen. The parent compound does not follow the H_0 acidity function (log$I$ vs. $-H_0$ slope is 0·8) and it is half protonated in 80 per cent sulphuric acid ($H_0^{(1/2)} = -7·35$). This means that it is less basic than benzophenone by about one pK unit. In 100 per cent sulphuric acid its ionization becomes complex (see Chapter 4).

There has been a great deal of interest in the structure of cations obtained by the protonation of carboxylic acids. For example, infrared, Raman and u.v. spectra of benzoic acid in sulphuric acid were studied (Hosoya and Nagakura, 1961; Casadevall et al., 1964; Hoshino et al., 1966) and all confirm that carbonyl protonation takes place, as was concluded from substituent effects on the basicity of benzoic acid (Stewart and Yates, 1959). Nuclear magnetic resonance spectroscopic studies of aliphatic carboxylic acids in superacid media at low temperature revealed, however, two resonances of the OH protons (Birchall and Gillespie, 1965), and this could be consistent with either OH or carbonyl protonation, accompanied by internal hydrogen bonding within the protonated carboxyl group (**3.41**). The position was clarified

(3.41)

by further studies of formic acid at low temperature in HF-BF$_3$ media (Hogeveen et al., 1966), which revealed three resonances for the OH protons, which could be explained by *cis* and *trans* positions for these protons, due to the partial double bond character of the CO bonds:

77 per cent *trans*   23 per cent *cis*

In contrast to formic acid, other acids show two peaks of equal area in the OH region in HSO_3F-SbF_5-SO_2 at $-60°C$ (Olah and White, 1967a) and thus occur only in the *trans* form. At higher temperature all these peaks collapse and at 0°C merge with the solvent peak. Dicarboxylic acids were studied under the same conditions (Olah and White, 1967b) and also the carbon-13 resonance of carboxylic acids (Olah and White, 1967c).

3.3.4.5. Esters

Esters undergo hydrolysis in sulphuric acid/water mixtures and therefore their basicities cannot be determined. In general, the basicities of simple alkyl esters would be expected to be closely similar to those of the corresponding carboxylic acids. Some esters hydrolyse sufficiently slowly in pure sulphuric acid to be recoverable in good yield on pouring the solutions into water after short storage (von Kothner, 1901; Oddo and Scandola, 1910; Kuhn and Corwin, 1948). Initial *i* factors have been found to be 2 for methyl and ethyl benzoate, and for ethyl acetate. The rates of hydrolysis of esters will be discussed in Chapter 5.

Information on the basicities of lactones is equally unobtainable because of hydrolysis, but some lactones have been studied cryoscopically in 100 per cent sulphuric acid, e.g. γ-valerolactone and coumarin (Oddo and Casalino 1917a). Coumarin is stable in pure sulphuric acid and forms a crystalline 1:1 addition compound with it with melting point 35·5°C (Kendall and Carpenter, 1914). Coumarin is undoubtedly more basic than other lactones because it is in fact an α-pyrone with an exceptionally resonance stabilized conjugate acid.

In spite of the cleavage of esters in acid solutions, the structures of their conjugate acids could be studied by n.m.r. spectroscopy in anhydrous superacid media at low temperature (Birchall and Gillespie, 1965; Olah *et al.*, 1967e). The results show that carbonyl protonation occurs and, consistent with this, conjugate acids of alkyl formates occur in the *cis* and the *trans* forms:

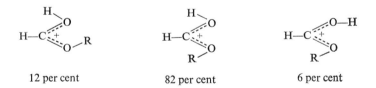

12 per cent   82 per cent   6 per cent

The given percentage distribution is for ethyl formate (Olah *et al.*, 1967e). The values of the coupling constants suggest a considerable delocalization of charge over the two oxygens and the carbon and, consequently, a reduced double bond character of the carbonyl bond. For other esters only one resonance for the proton on the oxygen was observed. This could not be assigned with certainty to either isomer, because of the absence of coupling of this proton to other hydrogens (Olah et al., 1967e).

3.3.4.6. Acid Chlorides

Acetyl chloride has for a long time been known to form stable mixtures with sulphuric acid, which evolve hydrogen chloride only at 100°C with the formation of acetylsulphuric acid (Aschan, 1913), and a crystalline addition compound between sulphuric acid and benzoyl chloride was reported by Bergmann and Radt (1921). A crystalline compound between acetyl chloride and sulphuric acid was more recently obtained by Tutundžić et al. (1955). A Raman spectroscopic investigation of the solutions of acetyl chloride in sulphuric acid showed conclusively that only simple protonation occurs (Casadevall et al., 1964b). It appeared from conductometric measurements on sulphuric acid/acetyl chloride mixtures that this protonation was incomplete (Tutundžić et al., 1955). This was confirmed and hence basic ionization constants were determined conductometrically for a number of aliphatic and aromatic acid chlorides both at 10°C and at 25°C (Liler, 1966b). The results are given in Table 3.17. The

TABLE 3.17

Basic ionization constants of acyl chlorides in 100 per cent sulphuric acid at 25°C (Liler, 1966b)

Acyl group	$K_b \times 10^2$	$-pK_{BH^+}$
Acetyl	3·3	10·90
Propionyl	4·9	10·82
n-Butyryl	5·4	10·78
iso-Butyryl	3·5	10·97
Chloroacetyl	2·4	11·13
Benzoyl	3·3	11·15
p-Toluoyl	3·1	11·18
p-Chlorobenzoyl	2·5	11·27
m-Nitrobenzoyl	2·1	11·35
p-Nitrobenzoyl	1·6	11·47
o-Chlorobenzoyl	2·8	11·21

basicities of substituted benzoyl chlorides correlate reasonably well with the σ^+ constants of the substituents, with a ρ value of 0·31. The spread of the basicities of aliphatic acid chlorides is also very small.

Oxalyl chloride and dichloroacetyl chloride behave as non-electrolytes, the latter also undergoing a slow reaction with sulphuric acid (Liler, 1966b).

3.3.4.7. *A Summary of the Behaviour of Carbonyl Bases*

The basicities of carbonyl compounds would be expected to be determined by the electron density on the carbonyl oxygen, which is itself related to the polarity of the carbonyl bond. As was discussed in Section 3.3.1.2, the carbonyl frequency is also primarily determined by the polarity of the bond. A correla-

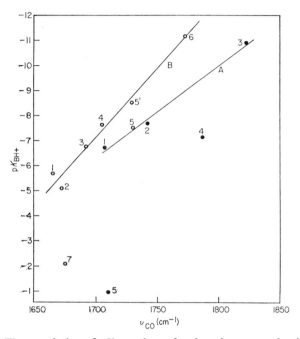

FIG. 3.12. The correlation of pK_{BH^+} values of carbonyl compounds with carbonyl frequencies (see text). (A) Acetyl compounds: 1, acetophenone; 2, acetone; 3, acetyl chloride; 4, propionic acid (in aqueous sulphuric acid, statistically corrected); 5, acetamide. (B) Benzoyl compounds: 1, benzophenone; 2, *trans*-chalcone; 3, acetophenone; 4, benzaldehyde; 5, benzoic acid (in aqueous sulphuric acid, statistically corrected); 5′, benzoic acid (in aqueous perchloric acid, statistically corrected); 6, benzoyl chloride; 7, benzamide.

tion would therefore be expected between the carbonyl frequency and the basicities of carbonyl compounds, and is indeed found (Liler, 1967). The data on the basicities of carbonyl compounds, reported in Sections 3.3.4.3 to 3.3.4.6, have been plotted against the carbonyl frequency in these compounds in Fig. 3.12. For aromatic compounds the carbonyl frequencies are those of Bellamy (1955) and Bellamy and Pace (1963)—they refer to dilute solutions in carbon tetrachloride and chloroform. The carbonyl frequencies for acetyl compounds refer to vapours (Hartwell *et al.*, 1948). For benzoic acid the pK_{BH^+} value

3. PROTONATION OF VERY WEAK BASES   135

obtained in perchloric acid is plotted, as well as the value obtained in sulphuric acid, which is not a thermodynamic pK (see Section 3.3.4.4). The basicity of propionic acid was plotted in preference to that of acetic acid, because it appears to be more reliable.

As can be seen from Fig. 3.12, there is a good linear correlation between the carbonyl frequency and the thermodynamic pK_{BH^+} values for all benzoyl compounds. Benzamide shows a striking deviation, which is consistent with its being an amino base and not a carbonyl base. The pK_{BH^+} value for benzoic acid obtained in sulphuric acid solution also deviates, which confirms the conclusion that this is not a thermodynamic pK value. There are fewer data for acetyl compounds, but it can be seen that a linear correlation exists for acetophenone, acetone and acetyl chloride. Apart from acetamide which,

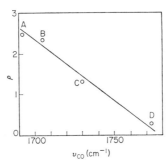

FIG. 3.13. The correlation of Hammett reaction constants for the protonation of carbonyl compounds with carbonyl frequencies: A, acetophenones; B, benzaldehydes; C, benzoic acids; D, benzoyl chlorides (Liler, 1965a).

being an amino base, deviates strikingly from the relationship, the pK_{BH^+} value for propionic acid also deviates like that for benzoic acid in sulphuric acid, only more so. This suggests that the pK_{BH^+} value for this compound obtained in sulphuric acid solution may also not be a thermodynamic pK value.

There is another interesting correlation involving the transmission of substituent effects on the pK_{BH^+} values in the series of benzoyl compounds. The protonation of all these compounds has been studied at the same temperature (25°C) in sulphuric acid/water mixtures, i.e. in media of high dielectric constant and closely similar structure throughout. The electronic requirements of proton addition at the reaction site are also the same for all protonation reactions. Therefore, according to a prediction of Hammett (1937), in such a series of closely structurally related compounds, all undergoing the same reaction under closely similar conditions, the reaction constants ρ should depend on the "displaceability of electrons on the reacting group", i.e. on the polarizability of the carbonyl group in this series of compounds. The polarizability of the group is determined by its electron density. It has been pointed out in Section 3.3.1.2 that the carbonyl frequency is a reflexion of the electron density on the carbonyl carbon (see Fig. 3.8). Therefore, a correlation of reaction constants for protonation, given in Sections 3.3.4.3 to 3.3.4.6, with carbonyl frequencies of benzoyl compounds would be expected. Fig. 3.13 shows that such a correlation exists (Liler, 1965a). Incidentally, the reaction constant for the protonation of benzamides deviates strongly from this relationship also.

3.3.4.8. *Pyrones, Benzopyrones and Related Compounds*

As was mentioned in Section 3.3.1, the capacity of pyrones to form salts with acids became well known at the turn of the century (Collie and Tickle, 1899), and soon afterwards Walden (1901) made an attempt to determine the base ionization constant of 2,6-dimethyl-γ-pyrone using several methods. His findings were confirmed by Rördam (1915), who determined the hydrolysis constant of the hydrochloride and hence obtained a pK_{BH^+} value of $-0\cdot28$. Indirect information on the basicities of several substituted γ-pyrones was also obtained by Johnson and Partington (1931). α-Pyrones are probably equally strong bases (e.g. coumarin), but no basicity determinations appear to be available. A number of benzo- and dibenzo-derivatives of pyrones and their thio- and seleno-analogues have been studied recently (Degani *et al.*, 1968a), as well as some flavones (Davis and Geissman, 1954). The results are given in Table 3.18. A good correlation of pK_{BH^+} values with σ_p substituent constants exists for 6-substituted thiochromones ($\rho = 1\cdot79$). Thiocoumarin has also

TABLE 3.18

Basicities of some derivatives of pyrones and their thio- and seleno-analogues at 25°C

Compound	pK_{BH^+}	References†
Chromone	$-2\cdot05$	a
Thiochromone	$-1\cdot20$	a
Selenochromone	$-1\cdot46$	a
Xanthone	$-4\cdot12$	a
Thioxanthone	$-3\cdot95$	a
Selenoxanthone	$-4\cdot36$	a
Flavone	$-1\cdot2$	b
Flavone, 4'-methoxy-	$-0\cdot8$	b
Flavone, 2',6'-dimethyl-4'-methoxy-	$-1\cdot7$	b
Flavone, 2',4',6'-trimethyl-	$-2\cdot05$	b
Flavone, 3',2',4',6'-tetramethoxy-	$-2\cdot45$	b

† *a*. Degani *et al.* (1968a); *b*. Davis and Geissman (1954).

been studied and found not to be a Hammett base (Degani et al., 1968b). The results for flavones (Table 3.18) show that the p-methoxy group enhances the basicity of the flavone by additional charge delocalization in the cation (**3.42**). This resonance requires coplanarity of the phenyl ring and the benzo-

(3.42)

pyrone ring. If coplanarity is impeded by the steric effect of *ortho* substituents, flavones become less basic even than the unsubstituted compound. This shows that charge delocalization into the phenyl ring is important in enhancing their basicity, regardless of the effect of other substituents.

3.3.4.9. *Sulphoxides, Sulphones and Related Compounds*

Although information on the basicities of organic sulphides is scanty and somewhat uncertain, sulphoxides appear to be more basic than the sulphides. The basicities of dimethyl sulphoxide and methylphenyl sulphoxide were evaluated by Haake and Cook (1968) by the n.m.r. chemical shift method in sulphuric acid ($pK_{BH^+} = -2·78$ and $-3·38$ respectively), but as the logarithms of the ionization ratios do not follow the H_0 acidity function the values obtained do not have thermodynamic significance. The H_A acidity function represents the protonation behaviour of sulphoxides more satisfactorily, and on this scale pK_{BH^+} values range from $-1·8$ for dimethyl sulphoxide to $-2·9$ for p-nitrophenylmethyl sulphoxide (Landini et al., 1969). Substituent effects on the basicity of the SO group are not large ($\rho = 0·85$).

The results of cryoscopic studies in 100 per cent sulphuric acid (Gillespie and Passerini, 1956) show that diphenyl sulphoxides with negative substituents (4-nitro- and 4,4′-dinitro-) behave as simple bases, but that diphenyl sulphoxide itself undergoes some reaction, probably sulphonation of the initially protonated compound. Phenyl benzyl sulphoxide and phenyl methyl sulphoxide are said to decompose in sulphuric acid (Oae et al., 1965a), but are also probably sulphonated. In addition it has been shown that protonated sulphoxides exchange their oxygen with the solvent (Oae et al., 1965b).

Sulphones are considerably less basic than sulphoxides. It was first shown by Gillespie (1950c) that diphenylsulphone behaves as a weak base in pure sulphuric acid. The introduction of p-nitro substituents into the aromatic rings causes slightly larger freezing-point depressions, probably owing to partial protonation of the nitrogroups themselves (Gillespie and Passerini, 1956). Doubt was cast on these conclusions by a conductometric study of these solutions (Hall and Robinson, 1964), which showed that these solutes depress the conductance of pure sulphuric acid. They were, therefore, classified as non-electrolytes. As was pointed out in Section 3.2.6, however, some weak bases do depress the conductance of pure sulphuric acid, but not to the same extent as a typical non-electrolyte does. Densities of the above solutions would be needed to decide this issue.

The basicities of several dialkyl sulphones have recently been determined in 100 per cent sulphuric acid by cryoscopy, conductance and conductometric titration (Hall and Robinson, 1964). The mean K_b values obtained are given in Table 3.19. Other compounds containing the sulphone group (CH_3SO_2F, CH_3SO_2Cl, $C_2H_5SO_2F$, $C_2H_5SO_2Cl$, and $CH_3SO_2OCH_3$) all behave as non-electrolytes (Hall and Robinson, 1964), whereas methane-sulphonic acid is either a non-electrolyte or a very weak base (Barr et al., 1961). 2:4-Dinitrobenzene-sulphonic acid and toluene-sulphonic acid behave as non-electrolytes (Hall and Robinson, 1964).

TABLE 3.19

The basicities of some dialkyl sulphones in 100 per cent sulphuric acid at 10°C

Sulphone	$K_b \times 10^2$	$-pK_{BH^+}$	References†
Dimethyl	1·48	12·27	a
Diethyl	1·20	12·37	a
Di-n-propyl	0·51	12·75	a
Di-n-butyl	0·40	12·85	a
Tetramethylene	0·38	12·88	a
Diphenyl	~0·5	~13	b

† a. Hall and Robinson, 1964; b. Gillespie, 1950c.

There is a linear correlation between the basicities of compounds containing the SO_2 group and the stretching frequencies of the SO bond (Hall and Robinson, 1964) from which the basicities of some other compounds could be predicted. For example, for sulphur trioxide $K_b = 2 \times 10^{-5}$ was estimated. For sulphamide a very low basicity is predicted from the SO stretching frequency assuming O-protonation, which is inconsistent with its observed behaviour as a strong base in pure sulphuric acid (Hall and Robinson, 1964). This means that sulphamide is N-protonated, and the situation is therefore analogous to that of carboxylic acid amides and other carbonyl compounds (see Fig. 3.12).

The lower basicity of sulphones compared with sulphoxides or, conversely, the much greater strength of their conjugate acids compared with the conjugate acids of sulphoxides (a difference of some 9 pK units) is a manifestation of the well-known phenomenon, that oxy-acids involving higher oxidation states of an element are always much stronger than those involving lower oxidation states. Because of the greater formal positive charge on the sulphur atom in sulphones than in sulphoxides, the sulphur-oxygen bonds are stronger in the former than in the latter.

3.3.5. HYDROCARBON BASES

In general unsaturated and aromatic hydrocarbons act as bases in acid solution and carbonium ions are produced by their protonation. Such species are very reactive and their transient formation is often inferred from the reactions that unsaturated hydrocarbons undergo in acid solution. However, in a certain number of instances sufficiently stable carbonium ions can be formed to allow a spectrophotometric study of protonation equilibria quite analogous to the study of other very weak bases in solutions of strong acids (sulphuric and perchloric). Otherwise solutions of aqueous strong acids are not the best media for the observation of carbonium ions, because of irreversible reactions that often take place (hydration, polymerization, sulphonation, oxidation). Other non-aqueous acidic media (e.g. HF-BF$_3$ and HSO$_3$F-SbF$_5$) have advantages in this respect, mainly because they do not contain any nucleophilic species. Readers interested in carbonium ions are referred for further information to more detailed texts on the subject (Bethell and Gold, 1967).

The reaction medium may be regarded as an external factor in the stabilization of carbonium ions. Internal stabilization of carbonium ions occurs by the spreading of charge from the carbonium centre on to substituent groups attached to it, or over a number of carbon atoms in a ring, or over a number of condensed aromatic rings. Examples of such carbonium ions are triphenylcarbonium (**3.43**), tropilium (**3.44**) and propenium (**3.45**), and anthracinium ions (**3.46**). The propenium ion is in fact known only if substituted by stabilizing substituents (alkyl or phenyl).

Protonation is not the only way of obtaining carbonium ions in aqueous acid solution, an equally common, or even more common way being by complex ionization of tertiary alcohols (carbinols). Triphenylcarbonium ions

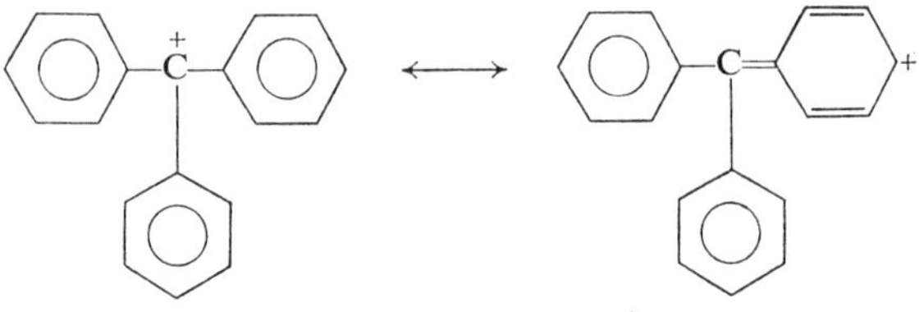

(3.43) ⟷ (3 structures + 3 structures with charge at *ortho* positions)

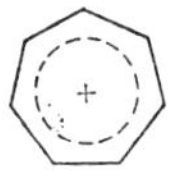

(3.44) 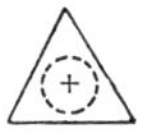   (3.45)

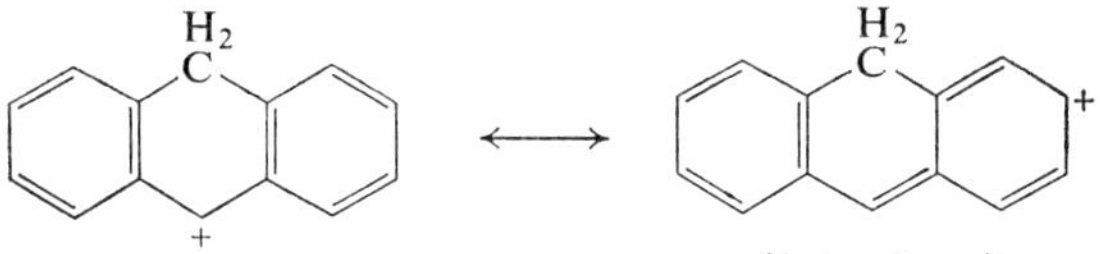

(3.46) ⟷ (4 structures)

are obtained in this way, for example. This type of reaction will be discussed in the following Chapter.

Two aryl substituents are sufficient to stabilize a carbonium centre in acid solution and protonation equilibria of 1,1-diaryl olefins have been systematically studied (Deno et al., 1959). Unsubstituted aliphatic olefins, whose basicities would be of considerable interest for the analysis of the kinetics of some of their very important reactions (hydration, polymerization, carbonylation) cannot be studied in this way because of their very high reactivity. In fact, one of the hopes of obtaining information on their basicities in these media is from correlations of kinetic data (Bethell and Gold, 1967). In other anhydrous acid media (mainly $HF-BF_3$) some determinations have been attempted using distribution measurements (Mackor et al., 1958) or other more indirect methods (Hogeveen and Bickel, 1967).

The protonation of 1,1-diarylolefins has already been referred to in Chapter 2 as the defining reaction of the H'_R acidity function. The direct spectrophotometric determination of this acidity function is difficult, because even diarylcarbonium ions are subject to complicating secondary reactions (in particular dimerization, which is especially troublesome when both the base and the conjugate acid are present in comparable concentrations). The pK'_{R+} values, based on the H'_R acidity scale, were obtained by Deno et al. (1959) for the 1,1-diphenylethyl cation and its *p*-methyl, *p*-chloro and *p*-methoxy derivatives, as well as for some closely related olefins (Table 3.20) The question of

TABLE 3.20
The pK'_{R+} values of some 1,1-diaryl olefins (Deno et al., 1959)

Cation	pK'_{R+}
1,1-Diphenylethyl	−10·4
1,1-Bis-(4′-methoxyphenyl)-ethyl	− 5·5
1,1-Bis-(4′-methylphenyl)-ethyl	−9·1
1,1-Bis-(4′-chlorophenyl)-ethyl	−11·4
1,1-Diphenylpropyl	−11·5
1,1-Diphenyl-2-methylpropyl	−12·8
1-Phenylindanyl	−10·1
1-Phenyl-3,4-dihydronaphthyl	−11·3

whether the equilibria may not be more complicated and involve the alcohol as well ($Ar_2\overset{+}{C}CH_3 + H_2O \rightarrow Ar_2CH(OH)CH_3 + H^+$) was examined. No alcohol was detected upon dilution of the acid solutions with water. Only the original olefin and some dimer were recovered.

A plot of the pK'_{R+} values for the four *p*-substituted compounds in the first part of Table 3.20 *vs.* σ^+ constants of the substituents is linear with $\rho = 8$, indicating strong conjugation of the substituents with the carbonium centre. It is interesting to compare the basicity of 1,1-diphenylethylene with the basicities of 1,1-diphenylpropylene and 1,1-diphenyl-2-methylpropylene, given in the second half of Table 3.20. It can be seen that the effect of additional methyl groups in the immediate vicinity of the carbonium centre is to destabilize the cation and make it a stronger Brønsted acid. This is contrary to expected stabilization by the inductive effect of these groups and must be due to their steric effect. Both the cation and the olefin contain the central carbon in the sp^2 state of hybridization, which would tend to minimize steric effects, but the cation is solvated and steric crowding reduces its stability by reducing its solvation energy (Deno et al., 1959). Ultraviolet spectra of some 1,1- and 1,3-diaryl carbonium ions in sulphuric and perchloric acid solution show changes with time, indicating instability of some kind (Bertoli and Plesch, 1966).

The protonation of azulene and some of its 1-substituted derivatives gives carbonium ions, which fall into the second category mentioned above, i.e. in which the charge is carried on a ring as a whole.

$$\text{azulene-R} + H^+ \rightleftharpoons \text{protonated azulenium-R-CH}_2 \quad (3.51)$$

The considerable basicity of azulenes was recognized by Plattner et al. (1949 and 1950), and it was established by n.m.r. spectroscopy that proton addition occurs at the position 3 (Schulze and Long, 1964a), whereas the resultant positive charge resides on the seven-membered ring, to which fact the high stability of the cation is due. The protonation behaviour of azulene and several 1-substituted derivatives in perchloric acid solution does not follow the H_0 acidity function, the slopes $-\mathrm{d}\log I/\mathrm{d}H_0$ being 1·6–1·9 (Long and Schulze, 1964). This suggests that these bases also follow the H'_R acidity function. Approximate pK'_{R+} values for azulene, its 1-methyl, 1-chloro and 1-cyano derivatives on this scale are $-1·7$, $-0·83$, $-3·25$ and $-8·41$, respectively (Long and Schulze, 1964; Reagan, 1969). The correlation between these pK'_{R+} values and the σ_m constants of the substituents is not very good ($\rho \simeq 10$). A spectrophotometric study of azulene in sulphuric acid/water mixtures (Deno et al., 1963c) gave $pK'_{R+} = -2·1$, in good agreement with the above estimate. Protonation occurs on the oxygen of the 1-substituent for 1-nitro, 1-formyl, 1-trifluoroacetyl and 1-carboxyl azulene (Schulze and Long, 1964a; Long and Schulze, 1964). These bases follow the H_0 acidity function. The protonation of cyclopenta(c)thiapyran and 2-phenyl-2-pyrindine gives cations which are iso-ū-electronic heteroanalogues of the azulinium ion (Anderson and Harrison, 1964).

Less complete protonation data on dienes and mixtures of dienes were reported by Deno et al. (1962a) and by Deno et al. (1963b) in terms of the percentage of sulphuric acid at half-ionization, because it is not clear which acidity function governs these reactions. Most frequently cyclic (allylic) carbonium ions result, as indicated by their n.m.r. spectra. For example, the

(3.47)   (3.48)

solution of cyclohexene (3.47) in pure sulphuric acid gives an i factor of 2, and the n.m.r. spectrum shows the presence of the cyclic cation (3.48). The equilibrium between the two forms (determined spectrophotometrically) is 1:1 in 50 per cent sulphuric acid (Deno et al., 1962a).

Protonations of some polycyclic hydrocarbons, studied spectrophotometrically in sulphuric acid/water mixtures, gave pK_{BH^+} values (on the Hammett H_0 scale) between -5 and -9 (Handa, 1955). The reliability of these data does not appear to be high, and the protonation sites are unknown. Molecular orbital calculations have been very successful in predicting the preferred site of protonation in such systems—for example, for anthracene as in formula (**3.46**) (Gold and Tye, 1952c)—as well as orders of basicities, which are in agreement with those found experimentally from distribution and vapour pressure measurements on solutions of hydrocarbons in anhydrous hydrogen fluoride (Mackor et al., 1958).

Structural information on numerous carbonium ions, which can be produced and stabilized in superacid media at low temperatures, has been accumulating fast in recent years, mainly owing to the application of the n.m.r. technique. More specialized works on carbonium ions give a fuller account of these findings than is possible within the scope of the present volume (see Bethell and Gold, 1967, and Olah and von Schleyer, 1968).

3.3.6. INORGANIC BASES

It has already been pointed out in Section 3.2 that metal hydrogen sulphates behave as strong bases in sulphuric acid by simply ionizing to provide the solvent anion. Other salts may also behave as strong bases by solvolysis, in which the weaker acid is displaced by sulphuric acid, and a metal hydrogen sulphate is formed in solution. It is the object of this brief section to mention a few inorganic compounds which are simply protonated by sulphuric acid and to discuss their basic strength. These compounds are mainly some inorganic oxides.

Enhanced solubility of compounds in acid solution is a reflexion of their tendency to form hydrogen bonds with the acid or to become protonated. Therefore information on the solubility of gaseous oxides is of interest in the present context. The solubility of carbon dioxide in sulphuric acid/water mixtures falls initially with increasing acid concentration to a minimum at 30 per cent acid, then rises to a maximum at 56 per cent acid, again falls to a minimum at 81 per cent acid and finally rises to a solubility in pure sulphuric acid about 20 per cent higher than that in water (Markham and Kobe, 1941). This final rise probably indicates some hydrogen bonding with sulphuric acid. Carbon dioxide does not appear to be protonated.

Sulphur dioxide is appreciably soluble in sulphuric acid. Absorption spectra of its solutions obey Beer's law over the concentration range 0·0005–0·005 M (Gold and Tye, 1950). A comparison of these absorption spectra with the absorption spectrum of the gas led to the conclusion that sulphur dioxide is present in solution chemically unchanged. The method used, however, does not rule out the possible presence of minute concentrations of the cation, HSO_2^+. Hall and Robinson (1964) estimated a base ionization constant of sulphur dioxide as $K_b = 1 \times 10^{-3}$ ($pK_{BH^+} = -13\cdot3$) from a correlation of the basicities of sulphonyl compounds with the SO stretching frequency. The basicity of sulphur trioxide estimated in the same way is 2×10^{-5}.

In recent years it has been shown that many transition metal/carbonyl complexes with strongly $\bar{u}$-bonding ligands exhibit basic properties with the formation of metal/hydrogen bonds (Davison *et al.*, 1962). Their span of basicities is large, and some are protonated only in concentrated sulphuric acid.

Hydrazoic acid was first shown to be protonated in concentrated sulphuric acid by Hantzsch (1930), but his cryoscopic work was apparently not very accurate, since it showed that a doubly positively charged species was obtained. A recent determination of the freezing-point depression by D. Jaques (quoted by Hoop and Tedder, 1961) gave an *i*-value of 4 for sodium azide. Hence the reaction is

$$NaN_3 + 2H_2SO_4 = Na^+ + 2HSO_4^- + H_2N_3^+ \qquad (3.52)$$

showing that hydrazoic acid is only monoprotonated in pure sulphuric acid. The basicity constant was earlier obtained by Back and Praestgaard (1957) by distribution experiments between chloroform and sulphuric acid. Half-protonation in 75·5 per cent sulphuric acid was established, corresponding to a pK_{BH^+} value of $-6\cdot8$ (on the new H_0 scale). Since hydrazoic acid decomposes in concentrated sulphuric acid at 40°C with nitrogen evolution ($\tau_{1/2} = 22$ min) (Briggs and Littleton, 1943), it is reasonable to assume that the structure of the cation is $H_2\overset{+}{N}-N\equiv N$, but it could also be $HN=\overset{+}{N}=NH$, a symmetrical structure, which might indeed be more stable and might be the dominant form.

CHAPTER 4

Complex Ionizations

4.1. Introduction

The fact that many substances undergo complex modes of ionization in pure sulphuric acid was made apparent by the pioneering work of Hantzsch (1908) who found that triaryl carbinols or ethyl nitrate, for example, give abnormally high freezing-point depressions. The most important subsequent discoveries of this kind of behaviour are due to Treffers and Hammett (1937), Newman (1941), Newman and Deno (1951), Gillespie et al. (1950b) and Leisten (1955).

The term "complex ionization" is often used to describe the behaviour of solutes which ionize like triphenylmethanol or nitric acid:

$$Ph_3COH + 2H_2SO_4 = Ph_3C^+ + H_3O^+ + 2HSO_4^- \quad (4.1)$$

$$HNO_3 + 2H_2SO_4 = NO_2^+ + H_3O^+ + 2HSO_4^- \quad (4.2)$$

Substances ionizing in this way are sometimes referred to as secondary bases. This kind of behaviour is not unexpected for some solutes, because the driving force for such a reaction may be ascribed, at least in part, to the high affinity of sulphuric acid for water. A rather unexpected mode of ionization was discovered in the more recent past in the case of acetic and benzoic anhydrides, which were found to abstract water from concentrated sulphuric acid (Leisten, 1955) and to ionize according to the equation

$$(RCO)_2O + 3H_2SO_4 = 2RC(OH)_2^+ + HS_2O_7^- + HSO_4^- \quad (4.3)$$

This reaction is in fact an acid catalysed hydrolysis, but the final state is reached rapidly, and will therefore be treated here as a complex ionization in a broader sense. Few examples of this kind of behaviour are known.

Apart from these two main types of complex ionization, a number of other reactions occur in concentrated sulphuric acid sufficiently rapidly to lead to high i factors immediately after dissolution of the solutes. These are sulphations, oxidations and polymerizations. Sulphonations, however, often occur slowly and lead to i factors increasing with time. This miscellany of behaviour, other than simple protonation, will here be regarded as "complex ionizations". Many organic, inorganic and metal/organic compounds show such variety of behaviour. Such ionizations have predominantly, but not exclusively, been observed in the concentrated sulphuric acid region and by methods applicable to this medium.

For many years the methods of study of complex ionizations in sulphuric acid have been exactly the same as the methods of study of simple protonations, i.e. classical cryoscopic and conductance measurements. Both these methods may not be sufficient to resolve the possible ambiguities of some complex ionizations. To this end the cryoscopic method has been considerably developed in recent years (Leisten and Wright, 1964) and the effectiveness of conductometric titrations for the same purpose has also recently been demonstrated (Robinson and Quadri, 1967a and 1967b).

These refinements of the methods used to study complex ionizations in pure sulphuric acid will be discussed first. A survey of the most important types of complex ionizations will then follow, classified according to the nature of the solutes (organic, inorganic, and organometallic).

4.2. Methods of Study of Complex Ionizations

4.2.1. THE CRYOSCOPIC METHOD

The classical cryoscopic method, the determination of van't Hoff's i factors, was the tool of discovery of the first complex ionizations by Hantzsch (1908) and of many more recent examples (Treffers and Hammett, 1937; Newman, 1941). The interpretation of large i factors is, however, often very difficult, and the cryoscopic results may be consistent with more than one reaction scheme (Leisten, 1955). Supplementation of cryoscopic results with a conductometric determination of the number of HSO_4^- ions produced sometimes removes the ambiguity (Leisten, 1955), but even the two methods used in conjunction cannot always decide between various possibilities. Two possible ionization schemes for acetone are an example of this kind (Leisten and Wright, 1964):

$$(CH_3)_2CO + H_2SO_4 = (CH_3)_2COH^+ + HSO_4^- \qquad (4.4)$$

$$2(CH_3)_2CO + 2H_2SO_4 = (CH_3)C\!\!=\!\!CH\!-\!\overset{+}{C}(OH)CH_3 + H_3O^+ + 2HSO_4^- \qquad (4.5)$$

The first is the simple protonation and the second is the formation of mesityl oxide. The cryoscopic i factor is 2 in both cases, and one HSO_4^- ion is generated per mole of acetone dissolved in both. This kind of ambiguity has led Leisten and Wright (1964) to develop four different cryoscopic mixtures in the solvent sulphuric acid, which can be used to establish whether H_3O^+ ions are formed in a complex ionization or not, how many HSO_4^- ions are formed, and whether or not water is abstracted from the solvent. These cryoscopic mixtures are dilute solutions in sulphuric acid of disulphuric acid (dilute oleum), disulphate ion, hydrogen sulphate ion, sulphuric acidium ion and hydronium ion in

4. COMPLEX IONIZATIONS

various combinations. The following four systems have been recommended and their characteristics determined: (1) sulphuric acid/disulphuric acid/disulphate; (2) sulphuric acid/disulphate/hydrogen sulphate; (3) sulphuric acid/sulphuric acidium/disulphuric acid; (4) sulphuric acid/sulphuric acidium/hydronium ion.

Mixture (1) is obtained by dissolving ammonium (or some other) sulphate in dilute oleum in such an amount as to leave an excess of disulphuric acid. The reaction

$$(NH_4)SO_4 + 2H_2S_2O_7 = 2NH_4^+ + 2HS_2O_7^- + H_2SO_4 \quad (4.6)$$

reduces the initial concentration of disulphuric acid and generates an equivalent amount of disulphate ion. If a solute undergoing complex ionization produces some water, this will react with disulphuric acid

$$H_2O + H_2S_2O_7 = 2H_2SO_4 \quad (4.7)$$

so that the net result would be an increase in the freezing point of the mixture, corresponding to $i = -1$. Hydrogen sulphate ions produced in the ionization of any kind of base will react as follows:

$$HSO_4^- + H_2S_2O_7 = HS_2O_7^- + H_2SO_4 \quad (4.8)$$

Thus each mole of HSO_4^- ions removes one mole of $H_2S_2O_7$ and replaces it by an equivalent amount of $HS_2O_7^-$ ions, the net effect on the freezing point of the solution being nil, i.e. $i = 0$ for HSO_4^- ions in mixture (1).

Mixture (2) is obtained like mixture (1), except that ammonium sulphate is added in excess relative to the amount of disulphuric acid present, so that the latter is fully converted to disulphate ions, the excess of ammonium sulphate yielding hydrogen sulphate ions. In this mixture any water generated in a complex ionization would be protonated

$$H_2O + H_2SO_4 = H_3O^+ + HSO_4^- \quad (i = 2) \quad (4.9)$$

and would further react according to the equation

$$H_3O^+ + HS_2O_7^- = 2H_2SO_4 \quad (i = -2) \quad (4.10)$$

the net effect on the freezing point of the mixture being nil. Any hydrogen sulphate ions produced would merely increase the concentration already present, and thus $i = 1$ for these ions.

It has been mentioned in Section 3.2.7, p. 83, that tetrahydrogenosulphatoboric acid is a fairly strong acid of the sulphuric acid solvent system. Therefore to generate the sulphuric acidium ion in mixture (3), this acid is produced by the reaction

$$H_3BO_3 + 3H_2S_2O_7 = H_3SO_4^+ + B(HSO_4)_4^- + H_2SO_4 \quad (4.11)$$

An excess of oleum must be present to provide the second component of the mixture. In this mixture water reacts according to equation (4.7) giving $i = -1$, and HSO_4^- ions behave according to the equation

$$HSO_4^- + H_3SO_4^+ = 2H_2SO_4 \quad (i = -1) \quad (4.12)$$

Mixture (4) is prepared like mixture (3), except that the amount of boric acid dissolved in oleum is between one half and one third of the molar concentration of disulphuric acid present. Under these conditions the following reaction occurs:

$$H_3BO_3 + 2H_2S_2O_7 = H_3O^+ + B(HSO_4)_4^- \quad (4.13)$$

in addition to reaction (4.11) so that the resultant solution contains some sulphuric acidium ion in addition to some hydronium ion. In this mixture HSO_4^- ions will have $i = -1$ through neutralization by $H_3SO_4^+$ ions, and water will be protonated

$$H_2O + H_3SO_4^+ = H_3O^+ + H_2SO_4 \quad (i = 0) \quad (4.14)$$

without any effect on the freezing point of the solution.

If the complex ionization of a solute is represented by the generalized equation

$$S + nH_2SO_4 \to xP + yHSO_4^- + zH_2O \quad (4.15)$$

and the i factors for HSO_4^- and H_2O are represented by i' and i'', then the observed freezing-point depression in any of the cryoscopic mixtures would be given by

$$i = x + yi' + zi'' \quad (4.16)$$

The values of i' and i'' in slightly aqueous sulphuric acid and in the four cryoscopic mixtures are summarized in Table 4.1 (Leisten and Walton, 1964). Experimentally found i factors are inserted into equations (4.16) with the values of i' and i'' appropriate to the solvent mixture used. Such equations are then solved for x, y and z. With five solvent mixtures available these numbers can be determined unambiguously. If the reaction involves abstraction of water from the solvent, z will be negative. It may be noted that for reactions that go to completion the procedure will yield an unambiguous equation for the reaction. For partial protonation, however, the results will depend on the solvent mixture used, because mixtures (3) and (4) are more acidic than pure sulphuric acid, and mixture (2) is more basic. So, for example, the results for acetonitrile, which is incompletely protonated in pure sulphuric acid (Liler

TABLE 4.1

The cryoscopic i factors in pure sulphuric acid and in cryoscopic mixtures (1)–(4) (Leisten and Walton, 1964)

Solvent	H_2SO_4	(1)	(2)	(3)	(4)
i' (HSO_4^-)	+1	0	1	−1	−1
i'' (H_2O)	+2	−1	0	−1	0

and Kosanović, 1959), indicate complete protonation in mixtures (3) and (4) (Leisten and Walton, 1964).

These cryoscopic mixtures were tested on several model solutes by Leisten and Walton (1964)—see Section 4.3. High accuracy of i factors is not required. With the help of this method, coupled with an examination of the solutions by n.m.r. spectroscopy, it should be possible to identify the most complicated modes of ionization in sulphuric acid.

4.2.2. THE CONDUCTOMETRIC METHOD

The use of conductances of solutions in sulphuric acid to determine the number of HSO_4^- ions, γ, produced by one solute molecule is essentially the same as the method for the determination of degrees of ionization of weak bases, described in Section 3.2.6, p. 71, except that complex ionizations which go to completion will produce one or more HSO_4^- ions per molecule of solute. The conductances of such solutions rise more steeply than the conductance of typical strong monoacid bases ($KHSO_4$ or other), and the ratio of concentrations of solutions having equal conductance gives

$$\gamma = c_s/c_x \tag{4.17}$$

where c_x is the molar concentration of the solute under investigation and c_s is the molar concentration of $KHSO_4$. For reactions that go to completion, γ is a whole number.

4.2.3. THE CONDUCTOMETRIC TITRATION METHOD

The use of conductometric titrations in establishing the mode of complex ionizations is also similar to their use in determining the strengths of weak bases (Section 3.2.7, p. 83). The application of this method to complex ionizations was recently elaborated by Robinson and Quadri (1967a) and compared with the method of cryoscopic mixtures (Section 4.2.1).

For the titration of an initially 0·1 M solution of disulphuric acid with the solution of a strong base ($KHSO_4$ or other), the conductance minimum occurs at a molar ratio of base to the initial number of moles of acid ($r = b/a$) equal to 0·56. In most complex ionizations water is either produced or removed from the solvent producing disulphuric acid. In addition, some electrolyte solutes (strong bases) or non-electrolytes may be produced.

If a solute reacts to give a molecule of water and a molecule of strong base according to

$$S \rightarrow P^+ + HSO_4^- + H_2O \tag{4.18}$$

then the addition of b moles of solute gives b moles of water, which remove an equivalent number of moles of $H_2S_2O_7$ from solution by reaction (4.7). Hence the number of moles of disulphuric acid remaining for neutralization of the base (HSO_4^-) is only $a - b$. The conductance minimum is now expected to occur at $b/(a-b) = 0.56$, i.e. at $b/a = 0.36$. If two moles of HSO_4^- ions are produced in the ionization, the ratio at minimum conductance is given by $2b/(a-b) = 0.56$, i.e. $b/a = 0.22$.

If a solute ionizes by abstracting water from the solvent, it will generate more disulphuric acid than was originally present. In general this type of ionization may be represented as follows:

$$S \to xP^+ + xHSO_4^- + yH_2S_2O_7 \qquad (4.19)$$

The conductance minimum should then occur at $xb/(a+yb) = 0.56$. The best-known example of this type of behaviour is the ionization of acetic anhydride (equation 4.3), which produces two moles of strong base, $CH_3C(OH)_2.HSO_4$, and a mole of $H_2S_2O_7$. Therefore the conductance minimum should occur at $2b/(a+b) = 0.56$, i.e. $b/a = 0.39$. This is clearly distinguishable from simple protonation, and from an alternative ionization scheme

$$(CH_3CO)_2O + 2H_2SO_4 = CH_3CO^+ + CH_3C(OH)_2^+ + 2HSO_4^- \qquad (4.20)$$

which would require a minimum at $2b/a = 0.56$ (since each mole of solute produces two moles of strong base). The experimentally observed position of minimum conductance is at $b/a = 0.38$, in agreement with prediction for the dehydration reaction (4.3).

The conductometric titration method is thus equally useful for elucidating more complex modes of ionization as the method of cryoscopic mixtures. It has the advantage of being experimentally simple and the results are unaffected by the non-ideality of solutions.

4.2.4. THE SPECTROPHOTOMETRIC METHOD

Many complex ionizations are accompanied by deep changes in the light absorption of the ionizing species and can therefore be studied by u.v. spectrophotometry. This is particularly true of one of the best-known types of complex ionization, that of triarylmethanols, which are well-known indicators. The absorption spectra of ions are themselves of considerable interest as sources of information about their structure.

4.3. Survey of Complex Ionizations

4.3.1. ORGANIC SOLUTES

Some of the material of this section was reviewed at an earlier date by Gillespie and Leisten (1954a). Emphasis will here be placed therefore on the more recent developments.

4.3.1.1. *Aliphatic Alcohols*

The basic facts about the behaviour of primary and secondary alcohols in sulphuric acid were correctly established by Hantzsch (1908), and Oddo and

Scandola (1909b). Their cryoscopic measurements showed that methanol and ethanol give three-fold freezing-point depressions, which can be accounted for by the formation of alkyl hydrogen sulphates according to the equation

$$ROH + 2H_2SO_4 = RHSO_4 + H_3O^+ + HSO_4^- \tag{4.21}$$

Accurate cryoscopic measurements by Gillespie (1950c) confirmed this, but also suggested that alkyl hydrogen sulphates may be slightly ionized further as bases:

$$ROSO_3H + H_2SO_4 \rightleftharpoons ROSO_3H_2^+ + HSO_4^- \tag{4.22}$$

Alternatively a slight ionization into a primary carbonium ion and HSO_4^- ion could account for the results, but this is unlikely in view of the known instability of primary aliphatic cations. The solutions of ethanol and methanol remain clear and colourless at ordinary temperatures. A reinvestigation of the ionization of methanol by Leisten and Wright (1964) using a cryoscopic mixture has fully confirmed reaction (4.21). Conductometric titrations also support this and offer no evidence for any further ionization of the alkyl hydrogen sulphate (Robinson and Quadri, 1967a).

The solutions of longer chain primary alcohols also give initially i factors close to three, but this value increases with time and simultaneously the pale yellow solutions turn red and brown. The complex reactions responsible for these changes lead finally to the separation of a hydrocarbon layer. Some sulphur dioxide is evolved also, indicating that some oxidation occurs. These reactions probably involve unstable carbonium ions as intermediates.

Cryoscopic experiments with branched chain primary and secondary alcohols initially give i factors between 2 and 3, which increase rapidly with time (Oddo and Scandola, 1909b). Thus it appears that the formation of alkyl hydrogen sulphates is incomplete at first. Observations of subsequent behaviour are similar to those of straight chain alcohols.

Tertiary alcohols, which give two-fold freezing-point depressions (Hantzsch, 1908; Oddo and Scandola, 1909b; Newman et al., 1949), were held for a long time to undergo simple protonation. The solutions, however, are not stable; their colour darkens and some sulphur dioxide is evolved. A recent reinvestigation of the ionization of t-butanol in cryoscopic mixtures (Leisten and Walton, 1964) has established unambiguously that this solute is immediately dehydrated and polymerized according to the equation

$$n\,t\text{-BuOH} + nH_2SO_4 \rightarrow \text{Polymer} + nH_3O^+ + nHSO_4^- \tag{4.23}$$

as already suggested by Craig et al. (1950). The polymer, polyisobutylene, is probably of fairly low molecular weight, because the results suggest that it also contributes to the freezing-point depression, which also changes slowly with time. In the light of this new information on the ionization of t-butanol, it seems likely that initial factors of less than 3 for branched chain and secondary alcohols are also due to some immediate dehydration and polymerization.

Also, in view of these results, some spectrophotometric observations on tertiary aliphatic and alicyclic carbinols in concentrated sulphuric acid (Lavrushin et al., 1955), ascribed to carbonium ions, are more likely to refer to polymerized products. The solutions are in fact not clear immediately after preparation.

Protonated alcohols and their cleavage to carbonium ions have been observed in superacid media (HSO_3F-SbF_5 diluted with SO_2) by Olah and Namanworth (1966) and by Olah et al. (1967b). Cleavage of protonated methanol could not be observed up to $+60°C$, whereas protonated n-propanol begins to cleave at $0°C$. Protonated secondary alcohols are in equilibrium with their corresponding secondary carbonium ions, which gradually rearrange into the more stable tertiary carbonium ions. Stable tertiary carbonium ions are produced by dissolving tertiary alcohols in the superacid medium.

4.3.1.2. *Aryl Alcohols*

The mode of ionization of triphenyl carbinol (equation 4.1), observed by Hantzsch (1908), was confirmed by a comparison of the absorption spectrum of the yellow solution with that of the solution of triphenylmethyl chloride in sulphur dioxide (Hantzsch, 1921), which showed electrical conductance owing to the dissociation into Ph_3C^+ and Cl^- ions. The spectra of many substituted triphenyl carbinols in pure sulphuric acid were studied by Newman and Deno (1951a), who found strong absorptions in the 400–500 nm range. All triarylcarbinols are recoverable upon dilution of their sulphuric acid solutions with water, or give ethers upon dilution with alcohols.

The possibility of using the ionization of these compounds in order to establish an acidity function scale (H_R) was recognized by Murray and Williams (1950), and such a scale was defined by Deno et al. (1955)—see Chapter 2. Hence the pK_{R^+} values for a number of triaryl carbonium indicators are well established. Further data were supplied more recently, and some indicators were studied over a range of temperature (Arnett and Bushick, 1964). A compilation of these data is given in Table 4.2. It can be seen that carbinol/carbonium ion equilibria cover the whole acidity range from pure water to concentrated sulphuric acid. The tris-*p*-amino-, tris-*p*-dimethylamino-, and bis-*p*-amino-derivatives were also studied cryoscopically (Newman and Deno, 1951a) and *i* factors of only 6 were found for them. This means that only two of the three amino groups are protonated, the third being involved in strong resonance interaction with the carbonium centre (Section 3.3.2, p. 95).

The question of the extent of resonance interaction of the phenyl rings with the carbonium centre is of considerable interest (Gillespie and Leisten, 1954a). It is clear from models that all three rings cannot be simultaneously coplanar with the carbonium centre, and therefore cannot all be involved in resonance. In the absence of resonance interaction with the carbonium centre, phenyl

TABLE 4.2

pK_{R^+} values for some triarylmethanols in aqueous sulphuric acid at 25°C

Substituents			pK_{R^+}	Reference†
4-OH	4-OH	4-OH	+1·97	a
4-OMe	4-OMe	4-OMe	+0·82	b
4-OMe	4-OMe	—	−0·89	c
4-OMe	—	—	−3·40	b
			−3·20	c
4-Me	4-Me	4-Me	−3·56	b
4-Me	—	—	−5·41	b
			−5·25	c
4-CD$_3$	—	—	−5·43	b
3-Me	3-Me	3-Me	−6·35	b
4-i-Pr	4-i-Pr	4-i-Pr	−6·54	b
H	H	H	−6·63	b
			−6·44	c
4-F	4-F	4-F	−6·05	d
4-Cl	4-Cl	4-Cl	−7·74	b
4-NO$_2$	—	—	−9·15	b
			−9·44	c
3-Cl	3-Cl	3-Cl	−11·03	b
4-NO$_2$	4-NO$_2$	—	−12·90	b
4-NMe$_3^+$	4-NMe$_3^+$	4-NMe$_3^+$	−15·16	c
4-NO$_2$	4-NO$_2$	4-NO$_2$	−16·27	b
			−16·35	e
F$_5$	F$_5$	F$_5$	−17·5	d

† a. Deno and Evans (1957); b. Deno et al. (1955); c. Arnett and Bushick (1964); d. Filler et al. (1967); e. Based on the ionization ratios of Murray and Williams (1950).

rings carrying electron-donating substituents would be expected to be rapidly sulphonated in sulphuric acid. The observed i factors do not bear out this expectation. The answer to this question emerges from X-ray crystallographic and infrared studies of some triphenylmethyl salts, which show that the three bonds around the carbonium centre are coplanar, but the rings are twisted from the plane into a propeller shape (Bethell and Gold, 1967). However, the angle of twist is only about 30°, so that some overlap of orbitals essential for resonance interaction is still possible.

Diarylmethanols behave in pure sulphuric acid like triarylmethanols ($i = 4$, recoverable on dilution), but on standing slow sulphonation may occur (Welch and Smith, 1950). Their ionization equilibria also obey the H_R acidity function (Deno et al., 1955). The pK_{R^+} values for some derivatives are given in Table 4.3. The figures show that the unsubstituted diphenylmethanol is some

TABLE 4.3

pK_{R^+} values for some diarylmethanols in aqueous sulphuric acid at 25°C

Substituents		pK_{R^+}	Reference†
4-OMe	4-OMe	−5·71	a
		−5·85	b
2,4,6-Me$_3$	2,4,6-Me$_3$	−6·6	a
4-OPh	4-OPh	−9·85	c
4-Me	4-Me	−10·31	b
		−10·4	a
2-Me	2-Me	−12·45	a
4-F	4-F	−13·03	c
4-t-Bu	4-t-Bu	−13·2	a
H	H	−13·25	b
		−13·3	a
4-Cl	4-Cl	−13·96	a, b
4-Br	4-Br	−14·16	c
4-I	4-I	−14·26	c

† a. Deno et al. (1955); b. Mocek and Stewart (1963); c. Deno and Evans (1957).

6·7 pK units less basic as a secondary base than triphenylmethanol (Table 4.2). This is a reflexion of the difference in the stability of the respective carbonium ions.

The basicities of several α-deuterated 4,4'-substituted diarylmethanols were compared with those of the corresponding protio compounds in aqueous sulphuric acid (Mocek and Stewart, 1963). The equilibrium constants for the equilibria

$$Ar_2CHOH + H^+ \rightleftarrows Ar_2CH^+ + H_2O \qquad (4.24)$$

$$Ar_2CDOH + H^+ \rightleftarrows Ar_2CD^+ + H_2O \qquad (4.25)$$

K_H and K_D respectively, give a ratio $K_H:K_D$ greater than unity, which seems to increase with the electronegativity of the *para* substituents. There is a change from tetragonal to trigonal symmetry at the central carbon atom in this ionization, and this is held to be responsible for the isotope effect (Mocek and Stewart, 1963).

Diphenylmethylmethanol also gives stable solutions in concentrated sulphuric acid, but upon dilution diphenylethylene is recovered (Gold and Tye, 1952a), which means that here the carbonium ion is in equilibrium with the olefin, rather than with the carbinol.

The behaviour of monoaryl alcohols in sulphuric acid solution has been less extensively studied, and there is conflicting information on this subject.

According to some earlier data, benzyl alcohol and *p*-chlorobenzyl alcohol give insoluble polymers in sulphuric acid (Cannizzaro, 1854). Heavily methylated benzyl alcohols (α,α,2,4,6-pentamethyl- and heptamethyl-) undergo rapid sulphonation, but their *i* factors extrapolated to zero time are close to four, whereas *p*-nitrobenzyl alcohol gives a constant *i* factor of 3·2 (Newman and Deno, 1951a). Most of these facts suggest considerable instability of monoaryl methanols in concentrated sulphuric acid. In contrast to this, dilute solutions in concentrated sulphuric acid can be prepared by dissolving the alcohols in glacial acetic acid first, and then diluting the solution with an excess of concentrated sulphuric acid (Williams, 1962). It appears that high dilution and prior dissolution in an acidic solvent provide milder conditions for the generation of carbonium ions and minimize secondary reactions.

A study of ionization equilibria for a number of monoaryl alcohols or olefins was attempted by Deno *et al.* (1960), but the solutions showed considerable instability. The estimated pK_{R^+} values of α,α,2,4,6-pentamethylbenzyl alcohol and for heptamethylbenzyl alcohol are $-12\cdot2$ and $-12\cdot4$ respectively. When this is compared with the value of $-6\cdot6$ for dimesitylmethanol (Table 4.3), it follows that monoaryl alcohols are some 6 pK units less basic as secondary bases than diaryl alcohols, i.e. there is an increase in secondary basicity of about 6 pK units for every additional phenyl group.

Some indirect information on the stability of monoaryl carbonium ions was obtained from their u.v. spectra, taken on solutions prepared by Williams (1962), which were sufficiently stable for meaningful spectra to be obtained. Following a suggestion of Grace and Symons (1959) that the absorptions at 400–450 nm of these carbonium ions are due to the transition

the effects of structure on the positions of absorption maxima (Table 4.4) can be interpreted in terms of the stabilization or destabilization of the ground state or the excited state. Apart from the inductive effect upon the stability of the ground state, which accounts for the increasing transition energies in part (a), the most notable effect is that of β-phenyl and γ-ethoxy groups in part (b), which also increase the transition energies by stabilizing the ground

(4.1)   (4.2)

state by a space polarization effect, as shown in (**4.1**) and (**4.2**). The effect of the lengthening of the aliphatic chains in dialkylphenyl carbonium ions—part (e)—may, on the other hand, be ascribed to the stabilization of the excited state (with the charge in the *para* position) by the alkyl groups overhanging the benzene ring. Angle strain can account for the results in part (f) (Williams, 1962). It would be interesting to know how these same effects would affect the pK_{R^+} values of these alcohols. The conversion to carbonium ions appears

TABLE 4.4

Absorption maxima ($\lambda > 360$ nm) of monoaryl carbonium ions

	Substituents			λ max
(a)	Ph	H	H	470†
	Ph	Me	H	435
	Ph	Et	H	420
	Ph	n-Pr	H	414
	Ph	n-Bu	H	408–414
	Ph	n-Hex	H	410–415
	Ph	iso-Bu	H	405–415
	Ph	t-Bu	H	395–400
(b)	Ph	Bz	H	457·5
	Ph	β-Phenylethyl	H	402·5
	Ph	β-Phenylpropyl	H	422·5
	Ph	γ-Ethoxypropyl	H	342·5
(c)	p-Me-phenyl	Me	H	421
				458
	p-Cl-phenyl	Me	H	421
				468
	p-Cl-phenyl	Me	H	442·5
	Mesityl	Me	H	407
(d)	Ph	Me	Me	395–400
	Ph	Me	Et	390–395
	Ph	Me	n-Pr	402
	Ph	Me	Benzyl	429
	Ph	Me	β-Phenylethyl	422·5
(e)	Ph	Et	Et	390–395
	Ph	Et	n-Pr	401
	Ph	n-Pr	n-Pr	403
	Ph	n-Bu	n-Bu	408
	Ph	n-Hex	n-Hex	420
(f)	Derived from 1-phenyl cyclopentanol			377
	Derived from 1-phenyl cyclohexanol			399
	Derived from 1-phenyl cycloheptanol			388

† Grace and Symons (1959). All others from Williams (1962).

to be complete in concentrated sulphuric acid for all the alcohols studied (Williams, 1962), and the results do not indicate any dehydration of alkyl and dialkyl derivatives into olefins, or any dimerization.

4.3.1.3. *Carboxylic Acids, Esters and Amides*

In addition to simple protonation, another type of ionization of carboxylic acids and their derivatives (esters and anilides) is known, which was discovered first by Treffers and Hammett (1937). Di-*ortho*-substituted benzoic acids give freezing-point depressions twice as large as benzoic acid itself, and are instantaneously esterified upon dilution of the solution with alcohols (Newman, 1941). Both these facts are consistent with an ionization, e.g. for mesitoic acid, according to the equation

$$Me_3C_6H_2COOH + 2H_2SO_4 = Me_3C_6H_2CO^+ + H_3O^+ + 2HSO_4^- \quad (4.26)$$

the observed i factor being 4. The existence of mesitoyl ions in sulphuric acid has also been confirmed by infrared spectroscopy (Oulevey and Susz, 1965).

Methyl esters of these carboxylic acids likewise ionize in a complex manner (Newman *et al.*, 1949)

$$RCOOCH_3 + 3H_2SO_4 = RCO^+ + H_3O^+ + CH_3HSO_4 + 2HSO_4^- \quad (4.27)$$

which leads to an i factor of 5. These esters are not recoverable upon pouring the sulphuric acid solution on ice; the carboxylic acids are obtained instead. The same type of ionization was reported for pentamethylbenzanilide (Newman and Deno, 1951b), on the basis of an eventually observed i factor of almost 6:

$$RCONHC_6H_5 + 3H_2SO_4 = RCO^+ + H_3O^+ + 3HSO_4^- + HO_3SC_6H_4NH_3^+ \quad (4.28)$$

There is no evidence for the alternative possibility of the formation of an iminocarbonium ion, which would lead to $i = 4$:

$$RCONHC_6H_5 + 2H_2SO_4 = R\overset{+}{-}C\!=\!N\!-\!C_6H_5 + H_3O^+ + 2HSO_4^- \quad (4.29)$$

Mesitylanilide and *o*-tolylanilide appear to react likewise according to equation (4.28), only more slowly (Newman and Deno, 1951b).

For many years this type of ionization was believed to be confined to sterically hindered acids and the effect was ascribed to steric crowding around the protonated carboxyl group. However, recently the same type of ionization was demonstrated by n.m.r. spectra for all acids in more strongly acidic media (e.g. for acetic acid in dilute oleum) (Deno *et al.*, 1964b). Thus steric crowding is only one of the factors, which destabilizes protonated carboxylic acids (an internal factor). Greater affinity for water in the more strongly acidic media is another (an external factor). The cations obtained in these ionizations have been variously called acylium ions, acyl cations, aroyl cations (for aromatic acids) and oxocarbonium ions. The latter term has gained wider acceptance recently.

Protonation is undoubtedly the first stage in the conversion of carboxylic acids to oxocarbonium ions in every case:

$$RCOOH + H_2SO_4 \rightleftharpoons RC(OH)_2^+ + HSO_4^- \qquad (4.30)$$

$$RC(OH)_2^+ + H_2SO_4 \rightleftharpoons RCO^+ + H_3O^+ + HSO_4^- \qquad (4.31)$$

Both stages of the conversion have been studied spectrophotometrically for mesitoic acid (Schubert et al., 1954) by using the fairly intense absorption of the oxocarbonium ion in the region 250–350 nm. The protonation of the acid occurs at lower acidities, whereas the conversion to the oxocarbonium ion occurs at >91 per cent acid, and is half completed in 97 per cent acid. Nuclear magnetic resonance studies of acetic, benzoic and mesitoic acids in sulphuric acid and oleum have shown that the conversion of the protonated acid to the oxocarbonium ion occurs for benzoic acid in oleum with >6 per cent SO_3, and that acetic acid is half converted in 15 per cent SO_3–85 per cent H_2SO_4 (Traficante and Maciel, 1966).

No systematic study of substituent effects on the conversion has been carried out, but some results suggest that charge delocalization in the oxocarbonium

(4.3)

ion (4.3) is an important factor. So, for example, dibromomesitoic acid gives an i factor of only 2–2·6 in pure sulphuric acid (Treffers and Hammett, 1937), whereas 2,4,6-trimethoxybenzoic acid is reported to be fully ionized to the oxocarbonium ion in only 64 per cent perchloric acid (Schubert et al., 1955). Thus electron-withdrawing substituents counter the conversion, whereas mesomerically electron-releasing substituents favour it.

In 90–100 per cent sulphuric acid benzoylbenzoic acid also undergoes complex ionization (Newman et al., 1945). Evidence has been adduced that this results in a lactol carbonium ion (Noyce and Kittle, 1965b).

The cleavage of protonated carboxylic acids to oxocarbonium ions was also studied in superacid media (Olah and White, 1967a and 1967c). The carbon-13 resonance of the oxocarbonium ion supports linear sp hybridization of the carbon electron orbitals in the cation.

4.3.1.4. *Acid Anhydrides*

The behaviour of carboxylic acid anhydrides is more complex than the early cryoscopic studies of Oddo and Casalino (1917b) have indicated. Accurate cryoscopic measurements by Gillespie (1950d) on acetic and benzoic anhydrides gave $i = 4$ for both solutes, and this was interpreted as indicating an ionization according to equation (4.20), with the formation of acetylium and benzoylium ions respectively. This suggestion proved, however, to be inconsistent with the conductometrically determined number of HSO_4^- ions produced by one molecule of solute, which is only one (Leisten, 1955). This is consistent with an ionization according to equation (4.3). The reaction is thus an extraction of water from pure sulphuric acid by the anhydride. Further conductometric studies fully confirmed this unusual type of complex ionization (Flowers et al., 1956; Robinson and Quadri, 1967a), and so did a study in cryoscopic mixtures (Leisten and Wright, 1964) and a Raman spectroscopic examination of the solutions (Casadevall et al., 1964a).

Some cyclic anhydrides were examined more recently (Leisten, 1961). The equilibrium

$$\begin{matrix} CO \\ | \\ CO \end{matrix}\!\!> O + H_3O^+ \rightleftharpoons \begin{matrix} C(OH)_2^+ \\ | \\ CO_2H \end{matrix} \qquad (4.32)$$

is critically balanced in the region of 100 per cent sulphuric acid for such anhydrides. For glutaric anhydride the hydrolysis is virtually complete even in dilute oleum. Succinic, phthalic and maleic anhydride show weakly basic behaviour, succinic anhydride being a stronger base than the latter two. Maleic anhydride is essentially a non-electrolyte according to Robinson and Quadri (1967b), who also found that succinic acidium ion is favoured even in dilute oleum for both the acid and its anhydride. For maleic and phthalic acids anhydride formation is favoured in 100 per cent sulphuric acid and oleum. Naphthalene-1,8-dicarboxylic anhydride is substantially monoprotonated.

For β-methylglutaconic anhydride an i factor of $3·3 \pm 0·07$ has been reported (Wiley and Moyer, 1954). It thus appears that this anhydride also partially hydrolyses in pure sulphuric acid.

4.3.1.5. *Miscellaneous Solutes*

A few more examples of complex types of ionization which have recently been elucidated will be mentioned here, in order to illustrate the complexity of behaviour that may be encountered and to demonstrate the power of a combined use of all cryoscopic and conductometric methods. Slowly changing freezing-point depressions are usually ascribed to sulphonation or oxidation of the solutes. In many such instances there is no certainty about the reactions occurring, and indeed if both sulphonation and oxidation occur the problem of interpreting cryoscopic observations may be an insoluble one. On the other hand if the final i factors are not too large, there is a good chance that by use of cryoscopic mixtures and conductance measurements an unambiguous identification of the reaction will be possible. Evolution of sulphur dioxide accompanies oxidations by sulphuric acid.

The ionization of trichloromethyl mesitylene in sulphuric acid has been the subject of some controversy recently. Hart and Fish (1958, 1960, 1961) interpreted the result of their cryoscopic study of this solute ($i = 5$) as meaning that a dipositive carbonium ion $(CH_3)_3C_6H_2CCl^{2+}$ is formed. A more detailed examination of both cryoscopic and conductometric results (Gillespie and Robinson, 1964) has shown that the ionization is

$$(CH_3)_3C_6H_2CCl_3 + H_2SO_4 = (CH_3)_3C_6H_2CCl_2^+ + HSO_3Cl + H_3O^+ + 2HSO_4^-$$

$$(i = 5, \gamma = 2) \quad (4.33)$$

giving a monopositive carbonium ion. These conclusions were confirmed by a study of the same solute in chlorosulphuric acid (Robinson and Ciruna, 1964). The rather unusual carbonium ion produced probably owes its stability to the back donation of the lone electron pairs of chlorine substituents to the carbonium centre. Mesitoyl chloride ionizes likewise to give the corresponding oxocarbonium ion (Gillespie and Robinson, 1964).

Hexamethylbenzene behaves as a Brønsted base in sulphuric acid/water mixtures and is half protonated in 90·5 per cent acid (Kilpatrick and Hyman, 1958) but in pure sulphuric acid it undergoes a displacement of methyl groups:

$$C_6(CH_3)_6 + 7H_2SO_4 = C_6(CH_3)_3(SO_3H)_3H^+ + 3CH_3HSO_4 + HSO_4^- + 3H_2O \quad (4.34)$$

This conclusion was reached by a study of this solute in several cryoscopic mixtures (Leisten and Walton, 1964).

The ionization of xanthen in sulphuric acid represents a case of simple sulphonation:

$$C_{13}H_{10}O + 2H_2SO_4 = C_{13}H_8O(SO_3H)_2 + 2H_2O \quad (4.35)$$

This was also deduced from a study of this solute in several cryoscopic mixtures (Leisten and Walton, 1964). This example shows that the use of cryoscopic mixtures can unambiguously establish the occurrence of sulphonation.

Scatole (3-methylindole) shows a more complicated type of behaviour, which could plausibly be ascribed to oxidative condensation (Leisten and Walton, 1964).

4.3.2. INORGANIC SOLUTES

A number of inorganic solutes which undergo complex ionization in sulphuric acid has been studied. The most interesting and most important of these from the point of view of reaction mechanisms in sulphuric acid media are nitric acid and nitrogen oxides, because their ionization results in the formation of reactive species. The ionization of iodine compounds is important for the same reason. The behaviour of only a few other solutes, which are of less direct interest, will also be mentioned.

4.3.2.1. Nitric Acid, Nitrous Acid and Nitrogen Oxides

The indications of the early cryoscopic work of Hantzsch (1908) that nitric acid ionizes in a complex manner in pure sulphuric acid and the subsequent Raman observations of Chédin (1935), showing the emergence of a new species in solutions of nitric acid in concentrated sulphuric acid, have been fully substantiated by extensive cryoscopic studies of nitric acid and nitrogen oxides in sulphuric acid by Gillespie et al. (1950b). Freezing-point depressions by nitric acid gave $v = 3 \cdot 77$ and this was ascribed to reaction (4.2). A four-fold freezing-point depression is not attained, because water as solute does not give an exactly two-fold depression (see Section 3.2.5, p. 67). The electrical conductivities of the solutions indicate the formation of two HSO_4^- ions per mole of acid (Gillespie and Wasif, 1953c).

Similarly, nitrogen oxides were found to ionize as follows:

$$N_2O_5 + 3H_2SO_4 = 2NO_2^+ + H_3O^+ + 3HSO_4^- \quad (4.36)$$
$$N_2O_4 + 3H_2SO_4 = NO_2^+ + NO^+ + H_3O^+ + 3HSO_4^- \quad (4.37)$$
$$N_2O_3 + 3H_2SO_4 = 2NO^+ + H_3O^+ + 3HSO_4^- \quad (4.38)$$

The electrical conductances of dinitrogen tetroxide in sulphuric acid are also consistent with equation (4.37)—Gillespie and Wasif (1953c) and Hetherington et al. (1955a).

Previously the maximum rate of nitration observed in 90 per cent sulphuric acid was ascribed to the full conversion of nitric acid to the active nitrating agent, NO_2^+ (Westheimer and Kharasch, 1946). The Raman spectra of nitric acid in several strong acids show a single characteristic line at 1400 cm^{-1} (Chédin, 1937; Ingold et al., 1950; Millen, 1950), which identifies the cation as NO_2^+. The presence of nitrogen in the cation was also demonstrated by transport measurements (Bennett et al., 1946). The NO_2^+ ion is considerably resonance stabilized (**4.4**):

$$\overset{+}{|\overline{O}}-N=\overline{O}| \leftrightarrow |\overline{O}=\overset{+}{N}=\overline{O}| \leftrightarrow |\overline{O}=N-\overline{O}|^{+}$$
(4.4)

The range of sulphuric acid/water concentrations in which the conversion of nitric acid to NO_2^+ occurs has been established more closely by u.v. spectrophotometry (Deno et al., 1961). The spectra indicate that HNO_3 still predominates in 85 per cent sulphuric acid. At higher concentrations marked changes occur in the absorption spectrum. In the 220–270 nm region, the extinction coefficients increase sharply to a maximum at 90 per cent acid, and then fall to low values in 99 per cent acid. On the basis of the Raman spectroscopic data of Chédin (1937) it can be estimated that HNO_3 is almost fully converted into NO_2^+ in about 90 per cent acid. Therefore the absorption between 220 and 310 nm must be due to a new spectroscopic species, which is tentatively considered to be nitronium hydrogen sulphate, O_2NOSO_3H. Its concentration falls with increasing acid concentration, because of the falling HSO_4^- concentration. The equilibrium constant for the reaction

$$NO_2^+ + HSO_4^- \rightleftharpoons O_2NOSO_3H \tag{4.39}$$

is estimated as only 0·01. Maximum concentration of nitronium hydrogen sulphate occurs in 90 per cent acid, in which about 9 per cent of the NO_2^+ may be present in this form (Deno et al., 1961).

The equilibrium between nitrate and nitric acid in sulphuric acid/water mixtures was also studied by u.v. spectroscopy (Deno et al., 1961). Nitrate is half converted to nitric acid in 44 per cent sulphuric acid. The logarithms of the ionization ratios follow the H_0 acidity function with a slope of 0·61.

Analogous information for nitrous acid has also been obtained spectrophotometrically by dissolving sodium nitrite in a series of sulphuric acid/water mixtures (Bayliss and Watts, 1955 and 1956). The rather weak band at 370 nm ascribed to HNO_2 disappears at concentrations >40 per cent, and is supplanted by a band at 250 nm, whose intensity reaches a maximum above 80 per cent acid. This is ascribed to the nitrosyl cation, NO^+, which also gives rise to the Raman line at 2290 cm^{-1} (Bayliss and Watts, 1956). Closely similar behaviour is found in perchloric acid solutions (Singer and Vamplew, 1956; Angus and Leckie, 1935). Not all nitrous acid can be accounted for in terms of these two species in the intermediate concentration range, and therefore the presence of protonated nitrous acid also is suggested. The conversion thus occurs in two stages:

$$HNO_2 + H_3O^+ \rightleftharpoons H_2NO_2^+ + H_2O \tag{4.40}$$

$$H_2NO_2^+ \rightleftharpoons NO^+ + H_2O \tag{4.41}$$

In the region 75–100 per cent sulphuric acid the absorption maximum at 250 nm changes to a maximum at 220 nm, with some evidence of an isosbestic point at 230 nm, indicating that a further chemical equilibrium is involved (Deschamps, 1957). These changes remain unexplained. The absorption of nitric oxide by these solutions has been studied (Seel and Sauer, 1957), and shown to arise from complex formation between the nitrosyl cation and NO:

$$NO + NO^+ \rightleftharpoons N_2O_2^+ \tag{4.42}$$

which gives the solutions a blue colour. In >97·5 per cent sulphuric acid the colour of these solutions fades, probably owing to oxidation of the complex cation by sulphuric acid.

NO_2^+ ions are also present in solutions of nitryl fluoride in concentrated sulphuric acid (Hetherington et al., 1955c). The freezing-point depression and conductance indicate that the following reaction occurs:

$$NO_2F + 3H_2SO_4 = NO_2^+ + HSO_3F + H_3O^+ + 2HSO_4^- \tag{4.43}$$

At higher concentrations (about 70 mol per cent of nitryl fluoride) colourless crystals separate from solution, believed to be nitronium hydrogen sulphate.

4.3.2.2. *Iodine Compounds*

Solutions of a number of iodine compounds in sulphuric acid have been studied in search of cations or oxycations containing iodine, because strongly acid media are known to stabilize such species, and iodine as the most electropositive halogen is the most likely to form them. Indications that such cations may exist arise from the existence of some crystalline compounds between iodine oxides and sulphur trioxide, such as the compound $I_2O_3 \cdot SO_3$ (Masson and Argument, 1938), which has frequently been formulated as iodosyl sulphate, $(IO)_2SO_4$. There is still a certain amount of controversy and uncertainty concerning the exact significance of some experimental results, but there is no doubt that some positive species of iodine and iodine oxides have been observed in sulphuric acid and oleum (Arotsky *et al.*, 1962; Gillespie and Senior, 1964a and 1964b).

Solutions of iodine in sulphuric acid on standing develop an absorption at 330 nm and a shoulder at 280 nm (Arotsky *et al.*, 1962). The band at 330 nm is ascribed to the I_5^+ cation, and that at 280 nm to sulphur dioxide, which are formed by the slow reaction

$$5I_2 + 5H_2SO_4 = 2I_5^+ + 2H_3O^+ + 4HSO_4^- + SO_2 \tag{4.44}$$

Iodine also dissolves in solutions of iodic acid in sulphuric acid, producing a steady increase in conductance, which stops at the stage required by the reaction

$$HIO_3 + 7I_2 + 8H_2SO_4 = 5I_3^+ + 3H_3O^+ + 8HSO_4^- \tag{4.45}$$

(Arotsky *et al.*, 1962). The spectrum of this solution in the region 400–600 nm is similar to that of iodine solutions in dilute oleum, in which I_3^+ cations are believed to be present (Arotsky *et al.*, 1961).

Good evidence for the existence of iodosyl cations has been presented by Gillespie and Senior (1964b), who have carried out a conductometric and cryoscopic study of solutions of iodosyl sulphate, $(IO)_2SO_4$, in sulphuric acid. The numbers of particles produced by its ionization, as well as the number of HSO_4^- ions, decrease with increasing concentration of the solution ($v = 3.22 \to 2.76$, $\gamma = 1.14 \to 0.75$ for $m = 0.02 - 0.07$ mol kg^{-1}) which can be explained by assuming that the iodosyl hydrogen sulphate formed according to the equation

$$(IO)_2SO_4 + H_2SO_4 = 2IOHSO_4 \tag{4.46}$$

is about 50 per cent ionized as a base:

$$IOHSO_4 \rightleftharpoons IO^+ + HSO_4^- \tag{4.46a}$$

This corresponds to a K_b value of 3.5×10^{-2} mol litre^{-1}.

Detailed cryoscopic and conductometric studies of the solutions of iodic acid (HIO_3) in sulphuric acid (Gillespie and Senior, 1964a) show that this solute is not simply protonated, as claimed by Arotsky et al. (1962), but ionizes to give a polymerized form of iodyl hydrogen sulphate:

$$nHIO_3 + 2nH_2SO_4 = (IO_2HSO_4)_n + nH_3O^+ + nHSO_4^- \qquad (4.47)$$

Both freezing-point depression and conductance suggest that this polymer may ionize slightly further according to the equation

$$(IO_2HSO_4)_n \rightleftharpoons (IO_2)_n(HSO_4)_{n-1}^+ + HSO_4^- \qquad (4.47a)$$

There is thus no evidence for the formation of the iodyl cation, IO_2^+, in sulphuric acid solutions.

4.3.2.3. Hydrogen Fluoride and Hydrogen Chloride

The conversion of hydrogen fluoride into fluorosulphuric acid by concentrated sulphuric acid was demonstrated by Traube and Reubke (1921) and the position of the equilibrium

$$H_2SO_4 + HF \rightleftharpoons HSO_3F + H_2O \qquad (4.48)$$

was studied by Lange (1933) by chemical analysis of the mixtures poured into strongly cooled alkali. Water has an exceptionally great effect on the position of this equilibrium and the simple mass law is not obeyed. The reaction is in fact more complicated, because of further ionization of both reaction products. In aqueous acid the following reaction occurs:

$$HSO_3F + HSO_4^- = H_2SO_4 + SO_3F^- \qquad (4.49)$$

in view of the fact that fluorosulphuric acid is stronger than sulphuric, and that the HSO_4^- ion is a strong base in sulphuric acid.

Conductometric and cryoscopic studies of solutions of hydrogen chloride in 100 per cent sulphuric acid show that it also is sulphonated, giving per mole almost one mole of hydrogen sulphate ion and almost three moles of particles (Gillespie and Robinson, 1964). This means that it almost quantitatively ionizes according to the equation

$$HCl + 2H_2SO_4 \rightleftharpoons HSO_3Cl + H_3O^+ + HSO_4^- \qquad (4.50)$$

The reversibility of the reaction is demonstrated by the fact that hydrogen chloride can be swept out of solution rapidly by dry nitrogen.

Both fluorosulphuric acid and chlorosulphuric acid are weak acids of the sulphuric acid solvent system with $K_a = 2 \cdot 3 \times 10^{-3}$ and 9×10^{-4} respectively (Barr et al., 1961).

4.3.3. ORGANOMETALLIC SOLUTES

4.3.3.1. Silicon Compounds

A relatively small number of organosilicon compounds have been studied in sulphuric acid, but the general pattern of behaviour appears well established. No siliconium ions have been observed. Alkyl–silicon bonds withstand treatment with the solvent, but aryl–silicon bonds undergo cleavage. Alkoxy groups attached to silicon are generally displaced by hydrogen sulphate groups. For example, hexamethyldisiloxane was repeatedly studied (Price, 1948; Newman *et al.*, 1949; Flowers *et al.*, 1963) and found to give a four-fold freezing-point depression and $\gamma = 1$. These findings are consistent with the following reaction:

$$[(CH_3)_3Si]_2O + 3H_2SO_4 = 2(CH_3)_3SiHSO_4 + H_3O^+ + HSO_4^- \qquad (4.51)$$

Thus silicon–oxygen bonds are cleaved and a hydrogen sulphate obtained.

The displacement of alkoxy groups on the silicon occurs as, for example, with dimethyldiethoxysilane:

$$(CH_3)_2Si(OEt)_2 + 6H_2SO_4 = (CH_3)_2Si(HSO_4)_2 + 2EtHSO_4 + 2H_3O^+ + 2HSO_4 \quad (4.52)$$

The behaviour of methyltriethoxysilane is more complicated and appears to involve the formation of polymers (Flowers *et al.*, 1963).

Apart from these alkylsilyl hydrogen sulphates, which have been postulated to explain cryoscopic and conductometric results, some trialkylsilyl sulphates have actually been prepared, e.g. trimethylsilyl sulphate, m.p. 56–8°C, from a solution of hexamethyldisiloxane in 20 per cent oleum (Sommer *et al.*, 1946).

Complete cleavage of Si—C bonds occurs when tetraphenylsilane is dissolved in sulphuric acid. Tetrabenzylsilane and hexaphenyldisiloxane are also cleaved to the corresponding sulphonic acid and partially sulphated polysilicic acid (Szmant *et al.*, 1951).

4.3.3.2. Tin Compounds

Virtually all information on these compounds in sulphuric acid is due to the work of Gillespie *et al.* (1966).

Tetramethyl tin reacts with sulphuric acid with methane evolution:

$$(CH_3)_4Sn + H_2SO_4 = (CH_3)_3SnHSO_4 + CH_4 \qquad (4.53)$$

The trimethyl tin hydrogen sulphate formed in solution behaves as a strong base:

$$(CH_3)_3SnSO_4H = (CH_3)_3Sn^+ + HSO_4^- \qquad (4.54)$$

Dialkyl tin hydrogen sulphates behave likewise. This difference in behaviour compared with trialkyl and dialkyl silicon hydrogen sulphates reflects the more highly electropositive character of tin.

As with silicon compounds, aryl–tin bonds undergo cleavage, giving hexa(hydrogenosulphato)tin anion and aryl-sulphonic acids.

4.3.3.3. Lead Compounds

Lead tetraacetate was shown by cryoscopy and conductance to ionize according to the equation

$$Pb(CH_3COO)_4 + 10H_2SO_4 = H_2Pb(HSO_4)_6 + 4CH_3C(OH)_2^+ + 4HSO_4^- \quad (4.55)$$

giving a yellow solution, containing hexa(hydrogenosulphato)plumbic acid (Gillespie and Robinson, 1957a). This acid is a weak acid of the sulphuric acid solvent system ($K_1 = 1 \cdot 1 \times 10^{-2}$ and $K_2 = 1 \cdot 8 \times 10^{-3}$ mol kg^{-1}). At higher concentrations its complex anions probably polymerize to give complex polymeric anions containing sulphate bridges.

CHAPTER 5

Reaction Mechanisms in Sulphuric Acid Solutions

5.1. Introduction

The most outstanding property of sulphuric acid/water mixtures as reaction media, which determines the types of reactions that occur, is their protonating ability. The conjugate acids of the substrates formed may be unstable in a variety of ways, or may be reactive towards other species present in solution. The high dielectric constant of the medium favours heterolytic splitting of the bonds with the formation of ionic products. The activity of water is the next important factor. It determines whether certain substrates are hydrated or dehydrated in the medium. This varies over a very wide range in sulphuric acid/water mixtures (see Chapter 1), and many reactions occurring in concentrated acid are due to low water activities in this medium. It is the combination of high acidity with low water activity that makes concentrated sulphuric acid an exceptionally suitable medium for some reactions. There is no other strong acid/aqueous medium that combines these two characteristics with chemical stability and ease of handling comparable to that of sulphuric acid/water mixtures. The concentrated acid also favours the formation of certain electrophiles, which attack otherwise stable substrates, and has therefore always been an important medium for electrophilic substitution reactions, especially nitration. The concentrated acid and dilute oleums themselves contain strong electrophiles involving sulphur trioxide and have the ability to sulphonate a variety of substrates.

Most reactions occurring in moderately concentrated sulphuric acid occur also, at closely similar rates, in solutions of other moderately concentrated strong acids. Such reactions are mostly hydrolyses and hydration reactions, which are acid-catalysed. A correlation of information on these reactions inevitably also involves reference to reactions occurring in weakly acidic aqueous solutions in the normal pH range. Reactions occurring in more concentrated sulphuric acid generally have less direct parallels with reactions in other aqueous acids.

All reactions occurring in sulphuric acid media and involving protons are not necessarily acid-catalysed, because the protons as strong electrophiles are in fact consumed in some of them. There is a tendency to regard such reactions as acid-catalysed, although they are in fact electrophilic substitutions by the hydrogen ion. They will be treated as such in the present Chapter.

Some of the most fruitful ideas underlying our understanding of reactions in strong acid solutions, involving the concept of the acidity function, were developed in the nineteen-thirties by L. P. Hammett and his school (Hammett, 1940). Later, mechanisms of reactions in strong acid solutions were reviewed in detail by Long and Paul (1957), and more recently a monograph on acid-catalysed reactions in organic chemistry has appeared (Willi, 1965). Ideas in this field, in particular on acidity functions, have, however, undergone considerable modification in recent years, as discussed in Chapter 2, and this has a bearing on the interpretation of the kinetics of acid-catalysed reactions. It is the object of this Chapter to present the current points of view and the problems still outstanding in this field.

In Section 5.2 the possible mechanisms of reactions in acid solutions will be outlined first, followed by a discussion of mechanistic criteria which have been developed in order to distinguish between kinetically indistinguishable alternatives. Kinetic information, however, remains basic to any attempt at elucidation of reaction mechanism, and therefore Section 5.3 is devoted to some methods of kinetic investigation, which are specific to concentrated sulphuric acid solutions. Kinetic methods used in the more aqueous acid are mostly conventional and are widely used in a variety of media. It will be assumed that these are familiar. Reactions occurring in sulphuric acid solutions will then be discussed in Section 5.4, classified largely according to general type and mechanism. In a few instances reactions classified on the basis of these two criteria might equally well have been placed in two different sections.

5.2. Mechanisms and Mechanistic Criteria in Strong Acid Solutions

Sulphuric acid/water mixtures may be regarded as a prototype of strongly acidic reaction media of high dielectric constant, and therefore much of the discussion that follows applies to acid-catalysed reactions in general.

5.2.1. MAIN TYPES OF REACTION MECHANISM IN STRONG ACID SOLUTIONS

The protonation of substrates in strongly acid media need not necessarily occur rapidly. Depending on the rate of this process, there are two possible mechanisms of acid catalysis which may be represented as follows:

$$\left.\begin{aligned} S + H^+\cdot\text{solvent} &\underset{\text{fast}}{\rightleftharpoons} SH^+ + \text{solvent} \\ SH^+ &\xrightarrow{\text{slow}} \text{products} + H^+ \end{aligned}\right\} \text{A-1}$$

5. REACTION MECHANISMS IN SULPHURIC ACID SOLUTIONS

$$\left. \begin{array}{l} S + H^+ \cdot \text{solvent} \xrightarrow{\text{slow}} SH^+ + \text{solvent} \\ SH^+ \xrightarrow{\text{fast}} \text{products} + H^+ \end{array} \right\} A\text{-}S_E 2$$

In these equations the solvated proton is represented in the most general form as $H^+ \cdot$ solvent, in order to ensure that they are valid at both low and high acid concentrations, where the solvating species are necessarily different. At low acid concentrations the proton will be exclusively solvated by water molecules (the primary H_3O^+ species being further solvated to $H_9O_4^+$), whereas at high acid concentrations the proton becomes increasingly available in the form of undissociated acid molecules. In dilute oleum it becomes available in the form of sulphuric acidium ion and disulphuric acid. Despite these changes in the solvating species, the activity of the protons increases uniformly from dilute sulphuric acid to dilute oleum, as studies of acidity functions show (Chapter 2).

Both mechanisms indicated above involve two steps. In the mechanism A-1 the first step is the rapid formation of the conjugate acid of the substrate, and is often described as a pre-equilibrium step. The conjugate acid formed then undergoes a slow unimolecular transformation into reaction products. One would expect this kind of mechanism to apply to substrates containing oxygen or nitrogen as the basic centre, because proton transfers to such sites are known to occur rapidly, with rate constants of $10^8 – 10^9$ litre mol^{-1} s^{-1} (Loewenstein and Connor, 1963).

The mechanism $A\text{-}S_E 2$ involves a slow transfer of the proton from an acidic species to the substrate, followed by a rapid further reaction of the conjugate acid. This kind of acid catalysis is in fact an electrophilic substitution, which involves the substrate and the proton in the rate-determining step. Such a mechanism would be expected to apply when the protons are transferred to a hydrocarbon base, because protonations of such bases are known to be slow. If the protonation step is slow, nothing can be learned from kinetic studies about the further reaction of the conjugate acid.

In dilute and moderately concentrated acid solutions many acid-catalysed reactions involve water in the overall equation for the reaction, and are variously described as hydrations and hydrolyses. These may proceed by the A-1 mechanism, since the conjugate acid may undergo a slow dissociation before reacting with water in a third fast step, or water as a nucleophile may attack the conjugate acid of the substrate. The mechanism of such a reaction may be formulated as follows:

$$\left. \begin{array}{l} S + H^+(H_2O)_n \rightleftharpoons SH^+ + nH_2O \\ SH^+ + H_2O \xrightarrow{\text{slow}} \text{product} + H^+ \end{array} \right\} A\text{-}2$$

In the more aqueous sulphuric acid/water mixtures, where water is present in high concentration, it is impossible to decide by kinetic studies alone whether water is involved in the rate-determining step or not. The two possibilities are kinetically indistinguishable, because simple kinetic data would just show in either case a first order reaction with respect to the conjugate acid, the order with respect to water not being obtainable in the largely aqueous medium. The unimolecular and the bimolecular hydrolysis mechanisms are also often denoted as S_N1 and S_N2 reactions, since they may be regarded as nucleophilic displacement reactions.

There is a further possibility of water acting on the conjugate acid of the substrate to remove a proton, which is different from the one added in the pre-equilibrium step. Examples of such reactions are enolizations of aldehydes and ketones.

The reaction mechanisms discussed so far involve the substrate and the acid reaction medium only, water being one of the species present in aqueous acid. In all of them the substrate is converted into a reactive conjugate acid. The enhanced reactivity of some conjugate acids may be exploited in reactions with other reagents, which themselves may remain unaffected by the acid medium. The mechanism of such reactions may be represented by the following two steps, either of which could be slow:

$$S + H^+ \cdot \text{solvent} \rightleftarrows SH^+ + \text{solvent}$$
$$SH^+ + X \rightarrow XS + H^+$$

Finally, when a reagent is introduced into an acid medium, there is a possibility that it will itself be converted into a more reactive conjugate acid form, or into a conjugate acid that will yield an electrophile by secondary reaction. The resulting electrophile could then react with a substrate in an electrophilic displacement reaction (S_E2). If the substrate itself exists in a protonation pre-equilibrium in the acid medium, in principle either the base or the conjugate acid of the acid/base pair could be attacked by the electrophile, the probability being weighted on purely electrostatic grounds in favour of the base.

5.2.2. MECHANISTIC CRITERIA

The basic question that must be answered concerning a reaction in acid solution is whether the reaction is acid-catalysed or not, and if it is, whether the substrate is involved in a rapid preprotonation step or is only slowly protonated. A number of other questions arise for various types of reactions, the most important ones concerning the exact nature of the reacting species, the identification of the bonds being broken (if there is more than one alternative), the molecularity of the rate-determining step, the composition of the activated complex, and finally the structure of the activated complex.

The possibility of arriving at the answers to these questions depends on the existence of some criteria, on the basis of which the various alternatives can be distinguished. The basic problem in the study of reaction mechanisms is therefore the development of such criteria. There has been a great deal of activity in this direction in recent years and some new approaches have been suggested and developed. The transition state theory has provided the most useful framework for these developments. The fundamentals of this theory, in its application to condensed phases, were developed by Wynne-Jones and Eyring (1935), and the reader is referred to the classic text on the subject by Glasstone, Laidler and Eyring (1941) for an introduction.

5.2.2.1. *Correlations of Rates with Acidity Functions*

The idea that the acidity function H_0 may be useful in identifying acid catalysis in strong acid solutions and in determining the rate of acid-catalysed reactions has been one of the mainsprings for the development of the concept by Hammett and Deyrup (1932). In an early review on this question, Hammett (1935) was able to quote a considerable number of reactions which show a parallelism between reaction rate and h_0, which is to be expected if the reaction mechanism involves a first order rate-determining reaction of the conjugate acid (A-1). For such a mechanism, i.e.

$$\left.\begin{array}{l} S + H^+ \cdot \text{solvent} \rightleftharpoons SH^+ + \text{solvent} \\ SH^+ \xrightarrow{k_1} X^{\ddagger} \rightarrow \text{products} \end{array}\right\} \text{A-1}$$

the rate of the reaction per unit volume of solution experimentally observed would be given by

$$-\frac{dc_s}{dt} = k_{obs} c_s = k_1[SH^+] \tag{5.1}$$

where k_{obs} is the observed first-order rate constant, k_1 the rate constant of the rate-determining step, c_s the stoichiometric concentration of the reactant, and $[SH^+]$ the concentration of the protonated substrate. In terms of the transition state theory the rate of the reaction is given by

$$-\frac{dc_s}{dt} = \frac{kT}{h}[X^{\ddagger}] \tag{5.2}$$

where k is the Boltzmann constant, T the absolute temperature, h the Planck constant, and $[X^{\ddagger}]$ is the concentration of the activated complexes. The equilibrium constant for the rapid pre-equilibrium formation of the conjugate acid may be written as

$$K_{SH^+} = \frac{[H^+][S]}{[SH^+]} \cdot \frac{f_{H^+} f_S}{f_{SH^+}} \tag{5.3}$$

where square brackets indicate concentrations and f's are activity coefficients. The transition state theory also assumes a quasi-equilibrium between the reactants and the activated complexes. In this case an equilibrium constant for this would be given by

$$K^{\ddagger} = \frac{[X^{\ddagger}]}{[SH^+]} \frac{f_{\ddagger}}{f_{SH^+}} \qquad (5.4)$$

If the concentration of the protonated substrate is only a very small fraction of the stoichiometric concentration ($[SH^+] \ll [S]$ and hence $[S] \simeq c_s$), then from the pre-equilibrium constant we obtain

$$[SH^+] = \frac{c_s}{K_{SH^+}} \cdot \frac{a_{H^+} f_S}{f_{SH^+}} \qquad (5.5)$$

On the other hand from (5.4) we have

$$[SH^+] = \frac{[X^{\ddagger}]}{K^{\ddagger}} \frac{f_{\ddagger}}{f_{SH^+}} \qquad (5.6)$$

and from (5.5) and (5.6) the concentration of the activated complexes may be obtained:

$$[X^{\ddagger}] = \frac{K^{\ddagger}}{K_{SH^+}} c_s \frac{a_{H^+} f_S}{f_{\ddagger}} \qquad (5.7)$$

Inserting this back into (5.2) and equating the rates as given by (5.1) and (5.2), we obtain the expression for the observed rate constant:

$$k_{obs} = \frac{kT}{h} \frac{K^{\ddagger}}{K_{SH^+}} \frac{a_{H^+} f_S}{f_{\ddagger}} \qquad (5.8)$$

It is now argued that the structure of the activated complex will be sufficiently similar to that of the reactant SH^+, so that their activity coefficients may be assumed to be virtually equal, i.e. $f_{SH^+} \simeq f_{\ddagger}$. If this is so, and if the protonation of the substrate parallels the protonation of simple Hammett bases, i.e. if $f_S/f_{SH^+} = f_B/f_{BH^+}$, then the term $a_{H^+} f_S / f_{\ddagger}$ may be replaced by the acidity of the medium, h_0, and consequently

$$\log k_{obs} = \log h_0 + \text{const.} \qquad (5.9)$$

or

$$\log k_{obs} = -H_0 + \text{const.} \qquad (5.10)$$

that is, the logarithms of the observed rate constants should be a linear function of the acidity function with a slope of -1. More commonly, $\log k_{obs}$ is plotted vs. $-H_0$ and lines with a positive unit slope are expected. A number of reactions in acid solution have been found to obey this relationship sufficiently closely to lend support to the assumptions made above.

If the fraction protonated is not small, then $c_s = [S] + [SH^+]$, and it can easily be shown from the expression for K_{SH^+} that

$$[SH^+] = c_s \frac{h_0}{K_{SH^+} + h_0} \qquad (5.11)$$

5. REACTION MECHANISMS IN SULPHURIC ACID SOLUTIONS

and the following expression for k_{obs} is then obtained:

$$k_{obs} = \frac{kT}{h} K^{\ddagger} \frac{h_0}{K_{SH^+} + h_0} \frac{f_{SH^+}}{f_{\ddagger}} \tag{5.12}$$

Making the same assumption as above that $f_{SH^+} \simeq f_{\ddagger}$, it can be seen that at sufficiently high acidities where $h_0 \gg K_{SH^+}$, the observed rate constant would be expected to reach a limiting value, independent of acidity. This type of expression was first derived by Schubert and Latourette (1952).

For the case of acid catalysis via a slow proton transfer step (A-S_E2), the rate-determining process may be represented as follows (neglecting the solvation of the proton):

$$S + H^+ \xrightarrow[k_2]{slow} (SH^+)^* \to SH^+$$

The rate in unit volume would then be given by

$$-\frac{dc_s}{dt} = k_2[S][H^+] = k_{obs} c_s \tag{5.13}$$

where square brackets indicate concentrations (there would be no distinction here between the stoichiometric concentration and the equilibrium concentration, i.e. $[S] = c_s$, since no equilibrium is assumed). In moderately concentrated acid, the concentration of hydrogen ions is high and would remain constant, since these ions are regenerated in the second step of the reaction. Experimentally the reaction would thus be first order in the reactant. In terms of the transition state theory the rate would be given by

$$-\frac{dc_s}{dt} = \frac{kT}{h} [(SH^+)^{\ddagger}] \tag{5.14}$$

where the square brackets indicate the concentration of the activated complexes, which may again be assumed to be closely similar to the protonated species, although not involving quite such a firm bonding of the proton with the substrate. From the equilibrium constant for the quasi-equilibrium between the reactants and the activated complexes we have

$$[(SH^+)^{\ddagger}] = K^{\ddagger}[S][H^+] \frac{f_{H^+} f_S}{f_{\ddagger}} \tag{5.15}$$

Inserting this into (5.14) we obtain

$$-\frac{dc_s}{dt} = \frac{kT}{h} K^{\ddagger}[S][H^+] \frac{f_{H^+} f_S}{f_{\ddagger}} = \frac{kT}{h} K^{\ddagger} c_s \frac{a_{H^+} f_S}{f_{\ddagger}} \tag{5.16}$$

and hence from (5.13) and (5.16)

$$k_{obs} = \frac{kT}{h} K^{\ddagger} \frac{a_{H^+} f_S}{f_{\ddagger}} \tag{5.17}$$

One may now introduce into this expression the definition of the acidity of the medium, $h_0 = a_{H^+} f_B / f_{BH^+}$, and obtain

$$k_{obs} = \frac{kT}{h} K^{\ddagger} h_0 \frac{f_{BH^+} f_S}{f_B f_{\ddagger}} \tag{5.18}$$

which, upon taking logarithms, leads to

$$\log k_{obs} = -H_0 + \log \frac{f_{BH^+} f_S}{f_B f_{\ddagger}} + \text{const.} \tag{5.19}$$

It can be seen that a linear correlation between $\log k_{obs}$ and H_0 can again be expected, if one is justified in assuming that $f_{\ddagger} \simeq f_{SH^+}$, and therefore that the activity coefficient term approaches unity. This is not an unreasonable assumption, and Long and Paul (1957) have argued that A-S_E2 reactions may well be expected to show a linear correlation between $\log k_{obs}$ and $-H_0$, similar to that found for the A-1 mechanism.

It thus follows that correlations of rate constants with acidity functions do not afford a means of distinguishing between A-1 and A-S_E2 mechanisms in strong acid solutions, where the solvated proton is the chief acid species present. The A-S_E2 mechanism implies general acid catalysis and can be identified in favourable cases by studying the rates of reaction in aqueous acid buffer solutions. In principle, in concentrated sulphuric acid both H_3O^+ and H_2SO_4 acid species could catalyse such reactions. It may be pointed out that no limiting rate would be expected for an A-S_E2 reaction at high acidities, in contrast to reactions proceeding by the A-1 mechanism.

Reactions occurring by the A-2 mechanism would be expected likewise to follow the H_0 acidity function and would therefore be indistinguishable in their acidity dependence from those occurring by the A-1 mechanism, so long as comparison concerns reactions in the dilute and moderately concentrated acid region, in which the concentration of water is large and remains virtually unchanged in the course of the reaction. At higher acid concentrations, where the activity of water falls sharply, the A-2 reactions would be expected to show a corresponding fall in rate.

In the foregoing discussion it has been assumed that the activity coefficient behaviour of the substrate and the protonated substrate is closely similar to that of Hammett bases, and this has led to an expectation of unit slopes in plots of $\log k_{obs}$ vs. $-H_0$ for both A-1 and A-S_E2 reactions. The question of the activity coefficient behaviour of very weak bases and the numerous deviations of the protonation behaviour of bases from the H_0 function have been discussed in detail in Chapter 2. Deviations from unit slopes in $\log k_{obs}$ vs. $-H_0$ plots should equally be a common phenomenon and for the same reason, i.e. because of the breakdown of the activity coefficient postulate for various substrates. Such deviations have indeed been observed, and before the widespread deviations of the protonation behaviour of bases from the H_0 function had become apparent they were thought to provide additional clues about the reaction mechanism, particularly regarding the distinction between A-1 and A-2 mechanisms in the moderately concentrated acid region. Several hypotheses were put forward in this respect, the best known being that by Zucker and Hammett (1939), which assumed that the rate constants of A-2 reactions should be linear with the stoichiometric concentration of the acid, rather than with h_0. A more recent treatment by Bunnett (1961) sought to exploit the deviations from unit slope in order to achieve a more precise classification of acid-catalysed reactions, according to the number of water molecules that take part in the activated complex. Both approaches are invalidated by the realization that activity coefficient behaviour of neutral bases varies from one class of compound to another and gives rise to acidity functions differing from H_0. It is therefore possible that $\log k_{obs}$ for some acid-catalysed reactions would be better correlated with some other acidity function, rather than H_0. However, as it is held (see Chapter 2) that acidity functions for various weak bases differ mainly owing to differences in the solvation of the base and the conjugate acid, adherence of the rates of various reactions to different acidity functions may also be taken to reflect differences in the solvation of reactants and transition states. In particular, those reactions that follow acid concentration rather than H_0, and which, according to the Zucker–Hammett criterion, were classed as involving a water molecule in the transition state, clearly follow an acidity scale which shows virtually no "excess acidity". This corresponds to strong solvation of the indicator conjugate acid in terms of ideas discussed in Chapter 2, and implies strong involvement of water molecules in the activated complex. Thus the Zucker–Hammett criterion, although based on a more specific and apparently false premise, often led to conclusions which are consistent with those which may be reached on the basis of current views on the factors determining differences in the acidity function behaviour of various bases.

A more recent attempt by Bunnett and Olsen (1966) to develop an alternative treatment of equilibria and rates in moderately concentrated mineral acids with reference to the acidity function H_0 by introducing a new parameter ϕ, supposed to measure the hydration change in going from substrate to conjugate acid or transition state, offers no advantage from the point of view of the understanding of mechanisms, since the mechanistic interpretation of the parameter ϕ remains uncertain.

Exactly analogous arguments to those developed above for simple pre-protonation equilibria predict that the rate constants of reactions which involve complex ionization of the substrate or of a reagent in a pre-equilibrium step will show linear correlations with the h_R function. Deviations from unit slopes are again a possibility, which may arise when the activity coefficient behaviour of the ionizing substance differs from that of the defining series of indicators.

Correlations with acidity functions play an important part in the study of reaction kinetics in strongly acid media. As the examples to be discussed in this Chapter will show, their main object of identifying acid catalysis can usually be achieved without difficulty, but the magnitudes of the slopes of linear correlations, or deviations from linearity, have as a rule less certain interpretations.

5.2.2.2. Activation Parameters

According to the transition state theory the rate of a reaction per unit volume of solution between two reactants S and Y is given by

$$\frac{\text{Rate}}{V} = \frac{kT}{h}[X^{\ddagger}] \equiv \frac{kT}{h} K^{\ddagger}[S][Y] \quad (5.20)$$

and hence the observed second-order rate constant is

$$k_2 = \frac{kT}{h} K^{\ddagger} \quad (5.21)$$

assuming that all activated complexes proceed to products. An example of such a reaction would be a slow protonation (A-S_E2). When the reaction involves a pre-equilibrium between S and Y (as in an A-1 reaction), the observed rate constant will be

$$k_1 = \frac{kT}{h} K^{\ddagger} K_e \quad (5.22)$$

where K_e is the formation constant of the unstable reactant formed in the pre-equilibrium (=$1/K_{SH^+}$ for the preprotonation reaction). For an acid-catalysed A-2 reaction the observed rate constant would be given by an expression analogous to (5.22), except that this would also involve the concentration of water molecules. If the equilibrium constants in equations (5.21) and (5.22) are expressed in terms of the enthalpy and the entropy of the reaction, the expressions for the above rate constants become

$$k_2 = \frac{kT}{h} \exp(\Delta S^{\ddagger}/R) \exp(-\Delta H^{\ddagger}/RT) \quad (5.23)$$

and

$$k_1 = \frac{kT}{h} \exp\left(\frac{\Delta S_e + \Delta S^{\ddagger}}{R}\right) \exp\left(-\frac{\Delta H_e + \Delta H^{\ddagger}}{RT}\right) \quad (5.24)$$

It can be seen that the usual procedure of plotting $\log k$ against $1/T$ for a series of temperatures would yield a pure enthalpy of activation only in the case of a rate-determining protonation, whereas for the pre-equilibrium case the overall enthalpy of activation will be the sum of the standard enthalpies for the pre-equilibrium and the activation process. Inserting the enthalpy terms so obtained into equations (5.23) and (5.24) and choosing a temperature within the range studied, one can obtain the respective entropy terms.

The theoretical calculation of entropies of reaction and of activation for reactions in solution is not possible at present, not only because of a lack of knowledge about the molecular properties of the activated complex, but also because of a lack of adequate description of molecular interactions between solute and solvent. This is particularly true of polar solvents and solvent mixtures, such as aqueous acids, because of the added complication of hydrogen bonding and association in the solvents themselves. Fortunately, however, the thermodynamics of acid/base reactions, which may be regarded as models for reactions involving preprotonation or slow protonation, has been studied for a considerable number of acid/base pairs. An extensive collection of such data has been published by Bell (1959). The possibility of using comparisons of entropies of activation with thermodynamic data for similar equilibria in order to draw mechanistic conclusions has been examined by Schaleger and Long (1963), mainly with reference to hydrolysis reactions, which present the difficult problem of distinguishing between the A-1 and A-2 mechanisms. In the latter case a water molecule would be bound in the activated complex. The loss of entropy in the freezing of water is 22·0 J K^{-1} mol^{-1}. In the dissociation of a carboxylic acid according to

$$RCOOH + 4H_2O \rightleftharpoons RCOO^- + H_9O_4^+ \qquad (5.25)$$

in which four water molecules may be regarded as losing their rotational and translational freedom, entropy changes of around −84 J K^{-1} mol^{-1} are observed. In reality, this loss of entropy may arise from the partial "freezing" of a greater number of water molecules involved in the primary and secondary solvation of the hydrogen ion and other species, and also from entropy differences between the carboxylic acid molecules and the carboxylate anions.

Analogously, entropies of activation must reflect differences in the entropies of the reactants and the activated complex, and the differences in the involvement of the solvent in the initial and the transition states. It is therefore not unreasonable, in considering the A-1 and the A-2 mechanisms of hydrolysis, to expect that the binding of a water molecule in the A-2 transition state should lead to a lower entropy of activation compared with the unimolecular case. Schaleger and Long (1963) collected a considerable amount of data, which show that this expectation is in fact fulfilled in many hydrolysis reactions. The A-1 hydrolyses have typically $\Delta S^{\ddagger}$ values between 0 and 42 J K^{-1} mol^{-1}, whereas A-2 hydrolyses have considerably more negative values (−60 to

-120 J K^{-1} mol^{-1}). However, intermediate values are also found in other S$_N$1 and S$_N$2 reactions. Hence the entropy criterion cannot be regarded as entirely certain and must be used with caution.

Volumes of activation (ΔV^+) and heat capacities of activation (ΔC_p^+) may also shed additional light on reaction mechanism. Neither is as readily accessible experimentally as entropies of activation. The interpretation of both is still more uncertain. These two criteria have recently been reviewed (Whalley, 1964; Kohnstam, 1967). It is argued that the preprotonation step should involve little change in volume, so that for A-1 reactions a positive volume of activation may be predicted, because the transition state should be greater than the initial state. For A-2 reactions the binding of a water molecule in the transition state should result in a decrease of volume, and activation volumes negative by 5–10 cm^3 mol^{-1} may be expected.

5.2.2.3. Solvent Isotope Effects

Since acid-catalysed reactions involve the transfer of a proton and alternative mechanisms are not always readily distinguishable, it is natural that soon after the discovery of hydrogen isotopes in the nineteen-thirties a number of studies of acid-catalysed reactions in both light and heavy water appeared in the literature. This early work was reviewed by Wiberg (1955). A detailed theoretical treatment of the effect of isotope substitution in the aqueous solvent on the rates of acid-catalysed reactions was developed more recently (Long and Bigeleisen, 1959). Mixed H$_2$O/D$_2$O solvents can also be useful in attempts to distinguish between the various mechanisms of acid catalysis (Gold and Kessick, 1965).

Considering first the A-1 mechanism, the rate of the slow unimolecular reaction of the conjugate acid would not be expected to be greatly affected by the change of solvent, since as a rule bonds other than the S—H$^+$ bond undergo rupture. Any solvent isotope effect on the rate of such a reaction must therefore be due to the difference in the extents of protonation of the substrate in the light and the heavy solvent. This point of view was first advanced by Bonhoeffer (1934) and by Bonhoeffer and Reitz (1937), who suggested that the higher rate in heavy water of most acid-catalysed reactions is due to deuterium oxide being less basic than ordinary water. Deuterium oxide was found to have an autoprotolysis constant only one fifth that of water (Wynne-Jones, 1936), which could be ascribed to its lower basicity. This suggestion is supported by pK determinations for a series of indicators in D$_2$O—DCl and D$_2$O—D$_2$SO$_4$, which show that protio conjugate acids are stronger than deuterio acids by a factor of 2·5–3·5 in most cases (Högfeldt and Bigeleisen, 1960). The ratio of the observed rate constants should on this basis be $k_{H_2O}/k_{D_2O} = 0·3$–$0·4$, as is in fact found for many hydrolysis reactions which follow the A-1 mechanism.

The solvent isotope effect on the rate of A-2 hydrolysis reactions will arise not only from the solvent effect on the equilibrium concentration of the conjugate acid, which is the same as for the A-1 mechanism, but also from the difference in the rate constant (k_2) of the slow nucleophilic attack by water molecules on the conjugate acid. The rate constant k_2 should reflect the difference in the nucleophilicity between H_2O and D_2O. Since deuterium oxide is a weaker base than protium oxide, the displacement reaction will be faster in the latter. This should lead to a larger ratio of observed rate constants, k_{H_2O}/k_{D_2O}, in the case of the A-2 mechanism than in the case of the A-1 mechanism. Experimental observations generally confirm this (Long and Pritchard, 1956). In the limiting case of a fully protonated substrate, the solvent isotope effect on k_2 becomes dominant, and this should lead to $k_{H_2O}/k_{D_2O} > 1$.

Finally, when proton transfer is rate determining (the A-S_E2 mechanism), statistical mechanical calculations lead to the prediction that the ratio k_{H_2O}/k_{D_2O} will be greater than unity, owing to the difference in zero-point energies of the bonds involving hydrogen and deuterium (Long and Bigeleisen, 1959). This is regarded as the normal isotope effect. Under certain conditions a reverse isotope effect may be expected (Bigeleisen, 1949, 1952). There is definite support for the normal solvent isotope effect in some reactions.

It was mentioned in Section 5.2.1 that A-1 and A-S_E2 mechanisms may also be distinguished by the fact that A-S_E2 reactions show general acid catalysis. It is not, however, always possible to measure the catalytic power of un-dissociated weak acids. When the Brønsted coefficient α is large, catalysis by undissociated acids may not be detectable in aqueous buffer solutions. For such a case Gold and Kessick (1965) have recently shown that studies of the kinetics of the reaction in mixed H_2O/D_2O solvents provide a means of determining the Brønsted coefficient α, and hence supporting the A-S_E2 mechanism.

5.2.2.4. *Other Criteria*

The criteria of reaction mechanism discussed so far are of considerable and special interest for reactions in strongly acid media. A number of other criteria, applicable more generally in aqueous and non-aqueous media, have also been applied to reactions in strong acid solutions.

Isotopic substitution in the substrate is generally useful in identifying the bonds broken in a reaction where two alternatives are possible, as in ester hydrolysis. This application of isotopes as tracers is probably their best known application in studies of reaction mechanism. The effect of isotopic substitution on the rates of reactions can also be informative. Isotopic substitution at the bond that undergoes rupture leads to a change in rate, and this phenomenon is known as the primary kinetic isotope effect. The theory of such effects, which are usually only 1–10 per cent for isotopes of elements other than hydrogen, has been developed by Bigeleisen (1949, 1952). The light molecule usually has a greater rate constant than the heavy molecule, mainly because of a mass ratio factor, which is always greater than unity, and the fact that atoms are in general more loosely bound in the activated complex than in the reacting molecule.

The exchange of the isotopic label in certain types of molecule may also provide useful information on the mechanism of some reactions. This is particularly true of ^{18}O-exchange of many oxygen containing compounds, e.g. derivatives of carboxylic acids, which undergo hydrolysis as well as oxygen exchange under the same conditions. The subject of oxygen isotope exchange reactions has been discussed recently by Samuel and Silver (1965). Such studies have provided evidence for the formation of tetrahedral intermediates in many hydrolysis reactions.

Another reaction that sheds light on the mechanisms of certain reactions is racemization. This is obviously useful only if the reaction centre is a tetrahedrally substituted atom. Certain tertiary alcohols in acid solution undergo this reaction by an A-1 mechanism:

$$R'R''R'''COH + H^+ \rightleftharpoons R'R''R'''COH_2^+ \rightleftharpoons R'R''R'''C^+ + H_2O$$

Both steps in this mechanism are reversible. If the carbonium ion formed has a finite lifetime, within which its planar configuration is fully established, the reverse reaction would lead to racemization. Studies of optically active substrates can thus provide evidence for the intermediacy of carbonium ions in many rearrangement reactions of alcohols.

Finally, substituent effects and linear free energy relationships are in many instances useful in drawing conclusions about reaction mechanisms and about the structure of activated complexes. Whether or not charge is generated at the reaction centre in the transition state of certain reactions may be decided on the basis of linear free energy correlations with closely related equilibria or rates of reactions of known mechanism. Also the behaviour of various groups in reactions such as migration and displacement is determined by their electron donor or acceptor characteristics within molecules, as measured by their substituent constants.

Substituent effects are virtually omnipresent in reactions of organic molecules and will be a constant theme in considering various types of reactions in this Chapter. The introduction on substituent effects in Chapter 3 is therefore highly relevant to this Chapter as well.

Steric effects of substituents near the reaction site may also be indicative of the nature of the slow step in acid-catalysed reactions. Bulky substituents should have little effect on the rate of A-1 reactions, whereas they should decrease the rate of A-2 reactions by hindering the approach of the attacking nucleophiles.

5.3. Methods of Kinetic Investigation

In this section some methods of kinetic investigation that have proved useful in moderately concentrated acid solutions and some that have been developed for use in concentrated sulphuric acid in particular will be described briefly. The advantages of physicochemical methods of continuously following the rate of reactions over the less accurate and more tedious sampling techniques have been the driving force for their development and application.

The spectrophotometric method, usually applicable to reactions involving aromatic and unsaturated compounds, is the most widely used. Other methods have to be used for aliphatic compounds. For reactions occurring at moderate rates in aqueous acid the dilatometric method may be used even when relatively minor changes in volume are involved (Bell et al., 1955). Capillaries of narrow radius must be used, and if difficulties arise owing to slow dissipation of heat through their walls, these may be minimized by using thin-walled tubes of small diameter. For very fast reactions the thermal maximum method was adapted (Bell and Clunie, 1952; Bell et al., 1954; Bell et al., 1955).

The cryoscopic method, so useful in elucidating the modes of ionization of solutes in pure or slightly aqueous sulphuric acid, has also been used to follow the rates of certain reactions (Leisten, 1956). The obvious requirement for its applicability is that there should be a change in the total number of solute particles in the course of the reaction, such as occurs for example in the hydrolysis of certain esters:

$$RCOOR' + H_2SO_4 = R-C{\overset{OH}{\underset{OR'}{+}}} + HSO_4^- \text{ (Initial } i = 2) \quad (5.26)$$

$$RCOOR' + 2H_2SO_4 = RC(OH)_2^+ + R'OSO_3 + HSO_4^- \text{ (Final } i = 3) \quad (5.27)$$

The reaction occurring at some higher temperature should proceed at a negligible rate at the freezing point of the sulphuric acid solution (5–10°C). The reaction mixture is rapidly cooled at specified times for freezing-point determination and subsequently equally rapidly heated to the reaction temperature. This is a slight inconvenience and introduces error, which may be comparable with that of some sampling methods. Nonetheless the i factors increase according to the first-order law for the above reaction (Leisten, 1956). The reactions amenable to study by this method must satisfy certain other conditions as well. The reaction product must be stable, i.e. further slow reactions must not occur. This is not a serious limitation, however, since i_∞ values do not have to be used in the evaluation of the rate constants. Slow secondary reactions would not affect seriously the rapid initial changes of the freezing point, and if the nature of the reaction is not in doubt, these may be used to evaluate the rate constants by the Guggenheim method (Guggenheim,

1926). The agreement between i_∞ values and the calculated final i values for the reaction under consideration is an extremely useful confirmation that the application of the method is sound.

The conductometric method was also shown to be applicable to the study of the kinetics of some reactions in the concentrated sulphuric acid region (Liler and Kosanović, 1959; Liler, 1963). If a reaction results in a change of the concentration of the highly conducting solvent self-ions, HSO_4^- and $H_3SO_4^+$, its course can be followed by measuring conductance. In most such reactions the number of kinetically distinct particles would also change and the cryoscopic method would also be applicable. The conductometric method has the advantage that the reaction mixture remains undisturbed in the conductance cell at the reaction temperature. If a change in the concentration of HSO_4^- ions takes place in the reaction, the treatment of the results requires a knowledge of the conductance of a strong base over a range of concentrations at the temperature of the kinetic experiments. A change in the concentration of HSO_4^- ions may occur when a non-electrolyte or a weak base generates a strong base in the reaction, as in the case of the decomposition of oxalic acid (Liler, 1963):

$$(COOH)_2 = CO + CO_2 + H_2O \tag{5.28}$$

Oxalic acid itself is only partly protonated, whereas the reaction product, water, is a strong base (the gaseous products escape from solution). The basic ionization constant of the weak base must be known, in order to enable a calculation of its contribution to the total HSO_4^- concentration at any time during the reaction.

Another possible type of reaction that may be studied conductometrically in 100 per cent sulphuric acid is hydrolysis, i.e. a reaction in which water is abstracted from the solvent (Liler and Kosanović, 1959). If the reactant behaves as a strong base, such a reaction may be represented in a generalized form by the equation

$$BH^+ + HSO_4^- + H_2SO_4 = B(H_2O)H^+ + HS_2O_7^- \tag{5.29}$$

As can be seen, the highly conducting HSO_4^- ions are replaced by the much less mobile $HS_2O_7^-$ ions, and the reaction results in a fall in conductance. This should be an approximately linear function of the decreasing concentration of HSO_4^- ions. Assuming that the reaction product is fully protonated in dilute oleum, the conductance at infinite time will be equal to that of a solution of the reaction product in an oleum of equivalent concentration of disulphuric acid. The concentration of $HS_2O_7^-$ ions generated in the reaction will not be equal to the fall in the concentration of the reactant, as might be expected on the basis of equation (5.29), but will be smaller because of the solvolysis reaction:

$$HS_2O_7^- + H_2SO_4 \rightleftharpoons H_2S_2O_7 + HSO_4^- \tag{5.30}$$

This regenerates a fraction of the HSO_4^- ions used up in the reaction.

5. REACTION MECHANISMS IN SULPHURIC ACID SOLUTIONS

Sulphonation reactions in pure sulphuric acid could also be studied conductometrically, because the water produced in the reaction should lead to an increase of conductance with time.

The n.m.r. method of following the rate at which the resonance lines of the reactants disappear, or those of the products appear, is also useful for following the rates of certain reactions in any acid concentration range (for example, see Olah et al., 1967e), although it has not been used very much so far. With improvements in measuring and thermostatting techniques accurate work at low concentrations and over wide temperature ranges should be possible. There are few limitations that one can foresee.

5.4. Mechanisms of Reactions in Sulphuric Acid Solutions

5.4.1. HYDROLYSES

Hydrolysis is a type of reaction that occurs for various substrates throughout the whole concentration range from very dilute acid to concentrated sulphuric acid and even dilute oleum. The variety of mechanisms encountered is also wide, especially for esters of carboxylic acids where, in addition to the question of whether the reactions occur by the A-1 or the A-2 mechanism, the more fundamental question of which bonds undergo fission arises. In view of the complexities in the hydrolysis of esters, some simpler hydrolysis reactions will be discussed first.

5.4.1.1. *Hydrolysis of Ethers, Cyclic Ethers and Epoxides*

Ethers are remarkably resistant to hydrolysis, but some early work reviewed by Burwell (1954) suggested that the reaction was acid catalysed. Long and Paul (1957) noted that entropies of activation, based on the measurements of Skrabal and Zahorka (1933), are positive, and on this basis the A-1 mechanism was favoured for primary and secondary ethers. The hydrolysis of diethyl ether, catalysed by mineral acids, has more recently been studied up to moderately high acid concentrations (5·65 M $HClO_4$) in the temperature range 120–160°C and at high pressures (Koskikallio and Whalley, 1959a). The rate of the reaction follows the H_0 acidity function and both the entropy of activation (-38 ± 10 J K^{-1} mol^{-1}) and the volume of activation (at 1 atm $-8·5 \pm 2$ cm^3 mol^{-1}) provide evidence that the slow step is bimolecular (A-2 mechanism).

One of the consequences of the A-2 mechanism of hydrolysis in dilute acid would be a fall in rate in more concentrated acid, where the activity of water decreases sharply. If a pK_{BH^+} value of $-3·59$ for diethyl ether is accepted (see Chapter 3), then its protonation should be complete in 70–80 per cent sulphuric acid. Jaques and Leisten (1964) found a levelling off in the rate of its hydrolysis at these concentrations, but no decrease was observed at still higher concentrations (up to 90 per cent). These kinetic results thus do not accord with the A-2 mechanism, although it seems possible that the A-2 mechanism gives way to the A-1 mechanism in more concentrated acid, before a decrease in the A-2 rate is observed.

Faster unimolecular hydrolysis of ethers in concentrated sulphuric acid has indeed been observed earlier (Jaques and Leisten, 1961). The rates could be followed cryoscopically, because the initial simple protonation of ethers gives an *i* factor of two, whereas solvolysis leads to approximately four-fold freezing-point depressions:

$$RR'O + 3H_2SO_4 = RHSO_4 + R'HSO_4 + H_3O^+ + HSO_4^- \quad (i = 4) \quad (5.31)$$

A number of dialkyl ethers and alkyl aryl ethers have been studied in this way. The results for aliphatic ethers are summarized in Table 5.1.

TABLE 5.1

Cryoscopic results for aliphatic ethers and rate constants for their solvolysis in 99·6 per cent sulphuric acid (Jaques and Leisten, 1961)

	R	R'	Initial *i*-factor	Final *i*-factor	Temperature (°C)	k_1 (10^{-3} min^{-1})
(a)	Me	Me	1·80	4·38	90	1·70†
	$CH_2CH_2SO_4H$	Me	1·4	3·7	65	1·69
	CH_2CH_2Cl	Me	1·79	4·28	65	3·4
	Et	Me	1·90	—	55	13·8
	n-Bu	Me	1·93	—	25	16
(b)	Et	Et	2·00	4·42	55	3·46†
	Et	Me	1·90	—	55	13·8
	Et	CH_2CH_2Cl	1·95	—	55	18·8
	Et	$CH_2CH_2SO_4H$	1·6	—	55	24·2
(c)	n-Bu	n-Bu	2·04	—	25	3†
	i-Pr	i-Pr	2·24	4·38	25	20†
	CH_2CH_2Cl	CH_2CH_2Cl	1·56	4·54	65	2·5†
(d)	CH_2CH_2Br	Me	1·82	3·66	25·8	27·2
	CH_2CH_2Br	Et	2·02	—	25·8	4·90

† Statistically corrected to allow for two identical bonds that may break.

In a unimolecular fission of the conjugate acid, i.e.

$$R\underset{\curvearrowleft}{\overset{H^+}{-O-}}R' \longrightarrow R^+ + R'OH$$

the breaking of the bond which leads to the formation of the more stable carbonium ion would occur.

Since β-hydrogen atoms stabilize carbonium ions by hyperconjugation (series (a), Table 5.1) it is the R—O bond that breaks rather than R'—O. Negative substituents, which oppose the dispersal of positive charge, slow down the reaction, so that the observed order is n-Bu > Et > 2-chloroethyl > 2-sulphatoethyl > methyl. The constitutional effects in R are very large (note differences in the reaction temperature). The effects of substituents in R' should be smaller and should be purely inductive. The results for series (b) fall in the expected inductive order. Series (c) demonstrates that di-isopropyl ether reacts faster than any other ether in Table 5.1, since it gives a secondary carbonium ion as intermediate. The results for bromo-ethers—series (d)—compared with those for the corresponding chloro-ethers, are anomalously high. The inductive substituent constants σ_I for chloromethyl and bromomethyl groups (+1·050 and +1·030 respectively) do not lead one to expect any detectable difference in the rates of hydrolysis of these ethers. Jaques and Leisten (1961) suggest some interaction of the bromine atom with the solvent as a possible explanation for the observed difference. This apart, the results in general support the assumption of a unimolecular rate-determining dissociation.

An examination of the effect of acid composition on the rate of these reactions led, however, to a surprising discovery (Jaques and Leisten, 1961). At least for fully protonated ethers, additions of water or $KHSO_4$ into the reaction mixtures in concentrated sulphuric acid should have virtually no effect on the rate. A substantial effect was observed, the reduction in rate being in fact larger for water as solute than for $KHSO_4$. This strongly suggests that sulphur trioxide is involved in the solvolysis of ethers, since water suppresses the dissociation $2H_2SO_4 \rightleftharpoons SO_3 + HSO_4^- + H_3O^+$ twice as effectively as HSO_4^- ions. The conclusion that the rate of the unimolecular fission of ethers in concentrated sulphuric acid is proportional to the concentration of sulphur trioxide may be accounted for in terms of the following mechanism:

$$RR'OH^+ + SO_3 \rightleftharpoons RR'SO_4H$$

$$R\text{—}\underset{\underset{SO_3H}{|}}{O^+}\text{—}R' \xrightarrow{\text{slow}} R^+ + R'HSO_4$$

$$R^+ + HSO_4^- \xrightarrow{\text{fast}} RHSO_4.$$

The concentration of the complex in the pre-equilibrium must be very small to be consistent with an i factor of 2. The heterolysis of the sulphur trioxide complex is faster than that of the conjugate acid of the ether, probably because it yields directly an alkyl hydrogen sulphate which is more stable in concentrated sulphuric acid than the alcohol. Alcohols are rapidly converted into alkyl hydrogen sulphates and are therefore less stable (Jaques and Leisten, 1961).

Aromatic ethers, except when heavily substituted by negative substituents, undergo sulphonation in pure sulphuric acid. 2,4-Dinitroanisole, 3,5-dinitroanisole, 2,4-dinitrophenetole and 2-methyl-5-nitrophenetole also undergo unimolecular fission (Jaques and Leisten, 1961), the phenetoles reacting faster than the anisoles. This suggests that the alkyl-oxygen bond is broken, and no evidence was found for the participation of sulphur trioxide in the reactive species.

Jaques and Leisten (1961) have also examined the question of the relative effectiveness of various concentrated mineral acids for the solvolysis of ethers. All acids are approximately equally effective at equal values of H_0 and at the same temperature. Concentrated sulphuric acid and oleum are extremely effective, owing to the formation of the reactive complexes with sulphur trioxide.

The rate of acid-catalysed hydrolysis of a strained cyclic ether, trimethylene oxide, is much faster than that of ordinary ethers (Long et al., 1957), presumably owing to steric strain in the ring. Hydrolysis occurs readily in <1 M acid at temperatures <40°C. The rate constants are linear functions of h_0, both in water and in 40 per cent aqueous dioxane (Pritchard and Long, 1958). The steric strain in the protonated ring suggests a probable unimolecular ring opening to yield a carbonium ion:

$$\underset{CH_2-CH_2-CH_2}{\overset{H}{\underset{O^+}{\diagup\diagdown}}} \longrightarrow \underset{CH_2-CH_2-CH_2{}^+}{\overset{OH}{|}}$$

This then reacts rapidly with a water molecule to give the dialcohol. All evidence, however, is not in favour of an A-1 mechanism. The deuterium isotope effect, measured in 1·5 M deuterosulphuric and sulphuric acid, is $k_H/k_D = 0·45$ (Pritchard and Long, 1958), and the entropy of activation is only slightly negative (Long et al., 1957). Both these criteria fail to place this reaction clearly into either the A-1 or the A-2 category, since the values are marginal and could be reconciled with either.

Epoxides hydrolyse even more rapidly in acid solutions than trimethylene oxide (Pritchard and Long, 1956; Long et al., 1957), so that even at 0°C only some negatively substituted derivatives (epichlorohydrin, epibromohydrin) can be studied dilatometrically in >1 M acid. Good correlations of rate constants with h_0 are found. The rate increases strongly with substitution of electron-releasing groups. This would be expected for both the A-1 and the A-2 mechanism in view of the pre-equilibrium involved, but these substituents would also stabilize the carbonium ion centre in an A-1 mechanism:

$$\underset{\underset{R}{|}}{\overset{H}{\underset{HC-CH_2}{\overset{O^+}{\diagup\diagdown}}}} \longrightarrow \underset{\underset{R}{|}}{\overset{OH}{\underset{H-\overset{+}{C}-CH_2}{|}}}$$

If this mechanism applies, the C—O bond whose fission results in the more stable carbonium ion would break. In accordance with this, studies of hydrolysis of asymmetric epoxides in ^{18}O-labelled water show incorporation of ^{18}O on the branched carbon (Long and Pritchard, 1956a). Substituent effects on rate are correlated satisfactorily with σ_I constants ($\rho = -1.95$)—Pritchard and Long (1956). Entropies of activation are slightly negative and again inconclusive (Long et al., 1957), but negative volumes of activation indicate that the reaction occurs by an A-2 mechanism (Koskikallio and Whalley, 1959c). Another important piece of evidence conflicts with the A-1 mechanism, namely hydrolysis of optically active epoxides has been found to occur with complete inversion of configuration (Winstein and Henderson, 1950). On the other hand, there is at least one case of acid-catalysed epoxide ring opening which occurs via a carbonium ion: the reaction of tetramethyl ethylene oxide results only 75 per cent in pinacol and 25 per cent in pinacolone (Pocker, 1959), which can only be explained by the partitioning of a carbonium ion intermediate between hydration and rearrangement. To reconcile this with a complete inversion of configuration, one must assume that one side of the carbonium ion is effectively "shielded", so that water molecules attack at the side opposite to the displaced group (Long and Paul, 1957). It thus appears that there is no clear-cut distinction between the A-1 and the A-2 mechanisms in this reaction, but rather that the carbonium ion character of the transition state becomes progressively more pronounced with increasing numbers of alkyl substituents on the ethylene carbons.

5.4.1.2. Hydrolysis of Acetals, Ketals and Glucosides

These compounds are characterized by the group —O—C—OR, and may be regarded as substituted ethers. They undergo hydrolysis much faster than similar dialkyl ethers (Brønsted and Wynne-Jones, 1929). Ingold (1953) estimated that the introduction of an ethoxy group into diethyl ether to give diethyl acetal increases the rate of hydrolysis by a factor of about 10^{11}. This is due to the ability of the alkoxy groups to stabilize an adjacent carbonium ion centre:

$$\text{RO—CH}_2\text{—}\overset{H}{\underset{}{O^+}}\text{—R'} \longrightarrow \text{RO}\overset{+}{\text{—CH}_2} + \text{HOR'}$$

Thus Ingold (1953) assumed that these compounds hydrolyse by an A-1 mechanism. Further support for this comes from the effect of substitution of methyl groups at the central carbon atom, which leads to enormous increases in rate in the series: formals < acetals < ketals. The effects of structural variation in the group R' are much smaller and compatible with the idea that they merely affect the extent of protonation. Optical activity in the group R' is preserved in hydrolysis (O'Gorman and Lucas, 1950) and hence no fission of the R'—O bond occurs. The reaction is faster in D_2O than in H_2O with $k_H/k_D = 0.37$ (Wiberg, 1955), consistent with the A-1 mechanism. Similar results were obtained by Kilpatrick (1963).

The rates of hydrolysis of dimethoxyethane in several mineral acids (perchloric, sulphuric, hydrochloric) show good linear correlations with h_0 (slope 1·15) up to moderately high concentrations (McIntyre and Long, 1954). The effect of substituents at the potential carbonium centre is correlated by the σ_I constants with a large negative value of ρ (−3·60), indicating that the transition state has some carbonium ion character (Kreevoy and Taft, 1955).

The activation parameters for the hydrolysis of a number of formals and acetals were recently analysed by Whalley (1964). The positive entropies of activation and small volumes of activation also support the A-1 mechanism.

Some cyclic acetals, 1,3-dioxolanes, hydrolyse rapidly in very dilute acid by the A-1 mechanism (Salomaa and Kankaanpera, 1961). Trioxane and paraldehyde undergo an acid-catalysed depolymerization, which is in fact also a hydrolysis, because the monomeric formaldehyde and acetaldehyde are completely or largely hydrated in aqueous acid solution. The rates of both reactions correlate satisfactorily with h_0 up to moderately high acid concentrations, with slopes close to unity (Paul, 1952; Bell and Brown, 1954; Bell et al., 1956). The fact that paraldehyde reacts much faster than trioxane is evidence that these reactions also occur by the A-1 mechanism, with slow formation of a reactive carbonium ion:

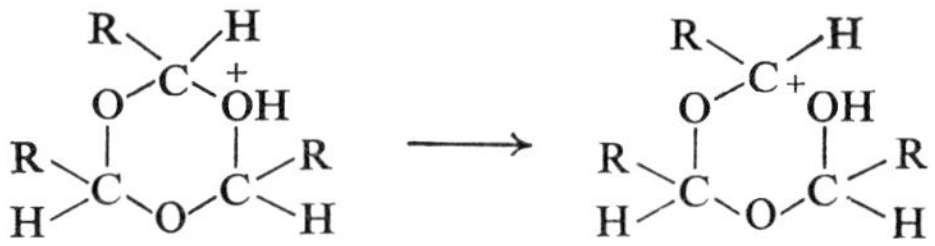

The extra methyl groups in paraldehyde increase the basicity of the compound, but their main accelerating effect is due to the stabilization of the carbonium ion centre. These reactions probably occur in three successive steps, the first being the slowest.

Sucrose and glucosides also involve groupings of the type —O—CH—OR and by analogy with acetals would be expected to hydrolyse by an A-1 mechanism. The inversion of sucrose has been repeatedly studied up to quite high acid concentrations and the rates show good correlation with h_0 with unit slopes (Long and Paul, 1957). The reaction is faster in D_2O than in H_2O with $k_H/k_D = 0.49$ (Wiberg, 1955). Hydrolyses of several glucopyranosides (α- and β-methyl- and α- and β-phenyl-) also show good correlations of $\log k_1$ vs. $-H_0$ with slopes 0·89–0·95 (Bunton et al., 1955b). Hydrolysis in ^{18}O-labelled solvent shows incorporation of ^{18}O in the glucose and not in the alcohol or phenol. Thus the hexose–oxygen bond undergoes fission, but whether this occurs in the rate-determining formation of the carbonium ion, or in the subsequent fast reaction of an open-chain carbonium ion is impossible to decide, because of the complexity of the structure involved and the similar probability that the ring oxygen or the OR oxygen may be protonated. An A-2 mechanism is *a priori* less likely for these complex structures, because of steric hindrance to the approach of a water molecule.

The considerable stabilization of alkoxycarbonium ions has enabled their direct observation by n.m.r. in liquid sulphur dioxide/SbF_5 mixtures (Olah and Bollinger, 1967).

Orthoesters hydrolyse much faster than acetals and have been studied in aqueous acetate buffer solutions (Brønsted and Wynne-Jones, 1929). The A-1 mechanism is also operative and the faster rate is due to the additional stabilization of the carbonium ion by one further alkoxy group. A small positive volume of activation determined more recently (Koskikallio and Whalley, 1959b) is consistent with this mechanism.

5.4.1.3. *Hydrolysis of Amides*

Some basic facts on the acid-catalysed hydrolysis of amides were firmly established many years ago (see Ingold, 1953). Kinetic indications that water is involved in the nucleophilic attack on the conjugate acid of the amide were obtained by Reid (1899, 1900), who showed that in dilute acid, where only small concentrations of the conjugate acid are present, polar substituent effects on the hydrolysis of benzamides are very small, whereas steric effects are large. Small polar effects are easily accounted for in terms of an A-2 mechanism:

$$R\text{—}CONH_2 + H_3O^+ \underset{fast}{\rightleftharpoons} R\text{—}CONH_3^+ + H_2O$$

$$R\text{—}CONH_3^+ + H_2O \xrightarrow{slow} RCOOH + NH_4^+$$

In this mechanism an electron-releasing substituent, which favours protonation and thus increases the concentration of the conjugate acid, at the same time decreases its reactivity to nucleophilic attack by water molecules by decreasing the partial positive charge on the carbonyl carbon. The reverse goes for electron-withdrawing substituents. Large steric effects are expected if steric hindrance to the attack by water molecules in the second, the rate-determining step, is involved. In addition it was known that the rate of hydrolysis of amides passes through a maximum in moderately concentrated acid (Krieble and Holst, 1938).

More recent work on both these aspects of amide hydrolysis has led to conclusions in agreement with the above mechanism. The basicities of substituted benzamides are now well established ($\rho = 0.92$)—see Chapter 3—and Leisten (1959) has argued that if hydrolysis is studied in sufficiently concentrated acid solutions for complete protonation of amides, the rates of hydrolysis should show the expected polar effect. This is indeed the case in

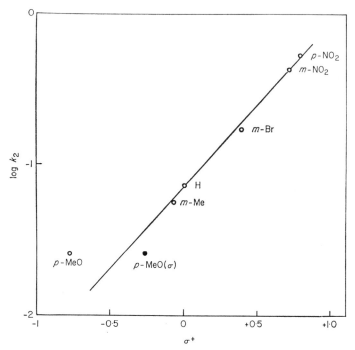

FIG. 5.1. Substituent effects on the rate of nucleophilic attack in the hydrolysis of benzamides (at $H_0 = -4.72$, 95°C).

8.54 M perchloric acid at 95°C (Leisten, 1959). The results are plotted against σ^+ substituent constants in Fig. 5.1. The hydrolysis step is accelerated by the same substituents which make the conjugate acid stronger ($\rho = 1.08$). On the basis of these results the small polar substituent effects on the hydrolysis of benzamides in dilute acid may now be quantitatively accounted for (Leisten, 1959). The rate of the reaction is given by the rate of the slow bimolecular step:

$$\text{Rate} = k_2[\text{RCONH}_3^+][\text{H}_2\text{O}] \tag{5.32}$$
$$= k_{\text{obs}}[\text{amide}]$$

Here k_{obs} is the experimental first-order constant at a given acid concentration. By substituting for the concentration of the conjugate acid, assuming only a small extent of protonation

$$[\text{RCONH}_3^+] = \frac{1}{K_{\text{BH}^+}}[\text{amide}][\text{H}_3\text{O}^+] \tag{5.33}$$

we obtain

$$\text{Rate} = \frac{k_2}{K_{\text{BH}^+}}[\text{amide}][\text{H}_3\text{O}^+][\text{H}_2\text{O}] \tag{5.34}$$

5. REACTION MECHANISMS IN SULPHURIC ACID SOLUTIONS

which means that at a given acid concentration $k_{obs} \propto k_2/K_{BH^+}$. Consequently, when the rates of hydrolysis of substituted benzamides in dilute acid are compared at the same acid concentration, the substituent effect on the observed rate constant will be related to the substituent effects on the pre-equilibrium and the rate-determining step by the equation

$$\log(k/k_0)_{obs} = \log(k/k_0)_2 - \log(K/K_0)_{BH^+} \qquad (5.35)$$

i.e.

$$\rho_{obs} = \rho_2 - \rho \qquad (5.36)$$

where ρ refers to the acid dissociation of the conjugate acids of amides. From the two ρ values mentioned (1·08 and 0·92) ρ_{obs} may be calculated by difference as 0·16, in satisfactory agreement with the value of 0·12 obtained from Reid's results (Jaffé, 1953). It can be seen that the substituent effect on the rate-determining step is slightly dominant, which is to be expected since the substituent effect on the reaction centre (the carbonyl carbon) is more direct than the effect on the protonation centre

It is worth mentioning that the plot of $\log k_2$ vs. σ constants (Leisten, 1959) showed a significant deviation of the point for the p-MeO-substituted compound, suggesting that an enhanced value of the substituent constant is needed. For this reason $\log k_2$ values were plotted vs. σ^+ constants in Fig. 5.1. It can be seen that the point for the p-MeO substituent deviates in this plot also, which means that in the transition state the conjugation of the p-MeO-substituent with the carbonyl carbon does not occur to the full extent (carbonyl frequencies of benzoyl compounds correlate with σ^+ constants of the *meta* and *para* substituents—Stewart and Yates, 1958). This is probably due to the partial loss of the carbonyl character in the transition state (**5.1**), owing to the partial formation of a bond between the nucleophile and the carbonyl

(**5.1**)

carbon. Alternatively it might be said that the carbonyl carbon acquires partial tetrahedral character in the transition state.

The rate of hydrolysis of amides in acid solution, given by equation (5.32), implies a dependence on acidity and a dependence on water activity, and thus accounts qualitatively for the observed rate profiles of the hydrolysis of amides in moderately concentrated acids, which show rate maxima. It has been shown by Moodie *et al.* (1963a) that the dependence on water activity is not first order, but that more likely three water molecules take part in the transition state. An analogous treatment was applied by Yates and Stevens (1965) to hydrolysis data on benzamide and *p*-nitrobenzamide in sulphuric acid, and by Yates and Riordan (1965) to literature data on the hydrolysis of acetamide (Rosenthal and Taylor, 1957; Edward *et al.*, 1955), propionamide (Rabinovich and Winkler, 1952), and benzamide (Edward and Meacock, 1957a). Since these amides are progressively protonated over the acid concentration range in which hydrolysis was studied, plots of $\log k_{obs} - \log h_A/(K_{BH^+} + h_A)$ vs. $\log a_{H_2O}$ were made and good linearity was found in every case. For the more weakly basic heterocyclic amides, nicotinamide (Jellinek and Gordon, 1949), picolinamide and isonicotinamide (Jellinek and Urwin, 1953), $\log k_{obs} + H_A$ vs. $\log a_{H_2O}$ was plotted. In all instances slopes greater than one were found, mostly in the range 3 ± 0.4. In view of this result water taking part in the transition state for the hydrolysis of amides should be regarded as hydrogen bonded to two

$$H_2O\cdots H \diagdown \atop H_2O\cdots H \diagup O\cdots \underset{\underset{O}{\|}}{\overset{\overset{R}{|}}{C}}-NH_3^+$$

(5.2)

further water molecules, as in (5.2). This hydrogen bonding enhances the negative charge on the attacking oxygen atom and hence facilitates the reaction. The nature of this transition state will be discussed further in Section 5.4.1.4.

Substituent effects on the hydrolysis of aliphatic amides have also been systematically studied. Aliphatic amides generally hydrolyse faster than benzamide, and this is reflected in the somewhat lower energies of activation for their hydrolysis (Rabinovich and Winkler, 1952). The order of decreasing reactivity is formamide, propionamide, acetamide, the difference between the latter two not being very large. As both formamide and propionamide are weaker bases than acetamide (see Chapter 3), their faster hydrolysis must be due to the faster nucleophilic attack in the rate-determining step. While for formamide this can readily be accounted for by reduced steric hindrance, the explanation for the faster hydrolysis of propionamide is less obvious. It has been suggested that this could be due to the smaller number of α-hydrogen atoms in propionamide as compared with acetamide, because these can hyperconjugate with the carbonyl carbon (Bolton, 1966). Such hyperconjugation can be formulated for the *N*-protonated cation as in (5.3):

$$H-\underset{\underset{H}{|}}{C}\overset{H}{\underset{}{\diagdown}}\overset{O}{\underset{}{\diagup}}C-NH_3^+$$

(5.3)

A linear free energy relationship holds for the hydrolysis of aliphatic amides, so that

$$\log k/k_0 = 0.858 E_s^c - 0.493(n-3)$$

where E_s^c is the steric substituent parameter (see Taft, 1956) and n is the number of α-hydrogen atoms in the acyl group (Bolton, 1966).

The effect of N-substituents on the rates of hydrolysis of amides has been studied less systematically. The rates of hydrolysis of N-methylformamide and N-methylacetamide in 1–3 M hydrochloric acid, studied by an n.m.r. method over a range of temperature (Saika, 1960) show that here also formamide reacts faster than acetamide. Acetylhydrazine hydrolyses more slowly than acetamide in dilute acid, but more rapidly at higher concentrations (Edward et al., 1955). One would expect the second protonation (i.e. that of the amide nitrogen) here to lead to hydrolysis, and the extent of this in dilute acid is certainly smaller than the extent of protonation of acetamide, hence the smaller rate, but the reasons for the inversion of rate at higher concentrations are not clear.

An exception to the described behaviour of amides is the hydrolysis of urea which is not acid catalysed (Shaw and Walker, 1958), and the hydrolysis of ethylurea which is almost independent of acid concentration between 35 and 90 per cent sulphuric acid (Armstrong et al., 1968a). The reasons for this are still not clearly understood.

The hydrolyses of some more complex amides and related compounds have also been studied. Acetylglycine shows a maximum rate in 5 M hydrochloric acid, whereas piperazine-2:5-dione and gelatine show a steady increase up to 10 M acid (Edward and Meacock, 1957b), which may be due to their smaller basicity. The hydrolysis of N-acylimidazolium ions was studied in dilute hydrochloric acid (Fee and Fife, 1966). The substrate is fully protonated in dilute acid, and increasing acid concentrations lead to a fall in rate (Marburg and Jencks, 1962). The solvent isotope effect, $k_H/k_D = 2.26-2.83$ (Marburg and Jencks, 1962; Fee and Fife, 1966), may therefore be due to the faster nucleophilic attack by H_2O than by D_2O on the conjugate acid. The hydrolysis of methyl benzimidate is considerably faster than the hydrolysis of benzamide and shows a fall in rate between $H_0 = 1$ and -2 (Edward and Meacock, 1957c).

All facts so far discussed on the rates of hydrolysis of simple amides in dilute and moderately concentrated acid are consistent with an A-2 mechanism. The solvent isotope effect, k_H/k_D, for the hydrolysis of acetamide has been reported as 0.68 in 0.1 N acid and as 1.16 in 4 N acid (Reitz, 1938). The result in dilute acid is consistent with a pre-equilibrium followed by a slow nucleophilic attack by water molecules, but the result in more concentrated acid appears to conflict with this. This may be explained, however, by assuming extensive protonation of the substrate in the more concentrated acid (Wiberg, 1955). As the pK_{BH^+} of acetamide is -0.93 (see Chapter 3), and the H_A function of

4 N hydrochloric acid is about $-1 \cdot 5$ (Yates and Riordan, 1965), this assumption is essentially correct.

Further important information on the nature of the rate-determining step in the hydrolysis of amides comes from studies of ^{18}O-exchange in the course of hydrolysis (Bender and Ginger, 1955). While exchange is observed in base-catalysed hydrolysis, there is virtually no exchange in acid hydrolysis ($k_{hydr}/k_{exch} \geqslant 374$). This means that the transition state for acid hydrolysis is adequately represented as that of a nucleophilic displacement reaction. There is no formation of a tetrahedral intermediate of any degree of stability. This question will further be discussed in Section 5.4.1.4.

The observed volumes of activation for the hydrolysis of acetamide and benzamide in dilute and moderately concentrated perchloric acid (0·3–4 M) are highly negative ($-9·4$ to -11.3 cm^3 mol^{-1} for acetamide and $-12·1$ to $-16·0$ cm^3 mol^{-1} for benzamide), i.e. typical of a reaction binding water molecules in the transition state (Osborn et al., 1961). The observed volumes of activation are the sum of the volumes of protonation and the volumes of activation of the rate-determining step. The considerable uncertainty in the volumes of protonation ($3·8 \pm 2·5$ cm^3 mol^{-1} for acetamide and 16 ± 8 cm^3 mol^{-1} for benzamide) as determined by Osborn et al. (1961) makes the calculation of the volumes of activation of the rate-determining step rather pointless, but it is clear that the values would be considerably negative.

The hydrolysis of thioacetamide, which occurs according to the equation

$$CH_3-C{\overset{S}{\underset{NH_2}{\diagup}}} + 2H_2O \rightleftharpoons CH_3COOH + NH_3 + H_2S \qquad (5.37)$$

shows a very similar acidity dependence as the hydrolysis of other amides in moderately concentrated acids, i.e. a maximum at about 4·5 M hydrochloric and at 4 M perchloric acid (Rosenthal and Taylor, 1957). A mechanism involving an A-2 displacement of NH$_3$ by water, which is operative with other amides, would lead here to the formation of thioacetic acid, which was in fact identified as an intermediate spectrophotometrically. Its subsequent reaction with water, analogous to the ^{18}O-exchange of carboxylic acids with water (see Samuel and Silver, 1965), leads to the formation of hydrogen sulphide.

Since the rates of the A-2 hydrolyses of amides decrease at sufficiently high acid concentrations, Duffy and Leisten (1960) examined the question of whether unimolecular amide hydrolysis may occur under anhydrous acid conditions. A number of N-substituted amides were studied, with N-substituents containing strongly electron-withdrawing groups, which would favour a heterolytic acyl–nitrogen bond fission:

$$R-\overset{O}{\overset{\|}{C}}-\overset{+}{N}H_2R' \longrightarrow R-\overset{+}{C}{\equiv}O + NH_2R' \qquad (A\text{-}1)$$

The complete reaction is

$$\text{RCO}\overset{+}{\text{N}}\text{H}_2\text{R}' + \text{HSO}_4^- + 2\text{H}_2\text{SO}_4 = \text{RC(OH)}_2^+ + \text{HSO}_4^- + \text{HS}_2\text{O}_7^- + \text{R}'\text{NH}_3^+ \quad (5.38)$$

Such reactions should lead to a change in the number of solute particles ($i = 2 \rightarrow 4$) and could thus be followed using the cryoscopic method. A number of amides that behave in this way were found. Some results are presented in Table 5.2. It can be seen from series (a) that electron-releasing substituents

TABLE 5.2

Effect of structure on the hydrolysis of amides in concentrated sulphuric acid
(Duffy and Leisten, 1960)

Amide R.CO.NHR'		Solvent (weight % H_2SO_4)	Temperature (°C)	k (10^{-3} min^{-1})
R	R'			
(a) CH$_3$	3,5-(NO$_2$)$_2$C$_6$H$_3$	100·0	60·0	8·56
CH$_3$CH$_2$	3,5-(NO$_2$)$_2$C$_6$H$_3$	100·0	60·0	22·2
CH$_2$Cl	3,5-(NO$_2$)$_2$C$_6$H$_3$	100·0	90·0	<0·2
(b) Ph	2,4-(NO$_2$)$_2$C$_6$H$_3$	100·0	13·3	22·0
p-NO$_2$.C$_6$H$_4$	2,4-(NO$_2$)$_2$C$_6$H$_3$	100·0	13·3	0·46
o-NO$_2$.C$_6$H$_4$	2,4-(NO$_2$)$_2$C$_6$H$_3$	100·0	13·3	20·7
(c) Ph	p-NO$_2$.C$_6$H$_4$	98·1	48·1	3·91
p-CH$_3$.C$_6$H$_4$	p-NO$_2$.C$_6$H$_4$	98·1	48·1	13·3
o-CH$_3$.C$_6$H$_4$	p-NO$_2$.C$_6$H$_4$	98·1	48·1	380
(d) CH$_3$	p-NO$_2$.C$_6$H$_4$	100·0	45·0	3·94
		100·0	55·0	14·8
		100·0	63·7	44·1

in R accelerate the reaction considerably, whereas electron-withdrawing substituents cause a sharp fall in rate, as expected for the A-1 mechanism. The same is observed in series (c) where methyl substitution in the phenyl ring increases the rate, especially effectively when in the *ortho* position. The introduction of a nitro-group into the phenyl ring in series (b) on the other hand causes a fall in rate, but not as effectively when in the *ortho* position as when in the *para* position, probably owing to steric hindrance to coplanarity of this rather bulky group. A comparison of series (b) and (c), however, shows that an increased number of nitro groups in the N-substituent considerably accelerates the reaction. The same is not true for series (a) and (d), where p-nitroacetanilide is seen to react faster than 3,5-dinitroacetanilide. The rates are, however, reversed at lower acid concentration, still in the region where the same mechanism applies. In general, the regions of hydrolysis by the A-1 and the A-2 mechanism are usually separated by a minimum in the rate/acidity profile curves.

The reaction constant ρ for substitution in the group R, obtainable from series (b) and (c) when the logarithms of the rate constants are plotted against σ^+ constants, because of the developing carbonium ion in the transition state, is $-2 \cdot 1$, in sharp contrast to the value of $+1 \cdot 08$ for the A-2 hydrolysis step for substituted benzamides (Fig. 5.1).

According to Duffy and Leisten (1960) the energies of activation for the A-1 hydrolysis of substituted benzamides (114–133 kJ mol^{-1}) are considerably higher than those for the A-2 hydrolysis of the same substrates at lower acid concentrations (94·1–98·7 kJ mol^{-1}). These figures show that nucleophilic attack by water molecules on the carbonyl carbon facilitates hydrolysis.

Addition of other acids into sulphuric acid should produce no change in rate, at least for the fully protonated amides. However, an effect was found in the hydrolysis of 3,5-dinitroacetanilide, following the order of acidity $H_2S_2O_7 > ClSO_3H > CH_3SO_3H$, but not proportional to the acidity function. This dependence of rate on acid composition suggests general acid catalysis. Therefore Duffy and Leisten (1960) believe that the A-1 mechanism of amide hydrolysis should be modified by assuming a proton transfer synchronous with the heterolysis of the CO—N bond, i.e. they suggest that proton addition to the nitrogen to form the ammonium ion commences before the transition state is fully formed.

5.4.1.4. *Hydrolysis of Esters and Lactones*

The hydrolysis of esters is probably one of the most extensively studied and discussed reactions of organic chemistry. Ingold and his collaborators made substantial contributions in elucidating and classifying the possible mechanisms (see Ingold, 1953). The subject was also discussed by Long and Paul (1957) and very extensively by Willi (1965).

There are two possibilities for the fission of carbon–oxygen bonds in an ester molecule, RCO.O.R', i.e. either the acyl–oxygen bond or the alkyl–oxygen bond may break. In acid solution this may occur either in a unimolecular reaction or in a bimolecular reaction with the solvent, so that four different mechanisms may be expected: $A_{Ac}1$, $A_{Al}1$, $A_{Ac}2$ and $A_{Al}2$, where the subscripts indicate which bonds are being broken. The first three of these mechanisms are known, the $A_{Ac}2$ mechanism, closely analogous to the bimolecular mechanism of amide hydrolysis, being the most common in dilute and moderately concentrated acid for esters of primary and secondary alcohols, in which there is no significant steric hindrance to the attack by water molecules. Evidence for acyl–oxygen fission was obtained by isotopic labelling with ^{18}O (Polanyi and Szabo, 1934), and the relative insensitivity of the rates to polar substituent effects in the hydrolysis of aromatic esters, similar to that in the hydrolysis of amides (Section 5.4.1.3), could be regarded as an indication of a two-step reaction mechanism:

$$R-\underset{OH}{\underset{|}{C}}-OR' + H^+ \rightleftharpoons R-\underset{OH}{\underset{|}{C^+}}-OR'$$

$$R-\underset{OH}{\underset{|}{C^+}}-OR' + H_2O \rightarrow R-\underset{OH}{\underset{|}{C^+}}-OH + HOR'$$

Here the substituent effects in the first step are compensated (indeed slightly overcompensated) in the second step, the nucleophilic attack by a water molecule on the conjugate acid ($\rho \simeq 0.3$ in the hydrolysis of m- and p-substituted ethylbenzoates, on the basis of the data of Timm and Hinshelwood, 1938). In the alkaline ester hydrolysis polar effects are large, because of the absence of the pre-equilibrium step. These differences in the polar substituent effects in the acid and the alkaline hydrolysis of esters have provided the basis for the definition of inductive substituent constants (see Chapter 3). Steric effects in ester hydrolysis are large.

Despite some analogies between the hydrolyses of esters and of amides, an important difference has been found by Bender (1951) in the fact that the acid hydrolysis of esters is accompanied by exchange of the carbonyl oxygen with the solvent, whereas this is hardly detectable in the acid hydrolysis of amides (Bender and Ginger, 1955). This can only be accounted for by assuming the reversible formation of an intermediate (**5.4**) in the reaction

$$R-\underset{}{\underset{|}{C^+}}\overset{^{18}OH}{\underset{}{}}-OR' + H_2O \rightleftharpoons R-\underset{OH}{\underset{|}{C}}\overset{^{18}OH}{\underset{|}{}}-OR' + H^+$$

(5.4)

rather than a transition state, which goes directly to products. Although in principle a transition state can also revert to reactants, transmission coefficients close to unity are most frequently assumed. The fact that either of the three C—O bonds may break implies that the adduct has a finite lifetime, and may be regarded as the hydrate of the ester. The difference between an intermediate and a transition state is merely in their degree of stability, or in the depth of the potential energy well, which is only very small for a transition state and larger for an intermediate. An intermediate must pass through transition states both on the way to reaction products and on the reverse way to reactants.

Oxygen exchange in the hydrolysis of some other derivatives of carboxylic acids (chlorides and anhydrides) has also been studied (Bunton et al., 1954) and the subject of O-exchange reactions has been reviewed by Samuel and Silver (1965). A brief summary of the results on the derivatives of benzoic acid is given in Table 5.3 for aqueous solutions wherever data are available.

TABLE 5.3

Oxygen exchange to hydrolysis data for the derivatives of benzoic acid

Derivative	Temperature (°C)	Solvent	$10^5 \, k_{obs}$	k_{hydr}/k_{exch}	Reference†
(1) $C_6H_5CONH_2$	109	0.1 M NaOH	263[OH$^-$]	0.21	a
(2) $C_6H_5COOC_2H_5$	25	0.1 M NaOH	—	4.8	b
		0.1 M HCl	—	5.2	
(3) $C_6H_5CO\!\!>\!\!O$, C_6H_5CO	62.6	75% dioxane (initially neutral)	0.0652	20	c
(4) C_6H_5COCl	25	75% dioxane (initially neutral)	8.2	25	c
(5) $C_6H_5CONH_3^+$	109	0.1 M HCl	—	≥374	a

† a. Bender and Ginger (1955); b. Bender (1951); c. Bunton *et al.* (1954).

It can be seen from this Table that, unlike the amide behaviour, the relative rates of hydrolysis and exchange for the ester are very much the same whether the reaction occurs in acid or in alkaline solution, although the rates of hydrolysis may be very different—the basic hydrolysis is faster by a factor of 10^4 under the same conditions of solvent and temperature (Bender *et al.*, 1958). This led Taft (1952a) to suggest that the same kind of intermediate is involved in both the acid and the alkaline hydrolysis. Also, the fact that oxygen exchange is observed in the hydrolysis of all acid derivatives in pure aqueous or partially aqueous solutions suggests that all acid derivatives (with the possible exception of amides in acid solution) hydrolyse by the same mechanism, which involves the reversible formation of a reaction intermediate with two structurally equivalent carbon to oxygen bonds. Such a mechanism may be represented by the following reaction scheme:

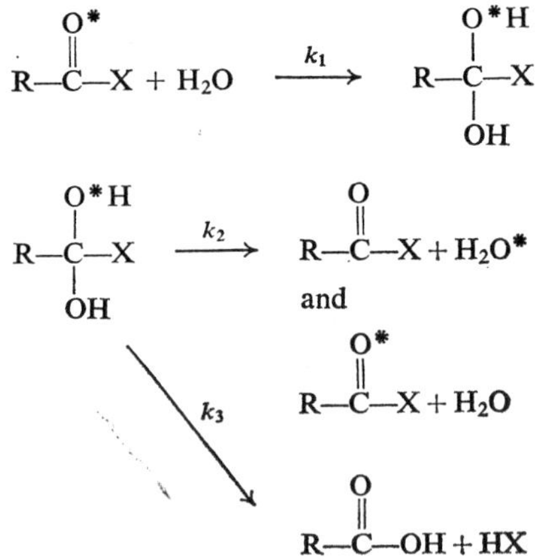

$$R-\overset{\overset{O^*}{\|}}{C}-X + H_2O \xrightarrow{k_1} R-\overset{\overset{O^*H}{|}}{\underset{\underset{OH}{|}}{C}}-X$$

$$R-\overset{\overset{O^*H}{|}}{\underset{\underset{OH}{|}}{C}}-X \xrightarrow{k_2} R-\overset{\overset{O}{\|}}{C}-X + H_2O^*$$

and

$$R-\overset{\overset{O^*}{\|}}{C}-X + H_2O$$

$$\xrightarrow{k_3} R-\overset{\overset{O}{\|}}{C}-OH + HX$$

In this reaction scheme the intermediate is partitioned between two reactions: the reverse reaction regenerates the reactants and the forward reaction gives reaction products (Bender et al., 1958). The first of these results in oxygen exchange. By applying the principle of the stationary state it can be shown that the observed rate constants for hydrolysis and exchange are given by

$$k_{\text{hydr}} = \frac{k_1 k_3}{k_2 + k_3} \quad (5.39)$$

$$k_{\text{exch}} = \frac{k_1 k_2}{2(k_2 + k_3)} \quad (5.40)$$

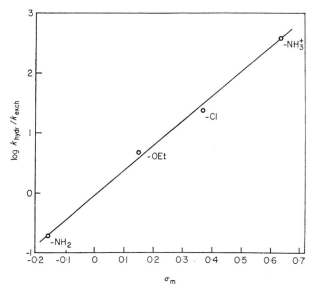

FIG. 5.2. The correlation of the relative rates of hydrolysis to oxygen exchange of the derivatives of benzoic acid (PhCOX) with σ_m constants of the leaving group (X).

(The factor 2 comes in the denominator of k_{exch} because only one half of the reverse processes leads to exchange.) The rates of hydrolysis and exchange are thus not determined by the rate of a single step, but by that of all the reactions involved in the scheme. Only in the limiting case, when the rate of the forward reaction is much greater than the rate of the reverse reaction ($k_3 \gg k_2$), the hydrolysis constant becomes $k_{\text{hydr}} \simeq k_1$, and the rate-determining step becomes the nucleophilic attack by a water molecule on the carbonyl carbon. This appears to be the case in acid-catalysed amide hydrolysis, in view of the almost complete absence of oxygen exchange. The experimentally observed ratio of the hydrolysis and the exchange rate constants follows from equations (5.39) and (5.40):

$$k_{hydr}/k_{exch} = 2k_3/k_2 \qquad (5.41)$$

Bunton et al. (1954) noted that the order of increasing k_{hydr}/k_{exch} ratios (Table 5.3) for various groups X is $NH_2 < OEt < OCOC_6H_5 < Cl < NH_3^+$, and is the same as the order of ease of ionization of X^- from C—X. This is also the order of increasing σ_m constants of these groups, and the relationship with σ_m is in fact quantitative (Fig. 5.2). This definitely means that the hydrolysis/exchange ratio gives a measure of the relative ease of breaking heterolytically the C—X and the C—OH bonds. The fact that the relationship is not with σ_I constants, but with σ_m constants, which have been shown to measure the ability of mesomerically electron-releasing groups X to interact with the sp^2 hybridized carbon (Liler, 1967)—see Chapter 3—means that in the transition state for the breaking up of the intermediate the group that has the greater ability to donate electrons to the sp^2 carbon will be preferentially retained. This is wholly consistent with the emergence of a carbonyl carbon in the reaction products, and the transition states should probably be formulated in general as follows:

(5.5)   or   (5.6)

Here the C—O or C—X bonds are being broken and the O—H or X—H bonds formed. The fact that the partitioning of the intermediate in both the acid and the alkaline ester hydrolysis is very much the same shows that the rôle of the proton is secondary in determining the partitioning ratio, because its presence assists to a similar extent and in the same manner the detachment of all of the three groups.

The fact that the hydrolysis/exchange ratio for amides in acid solution fits the relationship of Fig. 5.2 (on the assumption that $k_{hydr}/k_{exch} = 374$) probably means that the hydrolysis of amides in acid solution also occurs through an intermediate, but as double protonation is not possible the carbonyl oxygen probably merely forms hydrogen bonds with the solvent, which may easily lead to the equivalence of the two oxygens. The presence of three water molecules in the transition state for amide hydrolysis has been mentioned in

5. REACTION MECHANISMS IN SULPHURIC ACID SOLUTIONS

Section 5.4.1.3. One of these may be involved in hydrogen bonding with the carbonyl oxygen, as in (**5.7**):

$$\begin{array}{c} \text{HOH} \\ \vdots \\ \overset{\delta-}{\text{O}} \\ \| {\scriptstyle \delta+ \rightarrow +} \\ \text{R}-\text{C}\cdots\text{NH}_3 \\ \vdots \\ \text{O} \\ / \quad \backslash \\ \text{H} \quad \text{H}\cdots\text{OH}_2 \end{array}$$

(**5.7**)

The equivalence of the two oxygens in this structure is only a matter of rapid electronic rearrangement, and the intermediate is probably sufficiently long-lived for this. The large difference in the $k_{\text{hydr}}/k_{\text{exch}}$ ratio in the acid and the alkaline hydrolysis of amides (a factor of $>10^3$) leaves no doubt, however, that in acid hydrolysis the leaving group is NH_3, vs. NH_2 in alkaline hydrolysis. Only if this is so would the relationship in Fig. 5.2 be observed. The rapid addition of a proton onto the leaving ammonia molecule may commence in the transition state itself.

The existence of a pre-equilibrium in the acid hydrolysis of esters is supported by solvent isotope effects with $k_H/k_D = 0.43$–0.64 (Wiberg, 1955; Salomaa et al., 1964). The presence of a water molecule in the transition state could not be readily demonstrated in aqueous acid, but was demonstrated in "wet" acetone (Friedman and Elmore, 1941). The evidence for the formation of an intermediate discussed above necessarily means that at least one water molecule is involved in the transition states for either hydrolysis or exchange. The dependence of the rates of hydrolysis on the acidity of the solution at moderately high acid concentrations was also thought to provide evidence for this, because the rates were found to correlate more closely with the concentration of the acid, rather than with h_0 (Long and Paul, 1957). In terms of current ideas on acidity functions (Chapter 2) this merely means that the acid hydrolysis of esters follows an acidity function showing only slight "excess acidity". This implies that the transition state is probably highly hydrated, which is not unexpected in view of the presence of two OH groups in the intermediate. Negative volumes of activation and highly negative entropies of activation reflect also the strong hydration of the intermediate (Table 5.4).

Since at least one water molecule must be present in the intermediate of ester hydrolysis, one would expect the rate of hydrolysis to pass through a maximum at higher acid concentrations, as it does in the hydrolysis of amides. This has only recently been demonstrated for ethyl acetate (Lane, 1964). The interpretation of the curve requires knowledge of the pK_{BH^+} of the ester, which

TABLE 5.4

Entropies of activation and volumes of activation for the acid-catalysed hydrolysis of esters (Schaleger and Long, 1963; Whalley, 1964)

Ester	Mechanism	$\Delta S^{\ddagger}$ (J K^{-1} mol^{-1})	$\Delta V^{\ddagger}$ (cm^3 mol^{-1})
Methyl acetate	A-2	−89·1	−9·4 ± 0·7
Ethyl acetate	A-2	−96·2	−9·1 ± 0·7
t-Butyl acetate	A-1	+59·4	0·0 ± ∼1
β-Propiolactone	A-1	+39·3 ± 25	+2·5 ± ∼2
γ-Valerolactone	A-2	−102·9	—

is not available (see Chapter 3). A value could be estimated from the extinction coefficients at 190 nm at zero time, and from chemical shifts of acetyl protons. The equation $\log I = 0.645(-6.93 - H_0)$ was found to fit the data. It indicates half-protonation in 77 per cent sulphuric acid and shows that the ester follows an acidity function, which decreases with acid concentration much less steeply than H_0. By plotting $\log k_{obs}/[BH^+]$ vs. $\log a_{H_2O}$ a slope of 2·07 is obtained for hydrolysis and a slope of 1·84 for oxygen exchange (Lane, 1964). There are thus two water molecules and a proton in the transition states for both reactions. Taking this into account, Lane (1964) suggests transition states closely similar to (**5.5**) and (**5.6**), with the protons replaced by hydronium ions. This is implied in formulae (**5.5**) and (**5.6**). A transition state involving the ester, two water molecules and a proton has already been suggested by Laidler and Landskroener (1956).

The hydrolyses of several other acetates in sulphuric acid/water mixtures over a wide acid concentration range also follow an acidity function given by $H_s = 0.62\, H_0$ (Yates and McClelland, 1967), in close agreement with the slope found by Lane (1964). Rate maxima occur at about 50 per cent acid for several acetates, but plots of $(\log k_{obs} + H_s)$ vs. $\log a_{H_2O}$ are linear almost up to 80 per cent sulphuric acid, with slopes close to 2 for esters of primary alcohols. The same was found for some secondary alkyl acetates up to about 65–75 per cent acid (Fig. 5.3).

At still higher acid concentrations the rate of hydrolysis of these esters again shows an increase (Jaques, 1965; Yates and McClelland, 1967). The slopes of the lines $(\log k_{obs} + H_s)$ vs. $\log a_{H_2O}$ in this concentration region are very small for esters of primary alcohols (∼−0·2)—Fig. 5.3—indicating that no water molecules are involved in the rate-determining step. This is consistent with an A-1 mechanism, which has already been established cryoscopically for methyl, ethyl and *iso*-propyl benzoates (Leisten, 1956). For esters of

5. REACTION MECHANISMS IN SULPHURIC ACID SOLUTIONS

secondary alcohols the slopes of the above plots are larger and the breaks occur at lower acid concentrations (Fig. 5.3). This suggests that the A-1 mechanisms are not the same for esters of primary and secondary alcohols. This was shown to be the case for benzoate esters (Leisten, 1955). The rate of hydrolysis of alkyl benzoates in pure sulphuric acid varies as follows: methyl > ethyl < iso-propyl < t-butyl (Kuhn, 1949). Since esters are fully

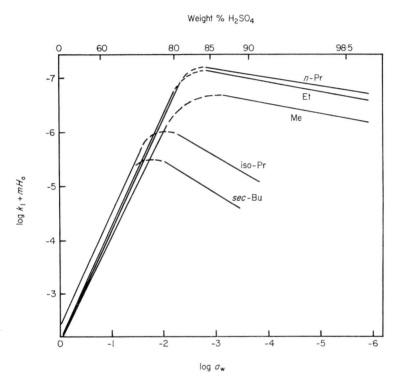

FIG. 5.3. $\log k_1 + mH_0$ vs. $\log a_w$ plots for the hydrolysis of some acetate esters in sulphuric acid/water mixtures (Yates and McClelland, 1967).

protonated in this medium, the effects of substituents may be interpreted simply in terms of the heterolysis of the conjugate acid. This has usually been done by considering esters protonated on the ether oxygen, for both alkyl–oxygen and acyl–oxygen fission (Ingold, 1953). As esters protonate on the carbonyl oxygen, without any evidence for ether–oxygen protonation even in superacid media down to −90°C (Olah et al., 1967e), it must be assumed either that the ether–oxygen-protonated species, present in undetectably small concentration, is the unstable form, or that the carbonyl–oxygen-protonated species decomposes. This poses no problem for the $A_{A1}1$ mechanism, which may be formulated as

$$R-C\overset{+}{\underset{O-R'}{\overset{OH}{\diagup}}} \longrightarrow R-C\overset{OH}{\underset{O}{\diagup}} + R'^{+}$$

but for the $A_{Ac}1$ decomposition a simultaneous proton transfer and acyl–oxygen fission have to be assumed (Olah et al., 1967e):

$$R-C\overset{+}{\underset{O-R'}{\overset{O-H}{\diagup}}} \longrightarrow R-\overset{+}{C}\!\!\equiv\!\!O + R'OH$$

The oxocarbonium ion, depending on its stability, may remain as such in strongly acid solution, or may react rapidly to give the protonated acid. Whatever the structure of the transition state in $A_{Ac}1$ hydrolysis, electron-withdrawing substituents in R should retard acyl–oxygen fission and accelerate alkyl–oxygen fission, and electron-donating substituents should act in the opposite way. The reverse should be true for substituents in R'. Substituent effects on the rate of hydrolysis of methyl benzoate for all four types of substitution are known and are consistent with the $A_{Ac}1$ mechanism (Ingold, 1953).

When substituents are present in the benzene ring which stabilize the oxocarbonium ion and in addition provide steric hindrance to the attack by water molecules, the $A_{Ac}1$ mechanism may be observed in moderately concentrated aqueous acids as, for example, for methyl mesitoate (Chmiel and Long, 1956; Bender et al., 1961). The rate of this hydrolysis follows the H_0 acidity function (slope 1.2). There is no exchange of carbonyl oxygen, and the entropy of activation is positive. The effect of *para* substituents on the rate of hydrolysis of methyl 2,6-dimethyl benzoates shows a correlation with σ^+ constants, with a highly negative slope ($\rho = -3 \cdot 22$), consistent with the formation of an oxocarbonium ion intermediate (Bender and Chen, 1963). Similar results were obtained for the hydrolysis of methyl and ethyl benzoates in pure sulphuric acid (Leisten, 1956; Kershaw and Leisten, 1960), except for ethyl benzoates containing *p*-nitro, 4-chloro-3-nitro and 3,5-dinitro substituents, where a change of mechanism to alkyl–oxygen fission occurs. The mechanism for negatively substituted methyl benzoates remains $A_{Ac}1$ even in solutions of tetra(hydrogenosulphate)boric acid in 100 per cent sulphuric acid (Hopkinson, 1969).

Alkyl–oxygen fission is facilitated for esters of secondary alcohols, and this explains the faster hydrolysis in pure sulphuric acid of isopropyl benzoate compared with ethyl benzoate (Leisten, 1956). The introduction of a *p*-nitro substituent leads here to a large increase in rate, as expected for the $A_{A1}1$

mechanism. Yates and McClelland (1967) postulate this mechanism also for acetate esters of secondary alcohols in concentrated sulphuric acid and interpret the larger negative slopes of the plots in Fig. 5.3 by a release of water molecules in the transition state, owing to the formation of an incipient carbonium ion which is not highly solvated. Substituent effects on the hydrolysis of isopropyl benzoates in 98·7 per cent sulphuric acid correlate with σ constants and a ρ value of +1·99 shows the absence of an oxocarbonium ion intermediate and confirms the $A_{A1}1$ mechanism (Hopkinson, 1969).

In superacid media (HSO_3F—SbF_5) at low temperature ($-80°C$) oxocarbonium ions are stable and methyl esters undergo only the first, rate-determining, step of unimolecular hydrolysis, i.e. acyl–oxygen fission, followed by rapid protonation of methanol (Olah et al., 1967e). Carboxylic acids also undergo cleavage under similar conditions, but considerably faster. This shows the great sensitivity of this reaction to the electron-releasing properties of the group R' ($CH_3 > H$). Under similar conditions secondary esters cleave exclusively by the $A_{A1}1$ mechanism to protonated acids and carbonium ions, which rearrange into the most stable t-butyl cation whenever possible (Olah et al., 1967e).

Esters of tertiary alcohols hydrolyse rapidly in dilute acid solution by the $A_{A1}1$ mechanism, as a result of the considerable stability of tertiary carbonium ions. The positive entropy of activation and the small volume of activation (Table 5.4) are consistent with this mechanism, and a host of other evidence has been reviewed recently by Willi (1965). In superacid media at $-80°C$ esters of tertiary alcohols cleave so rapidly that only the protonated acid and the tertiary carbonium ions are observed (Olah et al., 1967e).

Lactones, being inner esters of hydroxyacids, would be expected to behave like ordinary esters, if the rings formed are unstrained. This is so for γ-lactones. The hydrolysis and formation of γ-butyrolactone (the reaction is reversible), studied up to 4 M perchloric and hydrochloric acid, behave in the expected manner for an A-2 mechanism (Long et al., 1951). Also consistent with this mechanism is the entropy of activation for the hydrolysis of γ-valerolactone (Table 5.4). If the ring is strained, as in β-lactones, the acidity dependence suggests an A-1 mechanism (Long and Purchase, 1950), and the activation parameters (Table 5.4) are consistent with it.

5.4.1.5. *Hydrolysis of Nitriles* (*Cyanides*)

The hydrolysis of nitriles has been studied less extensively than the hydrolysis of other acid derivatives, but some basic features are reasonably well established. Hammett and Paul (1934) showed that the rates of hydrolysis of cyanamide in aqueous nitric acid (up to 5 N) follow the H_0 acidity function with a unit slope. The reaction is thus acid-catalysed and the conjugate acid undergoes hydrolysis. A preprotonation equilibrium is also confirmed by the solvent isotope effect on the hydrolysis of acetonitrile ($k_H/k_D < 0.77$—Reitz, 1938), but the magnitude of the effect indicates a two-step mechanism (A-2), the second step being the nucleophilic attack by water molecules on the conjugate acid:

$$RCNH^+ + H_2O \xrightarrow{k_1} RCONH_3^+$$

The nitriles are, however, incompletely protonated even in 100 per cent sulphuric acid (see Chapter 3).

In dilute acid solutions the hydrolysis of amides is much faster than that of nitriles ($k_2 \gg k_1$), so that the ammonia formed in the subsequent reaction

$$RCONH_3^+ + H_2O \xrightarrow{k_2} RCOOH + NH_4^+$$

may be used to follow the rate of nitrile hydrolysis. In this way Krieble and Noll (1939) studied the hydrolysis of several nitriles in sulphuric, hydrochloric and hydrobromic acids (up to 6–8 M). The three acids differ considerably in their catalytic power, sulphuric acid being the poorest catalyst, especially in the hydrolysis of hydrogen cyanide. The plots of $\log k$ vs. $-H_0$ are satisfactorily linear for hydrogen cyanide as well as for acetonitrile and cyanoacetic acid (Krieble and Noll, 1939), with slopes of only 0.6–0.7. These are consistent with a more highly hydrated transition state compared with the protonated reactant, which can be expected if a water molecule enters into it, as in (5.8),

(5.8)

with the possibility of further hydrogen bonding with solvent molecules. The orders of reactivity are $HCN > CH_3CH(OH)CN > CH_3CH_2CN > CH_3CN$ and $CH_2OH.CH_2CN > COOH.CH_2CN$ (Krieble and Noll, 1939) and are generally consistent with the idea that electron-attracting substituents accelerate the reaction, which means that their effect on the pre-equilibrium is less than their effect on the nucleophilic attack step. The second sequence probably arises from the protonation of the OH group, i.e. $H_2\overset{+}{O}CH_2CH_2CN > COOHCH_2CN$. The greater reactivity of propionitrile compared with acetonitrile is probably due to the smaller number of α-hydrogen atoms and thus to a smaller hyperconjugative delocalization of the partial positive charge on the nitrile carbon, which is essential for a fast nucleophilic attack by water molecules.

An A-2 mechanism involving attack by water molecules in the rate-determining step has been shown to lead to rate maxima at moderately high acid concentrations both in the hydrolysis of amides (Section 5.4.1.3) and of esters

(Section 5.4.1.4). A search for such a maximum in the hydrolysis of nitriles was unsuccessful: the rate was found to increase steadily up to 10 N hydrochloric acid (Rabinovich et al., 1942) and up to 18·86 N sulphuric acid (McLean et al., 1942). Since the rates of hydrolysis of amides pass through a maximum at 4–6 N acid concentration, the rate of hydrolysis of nitriles cannot be followed at higher concentrations by determining ammonia alone. The hydrolysis becomes a typical case of consecutive reactions, i.e.

$$\text{RCN} \xrightarrow{k_1} \text{RCONH}_2 \xrightarrow{k_2} \text{RCOOH} + \text{NH}_3$$

with $k_2 < k_1$, requiring determination of both the amide and ammonia in order to calculate the rate of hydrolysis of the nitrile (Rabinovich et al., 1942). In sulphuric acid > 20 N k_1 becomes much greater than k_2 for propionitrile (Rabinovich and Winkler, 1942), in agreement with the well-known fact that amides may be isolated quantitatively when nitriles are dissolved in concentrated sulphuric acid.

The absence of a rate maximum in the hydrolysis of nitriles up to high acid concentrations is consistent with a steadily increasing proportion of the protonated nitrile. Extensive protonation of nitriles occurs only in pure sulphuric acid (see Chapter 3), and even in that medium nitriles hydrolyse according to the equation

$$\text{RCNH}^+ + \text{HSO}_4^- + \text{H}_2\text{SO}_4 = \text{RCONH}_3^+ + \text{HS}_2\text{O}_7^- \qquad (5.42)$$

so that the reaction can be followed conductometrically. The hydrolysis of acetonitrile and benzonitrile (at 25°C) studied in this way (Liler and Kosanović, 1959) obeys a second-order law, which indicates that HSO_4^- ions may be acting as nucleophiles. Whether there is a maximum or any levelling off in the rate of hydrolysis of nitriles between 18·86 N and pure sulphuric acid is not known. Deno et al. (1957) report that the hydrolysis of acrylonitrile to acrylamide between 25 and 85 per cent sulphuric acid obeys approximately the H_0 function.

In pure sulphuric acid benzonitrile, although protonated to a smaller extent, reacts faster than acetonitrile (Liler and Kosanović, 1958), which is in agreement with the greater electron withdrawal by the phenyl group (compared with methyl) resulting in a greater rate of nucleophilic attack on the nitrile carbon.

Complete decomposition of hydrogen cyanide to yield carbon monoxide occurs at maximum rate in 78–79 per cent sulphuric acid or in 100 per cent phosphoric acid (Cobb and Walton, 1937), undoubtedly via the final product of the hydrolysis, formic acid. In concentrated sulphuric acid, in which hydrolysis goes only as far as the amide, no decomposition is observed. Little or no hydrogen cyanide can be generated by acting with concentrated sulphuric acid upon sodium cyanide (Moore, 1932). A crystalline adduct between hydrogen cyanide and sulphuric acid has in fact been described (Cobb and Walton, 1937) and its rate of "decomposition", i.e. hydrolysis, studied in 96·2 per cent and in 100 per cent acid. The rate in 96·2 per cent acid is one-third that in 100 per cent acid, a relatively small decrease compared with other reactions governed by H_0. This may be an indication that there is a levelling off in the rate of hydrolysis of nitriles in concentrated sulphuric acid.

Aldehyde cyanohydrins and ketone cyanohydrins are also hydrolysed to amides in concentrated sulphuric acid via the intermediate formation of sulphates, $R_2C(OSO_3H)CONH_2$ (Verhulst, 1931; Terada, 1959). The intermediates have been prepared for R = H, Me, Et, Pr, Me_2CH, Bu, and *iso*-Bu, and the kinetics of their acid hydrolysis studied (Verhulst, 1931).

Thiocyanic acid also hydrolyses in acid solution according to the equation

$$HNCS + H^+ + 2H_2O \rightleftharpoons H_2S + CO_2 + NH_4^+ \qquad (5.43)$$

and the rate follows the H_0 acidity function in sulphuric, phosphoric and hydrochloric acids up to 11 M, with a unit slope above $H_0 = -2$, and 1·4 slope in more dilute acid (Crowell and Haukins, 1969). Thiocyanic acid is itself fairly strong ($pK_a = -2$ to -2.3) and it is suggested that two decomposition paths are followed, one involving the undissociated acid and one involving the conjugate acid, $H_2\overset{+}{N}=C=S$, but the detailed mechanism is still not clear.

5.4.1.6. *Hydrolysis of Acid Anhydrides*

Acid anhydrides are hydrolysed with ease in aqueous acid solutions and the rate of the reaction can only be followed at low temperature. Gold and Hilton (1955) studied the hydrolysis of acetic anhydride in aqueous hydrochloric, perchloric, sulphuric and phosphoric acids at 0°C by dilatometry and the thermal maximum method up to H_0 values of -3.5. While there are small differences between different acids, all log$k$ data fall close to a line of unit slope in a plot $vs. -H_0$ (Long and Paul, 1957). The reaction is thus acid catalysed and Gold and Hilton (1955) propose the following mechanism:

$$Ac_2O + H^+ \rightleftharpoons Ac_2OH^+$$

$$Ac_2OH^+ \xrightarrow[\text{slow}]{} Ac^+ + AcOH$$

$$Ac^+ + H_2O \xrightarrow[\text{fast}]{} AcOH + H^+$$

This mechanism involves a unimolecular decomposition of the conjugate acid as the rate-determining step (A-1). However, Bunton *et al.* (1954) have established that in the hydrolysis of benzoic anhydride in an initially neutral solution in 75 per cent dioxane carbonyl oxygen exchange occurs in parallel with the hydrolysis reaction, the ratio k_{hydr}/k_{exch} being 20. This suggests that the hydrolysis of acid anhydrides occurs in the same manner as that of other acid derivatives, i.e. via the formation of a tetrahedral intermediate. This mechanism was discussed in detail in Section 5.4.1.4.

The unimolecular mechanism is, however, probably operative in the pure sulphuric acid region and in superacid media. The equilibria

$$R_2(CO)_2O + 3H_2SO_4 \rightleftharpoons 2RC(OH)_2^+ + HS_2O_7^- + HSO_4^- \qquad (5.44)$$

are rapidly established, and their position depends on the nature of the anhydride (see Chapter 4).

The anhydrides of acetic, propionic, succinic, and glutaric acid undergo complete cleavage in superacid media (HSO_3F—SbF_5—SO_2) even when prepared at $-80°C$. The n.m.r. spectra of the solutions correspond to the mono-oxocarbonium-monocarboxonium ions, which are also obtained from protonated dicarboxylic acids (Olah and White, 1967b).

5.4.1.7. *Hydrolysis of Oximes*

The hydrolyses of cyclohexanone and cyclopentanone oximes (Vinnik and Zarakhani, 1960; Zarakhani *et al.*, 1965) show rate–acidity profiles very similar to those for the hydrolysis of methyl benzimidate (Edward and Meacock, 1957c), namely a rate maximum in very dilute acid followed by a steady decrease with increasingly negative values of H_0 in both hydrochloric acid and sulphuric acid solutions (up to 80 per cent acid). Vinnik and Zarakhani (1960) put forward the hypothesis that the rate decreases at higher acid concentrations because of the formation of ion pairs between the protonated oxime (which they assume to be the reactive species) and the anions of the acid. From the analysis of their data they obtain a pK_a for the oximes of 3·3 (cf. Chapter 3). The basic idea of ion pair formation is not easily acceptable, however, in view of the high dielectric constant of aqueous acids, and the decreasing rate of oxime hydrolysis with increasing acid concentration is probably due to the decreasing activity of water.

5.4.1.8. *Hydrolyses of Some Halogenated Compounds*

Earlier reports in the literature (C.A. **41**, 2073e) that halogenated compounds of the type $RR'CHCl_3$ or $RR'C{=}CCl_2$ hydrolyse in 85–90 per cent sulphuric acid at $<125°C$ (the first apparently through the intermediate formation of a dichloro-olefin and its subsequent hydration) and that benzylidyne trifluoride ($PhCF_3$) likewise hydrolyses in 100 per cent sulphuric acid (Le Fave, 1949) have recently led to a systematic study of the kinetics of this latter reaction over the concentration range 80–100 per cent acid (Coombes *et al.*, 1969). The rate constant for the hydrolysis is proportional to h_0 in the range 81–92 per cent sulphuric acid, but increases less rapidly at higher concentrations. Electron-withdrawing substituents reduce the rate considerably. Two reaction mechanisms were suggested to account for these observations. The first involves a preprotonation step:

$$PhCF_3 + H^+ \underset{fast}{\rightleftharpoons} PhCF_3H^+$$

$$PhCF_3H^+ \underset{slow}{\longrightarrow} PhCF_2^+ + HF$$

$$PhCF_2^+ + H_2O \underset{fast}{\longrightarrow} PhCF_2OH + H^+$$

The alcohol is further rapidly converted to the carboxylic acid. The second mechanism involves a slow proton transfer. There is no compelling reason to assume the unimolecular formation of $PhCF_2^+$ as intermediate, and nucleophilic attack by water on the conjugate acid could equally account for the experimental facts (Coombes et al., 1969). The mechanism is thus not yet established.

5.4.2. HYDRATION AND DEHYDRATION REACTIONS

Hydration and dehydration reactions are often reversible, and by the principle of microscopic reversibility must occur in both directions through the same set of steps and intermediates. For practical reasons the study of a hydration may be easier than that of its reverse, dehydration, or vice versa, especially when the position of equilibrium is far to one side, but the kinetic data obtained for either reaction provide information about the same mechanism of interconversion. For this reason the hydration and the dehydration reactions will be discussed together in this section.

The most extensively studied reactions of this type are the hydration of multiple bonds and its reverse, the dehydration of alcohols. They have been discussed recently by Willi (1965), but have since been further elucidated. They will be discussed first. Concentrated sulphuric acid has long been known as a dehydrating agent. Of the many reactions of this type a certain number have been studied kinetically in the more recent past and their mechanisms formulated. These will be discussed in the latter part of the section.

5.4.2.1. Interconversion of Olefins and Alcohols

The hydration of olefins in acid solution may be accompanied by a number of other reactions, such as polymerizations, rearrangements and the formation of sulphuric esters, but in moderately concentrated acid and at low olefin concentrations the rates of esterification and polymerization may be made small. All these reactions were believed for a long time to involve carbonium ion intermediates (Hammett, 1940), and recent studies fully confirm this view. Systematic work on this reaction was initiated by Hammett and his collaborators (Levy et al., 1951; Taft et al., 1952) and has been further developed mainly by R. W. Taft, Jr. In more recent years several groups of investigators have made further significant contributions.

The fact that isobutene and trimethylethylene undergo hydration to t-butanol and t-amyl alcohol in dilute acid (0·1–1 M) at room temperature, whereas propylene reacts at a measurable rate in 8 M perchloric acid and ethylene only in 12 M acid (Purlee and Taft, 1956) is strongly suggestive of the existence of a carbonium ion intermediate in the reaction, because the stability of carbonium ions resulting from the above olefins should decrease in the order tertiary > secondary > primary (see Chapter 3). Whether the carbonium ions are formed in a pre-equilibrium or in a slow proton transfer step has been a matter of some uncertainty. Either possibility is consistent with acid catalysis, which was demonstrated for several aliphatic olefins for which slopes of $\log k_{hydr}$ vs. $-H_0$ close to unity were observed (Table 5.5). A

TABLE 5.5

Equilibrium and rate data for the hydration of some olefins

Olefin	$K = \dfrac{[\text{alcohol}]}{[\text{olefin}]}$ †	k_{rel} (30°C)‡	Slope $\log k_{hydr}$ vs. $-H_0$	Reference§
Isobutene	7·46 × 10³	1	1·07	a
2-Methyl-2-butene (trimethylethylene)		0·67	0·98	b
2-Methyl-1-butene		1·30		
2,3,3-Trimethyl-1-butene		1·20	0·99	b
Cyclohexene	~4			c
1-Methyl-1-cyclopentene	54·5	2·29		
1-Methyl-1-cyclobutene	>2 × 10⁵	0·1	1·25	b
Methylenecyclobutane	>1·5 × 10⁶	0·60	1·11	b

† The value for isobutene is that of Eberz and Lucas (1934). Most other values have been obtained from this value and the estimates of equilibrium constants relative to that system as summarized by Riesz et al. (1957).
‡ Data based largely on the values collected by Riesz et al. (1957).
§ a. Taft (1952); b. Taft et al. (1955); c. Gold and Satchell (1963).

slow proton transfer step requires that the reaction should be general acid catalysed. Only recently Schubert et al. (1964) have succeeded in demonstrating that the hydration of styrene is general acid catalysed. The detection of general acid catalysis may be difficult when the Brønsted coefficient α is large. Gold and Kessick (1965) determined this coefficient for the hydration of isobutene from a study of this reaction in H_2O/D_2O mixtures and found $\alpha = 0.85 \pm 0.1$. Such a high value of α indicates that proton transfer is far advanced in the transition state, but not complete.

A further requirement of a slow proton transfer step is that the reaction should be faster in ordinary water than in heavy water. This is the case for isobutene with $k_H/k_D = 1\cdot45 \pm 0\cdot1$ (Gold and Kessick, 1965). Propene is also more rapidly hydrated in 90 per cent aqueous formic acid than in tritiated formic acid ($k_H/k_T \simeq 4\cdot5$)—Coe and Gold (1960). The solvent isotope effect in the hydration of several styrenes, $k_H/k_D = 2$–4, also indicates a slow proton transfer step (Schubert et al., 1964).

All this evidence points to a slow formation of a carbonium ion as the first step in the hydration of olefins. For instance, for isobutene we may write

$$CH_3-\underset{\underset{CH_3}{|}}{C}=CH_2 + H_3O^+ \xrightarrow{slow} CH_3-\underset{\underset{CH_3}{|}}{C^+}-CH_3 + H_2O$$

with a transition state representing an incompletely formed carbonium ion in which the developing carbonium centre may be solvated by water molecules, as in (5.9). The fact that the rates of hydration of olefins follow the H_0 rather

$$\begin{array}{c} CH_3 \\ \diagdown{}^{\delta+} \\ C-CH_2 \\ CH_3\diagup \vdots \vdots \\ H_2\ddot{O}\cdots H^{\delta+} \end{array}$$

(5.9)

than the H_R acidity function (Table 5.5), led Boyd et al. (1960) to suggest that the carbonium ion formed may not be "free", but may remain in interaction with the leaving water molecule. This does not necessarily mean that the addition of the proton and the water molecule occur simultaneously in the hydration reaction. Studies of the dehydration and oxygen exchange of alcohols, to be discussed below, provide proof that this is not so, but that the reaction of the carbonium ion with water is another slow step, although much faster than the proton transfer to the olefin.

There are some indirect indications that the carbonium ion is formed as an intermediate in the reaction, through a transition state which bears no resemblance to the reaction product—the alcohol. It can be seen from Table 5.5, for example, that the equilibrium strongly favours the alcohol for all aliphatic olefins, but especially for 1-methyl-1-cyclobutene and methylenecyclobutane, where the formation of the alcohol leads to a considerable reduction in ring strain, owing to a change from sp^2 hybridization to sp^3. The relative reaction rates do not reflect a similar effect, as both compounds react more slowly than isobutene. This is consistent with the formation of an intermediate in which ring strain is not released (the hybridization in the carbonium ion remains sp^2). Also the entropies of activation for the hydration of isobutene and other olefins are mostly slightly negative, $\Delta S^{\ne} = +4\cdot2$ to $-33\cdot5$ J K^{-1} mol^{-1} (Taft and Riesz, 1955), whereas the entropy changes for

5. REACTION MECHANISMS IN SULPHURIC ACID SOLUTIONS   213

the overall reaction are much more negative, with $\Delta S \simeq -84$ J K^{-1} mol^{-1} (Lucas and Eberz, 1934; Lucas and Liu, 1934), also showing that the transition state is not closely similar to the final product.

Some data on the hydration of styrenes, which have been studied more recently, are collected in Table 5.6. All styrenes are hydrated in moderately concentrated acid, although they are capable of forming more stable carbonium ions than aliphatic olefins from Table 5.5 which are readily hydrated in dilute acid. This is probably due to the greater stability of the styrenes themselves in which the olefinic bond is conjugated with the benzene ring. This also accounts for the fact that the alcohol is not so strongly favoured at equilibrium in these systems (the styrene is indeed favoured in compounds 5 and 6, Table 5.6). Rates of hydration of ring substituted styrenes and α-methyl styrenes correlate with σ^+ constants of the substituents, giving reaction constants, ρ, respectively, $-3\cdot42$ (Schubert et al., 1964) or $-4\cdot00$ (Durand et al., 1966), and $-3\cdot21$ (Deno et al., 1965) or $-3\cdot27$ (Durand et al., 1966). The large negative values of ρ suggest a large degree of carbonium ion character in the transition state. Also the slope of $\log k_{\text{hydr}}$ vs. $-H_0$ for styrene is considerably greater than one, owing to the carbonium ion character of the transition state (Schubert et al., 1964). The correlation with H_R' is almost as good (slope 0·71). Such behaviour is not found in styrenes containing the carboxyl group (Table 5.6), possibly owing to the hydration of the carboxyl group, which need not be the same in the initial and the transition states.

It was mentioned in the introduction that rearrangements may accompany the hydration of olefins, but that these reactions are slow. Several demonstrations that rearrangements are either absent or extremely slow lend strong support to the protonation of olefins as the slow step in their hydration, because the reverse reaction, the loss of a proton, must be slow if there are no rearrangements in cases where more than one olefin may be formed from a given carbonium ion. For example, both 2-methyl-2-butene and 2-methyl-1-butene form the same carbonium ion on protonation, but when one of them is hydrated, the presence of the other isomer cannot be detected up to 50 per cent conversion into the alcohol (Levy et al., 1953). Likewise, when hydration is studied in a deuterated solvent, any reformation of the olefin in the course of the reaction should result in some incorporation of deuterium in the olefin, e.g.

$$\underset{CH_3}{\overset{H}{>}}C=C\underset{CH_3}{\overset{CH_3}{<}} \xrightarrow{D^+} H-\underset{CH_3}{\overset{D}{\underset{|}{C}}}-\overset{+}{C}\underset{CH_3}{\overset{CH_3}{<}} \xrightarrow{-H^+} \underset{CH_3}{\overset{D}{>}}C=C\underset{CH_3}{\overset{CH_3}{<}}$$

but this is not found after the first half-life of the hydration (Purlee and Taft, 1956). Also the hydration of optically active 3-p-menthene is not accompanied

TABLE 5.6

Equilibrium data and acidity dependence for the hydration of styrenes at 25°C

Styrene	Alcohol	$K_e = \dfrac{[\text{Styrene}]}{[\text{Alcohol}]}$	Slope $\log k_{\text{hydr}}$ vs. $-H_0$	Acid concentration range	References§
(1) $\underset{\text{H}}{\overset{C_6H_5}{>}}C=CH_2$	$C_6H_5\underset{\text{H}}{\overset{CH_3}{-}}C-OH$	$(2\cdot3-2\cdot6) \times 10^{-2}$	1·23	4·6–7·4 M $HClO_4$	a
(2) $\underset{\text{D}}{\overset{C_6H_5}{>}}C=CH_2$	$C_6H_5\underset{\text{D}}{\overset{CH_3}{-}}C-OH$	$1\cdot8 \times 10^{-2}$		4·6–7·4 M $HClO_4$	a
(3) $\underset{CH_3}{\overset{C_6H_5}{>}}C=CH_2$	$C_6H_5\underset{CH_3}{\overset{CH_3}{-}}C-OH$	$\leqslant 0\cdot1$	1	5–40% H_2SO_4	b
(4) $\underset{H}{\overset{C_6H_5}{>}}C=\underset{H}{\overset{COOH}{<}}$	$C_6H_5\underset{H}{\overset{CH_2COOH}{-}}C-OH$	†	1	46–76% H_2SO_4	c

5. REACTION MECHANISMS IN SULPHURIC ACID SOLUTIONS

	Alkene	Alcohol	Ratio	H_2SO_4	Ref.
(5)	C_6H_5-CH=CH-COOH (H, H trans)	C_6H_5-C(OH)(H)-CH$_2$COOH	58	‡	52% H_2SO_4	d
(6)	C_6H_5-C(CH$_3$)=CH-COOH (H trans)	C_6H_5-C(OH)(CH$_3$)-CH$_2$COOH	2·2	1	47–60% H_2SO_4	e
(7)	C_6H_5-C(CH$_2$COOH)=CH$_2$	C_6H_5-C(OH)(CH$_3$)-CH$_2$COOH	0·142	1	39–49% H_2SO_4	e

† The alcohol in this system undergoes fast dehydration to *trans*-cinnamic acid (Noyce and Lane, 1962), so that the observed rate of isomerization equals the rate of hydration.
‡ Hydration not studied owing to the unfavourable position of equilibrium.
§ *a.* Schubert *et al.* (1964); *b.* Deno *et al.* (1965); *c.* Noyce *et al.* (1962a); *d.* Noyce and Lane (1962); *e.* Noyce and Heller (1965).

by racemization, which should be the case, if there was a reformation of the olefin from the carbonium ion (Kwart and Weisfeld, 1958). All these results indicate that the carbonium ion once formed rapidly reacts to produce the alcohol. The fact that the whole reaction is reversible, i.e. that alcohols are dehydrated to olefins, is proof that the conversion of carbonium ion to olefin does occur, but in a slow step, which is the reversal of the slow protonation of the olefin.

As the reaction steps which follow the slow formation of the carbonium ion are fast by comparison, information about them can only be obtained from studies of the reverse reaction, the dehydration of alcohols. Considerable advance in the understanding of this reaction was achieved when oxygen isotope exchange and racemization were studied in the course of the dehydration reaction (Samuel and Silver, 1965). Both oxygen exchange and racemization are acid-catalysed, i.e. they occur via the conjugate acid of the alcohol, and therefore involve the displacement of one water molecule by another, according to the equation

$$R_3COH_2^+ + H_2O^* \rightleftharpoons R_3C\overset{*}{O}H_2^+ + H_2O \tag{5.45}$$

The water molecules act here as nucleophiles and the displacement may be imagined to occur either as a unimolecular reaction (S_N1) in which the conjugate acid dissociates to give a carbonium ion and the leaving water molecule equilibrates with the solvent before recombination occurs, or as a bimolecular nucleophilic attack (S_N2) in which one water molecule displaces another in a single step. The two mechanisms lead to different expectations for the relative rates of oxygen exchange and racemization for optically active alcohols. In an S_N1 displacement the recombining water molecule may enter from either side of the carbonium ion with equal probability, and hence $k_{exch} = k_{rac}$. In an S_N2 displacement the attacking water molecule enters on the side opposite to the leaving water molecule, and hence inversion of configuration occurs at each exchange step, which means that $k_{rac} = 2k_{exch}$. Some experimentally found values are given in Table 5.7. All the reactions follow the H_0 acidity function at higher acidity. The k_{rac}/k_{exch} result for s-butyl alcohol is consistent either with an S_N2 displacement, or with an S_N1 reaction in which the leaving water molecule maintains an interaction with the carbonium ion formed, thus "shielding" it from one side, so that the attacking water molecule can only enter from the other side. The carbonium ion formed in such a reaction could not be regarded as "free", but is envisaged as (5.10)

$$\begin{array}{c} OH_2 \\ CH_3 \diagdown \vdots \\ C^{\pm}\!\!-\!\!H \\ C_2H_5 \diagup \vdots \\ OH_2 \end{array}$$

(5.10)

TABLE 5.7

Relative rates of racemization and oxygen exchange of some optically active alcohols in acid solution

Alcohol	k_{rac}/k_{exch}	Slopes log k vs. $-H_0$	Reference†
(1) $\begin{array}{c}CH_3\\C_2H_5\end{array}\!\!\!>\!CHOH$ (s-butyl alcohol)	2·0	1	a
(2) $\begin{array}{c}CH_3\\C_6H_5\end{array}\!\!\!>\!CHOH$ (1-phenylethanol)	1·22		b
(3) $\begin{array}{c}C_6H_5\\p\text{-MeOC}_6H_4\end{array}\!\!\!>\!CHOH$ (p-methoxy benzhydrol)	1·00	1·3	c

† a. Bunton et al. (1955a); b. Grunwald et al. (1957); c. Bunton et al. (1958c).

and has been variously described as "shielded" or "encumbered". It should be noted that the stability of the carbonium ions formed from the alcohols in Table 5.7 increases down the series, and the idea of a "shielded" carbonium ion can account for all the results if it is assumed that the lifetime of the carbonium ion increases in that order. The longer the lifetime of the carbonium ion, the more complete the equilibration of the water molecules solvating it with the bulk of the solvent, and the smaller the probability of a recapture of the leaving water molecule. The straight S_N2 substitution is unlikely on steric grounds.

The possibility that oxygen exchange of alcohols may occur by a dehydration-rehydration mechanism has been ruled out by a demonstration that oxygen exchange for t-butanol is 20–30 times faster than dehydration to isobutene (Dostrowsky and Klein, 1955). Similar results have been found for other alcohols. For tertiary alcohols both the rates of exchange and of dehydration follow the H_0 acidity function with a slope of 1·2 (Boyd et al., 1960), suggesting that both reactions occur through the same carbonium ion intermediate.

The direct study of the dehydration of t-butyl alcohol is difficult because the reaction proceeds only to a very slight extent as can be seen from Table

5.5 (Taft et al., 1955), but that of t-amyl alcohol could be studied directly and was found to lead to two olefins:

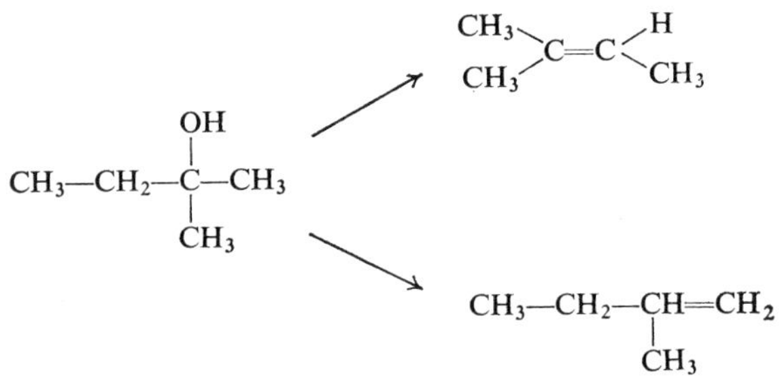

$$CH_3-CH_2-\underset{\underset{CH_3}{|}}{\overset{\overset{OH}{|}}{C}}-CH_3 \quad\begin{array}{l}\nearrow \quad \overset{CH_3}{\underset{CH_3}{>}}C=C\overset{H}{\underset{CH_3}{<}} \\ \\ \searrow \quad CH_3-CH_2-\underset{\underset{CH_3}{|}}{CH}=CH_2\end{array}$$

These form in the relative amounts 5:1 (Riesz et al., 1957; Boyd et al., 1960). Both these olefins give the same carbonium ion on protonation and may be assumed to be formed through such an intermediate.

An ingenious indirect method for the determination of the rate of dehydration of aliphatic alcohols has recently been developed by Gold and Satchell (1963) in an attempt to explain the mechanism of CH exchange of alcohols. It has been suggested that exchange may take place in the carbonium ion (Setkina and Kursanov, 1958), but Gold and Satchell (1963) and Gold and Gruen (1966) have shown that it takes place by a dehydration-rehydration mechanism via the olefin. As the position of equilibrium (Table 5.5) indicates that the hydration of isobutene is much faster than the dehydration of t-butanol, the rate of tritium incorporation gives a measure of the rate of dehydration. The $\log k$ vs. $-H_0$ plot was found to be linear with a slope of 1·35 (Gold and Gruen, 1966).

On the basis of all this information the mechanism for the interconversion of alcohols and olefins may be represented by the following scheme:

$$R_3COH + H^+ \underset{fast}{\rightleftharpoons} R_3COH_2^+ \text{ (pre-equilibrium)}$$

$$R_3COH_2^+ \underset{k_{-1}}{\overset{k_1}{\rightleftharpoons}} R_3C^+ + H_2O \text{ (ionization)}$$

$$R_3C^+ + H_2O \underset{k_{-2}}{\overset{k_2}{\rightleftharpoons}} \text{Olefin} + H_3O^+ \text{ (elimination)}$$

The application of the steady state treatment to the carbonium ion intermediate gives its steady state concentration as

$$[R_3C^+] = \frac{k_1[R_3COH_2^+]}{(k_{-1} + k_2)[H_2O]} \tag{5.46}$$

5. REACTION MECHANISMS IN SULPHURIC ACID SOLUTIONS

From the expression for the equilibrium constant for the preprotonation (K_{BH^+}) we obtain

$$[R_3COH_2^+] = \frac{h_0}{K_{BH^+}}[R_3COH] \tag{5.47}$$

Assuming a small degree of conversion of the alcohol to its conjugate acid, its concentration may be replaced by its stoichiometric concentration, c_s:

$$[R_3COH_2^+] = \frac{h_0}{K_{BH^+}} c_s \tag{5.48}$$

The carbonium ion concentration then becomes

$$[R_3C^+] = \frac{k_1}{(k_{-1}+k_2)} \cdot \frac{h_0}{K_{BH^+}} \frac{c_s}{[H_2O]} \tag{5.49}$$

According to the above reaction scheme, the carbonium ion is partitioned between a faster recombination reaction (k_{-1}), which leads to oxygen exchange and racemization, and a slower elimination reaction (k_2), which leads to the formation of the olefin. The rate of olefin formation is given by

$$\frac{d[\text{Olefin}]}{dt} = k_2[R_3C^+][H_2O] = k_{\text{dehydr}} c_s \tag{5.50}$$

and by substituting (5.49) for the carbonium ion concentration we obtain

$$k_{\text{dehydr}} = \frac{k_1 k_2}{(k_{-1}+k_2)} \frac{h_0}{K_{BH^+}} \tag{5.51}$$

If $k_{-1} \gg k_2$, this simplifies to

$$k_{\text{dehydr}} = \frac{k_1 k_2}{k_{-1}} \frac{h_0}{K_{BH^+}} \tag{5.52}$$

showing that in such a case the ionization step may also be regarded as a pre-equilibrium.

The rate of oxygen exchange is given by

$$\frac{d[R_3C\overset{*}{O}H]}{dt} = k_{-1}[R_3C^+][H_2O] = k_{\text{exch}} c_s \tag{5.53}$$

and again by substituting for the carbonium ion concentration, this leads to

$$k_{\text{exch}} = \frac{k_{-1} k_1}{(k_{-1}+k_2)} \frac{h_0}{K_{BH^+}} \tag{5.54}$$

and finally for $k_{-1} \gg k_2$ we obtain

$$k_{\text{exch}} = k_1 \frac{h_0}{K_{BH^+}} \tag{5.55}$$

so that the rate of exchange is primarily determined by the rate of ionization of the conjugate acid to give the unstable carbonium ion.

The expressions for the rate of dehydration and the rate of exchange show first of all that these reactions follow the h_0 acidity scale simply because of the preprotonation step, and slopes of log k vs. $-H_0$ close to unity found experimentally suggest therefore that alcohols behave as Hammett bases. Unfortunately this point could not be established with certainty from protonation studies (see Chapter 3). The properties of the transition states in these two reactions do not affect the slopes. Secondly, the relative rate of the two reactions is given by

$$k_{\text{exch}}/k_{\text{dehydr}} = k_{-1}/k_2 \qquad (5.56)$$

i.e. it is determined by the rate of recombination (k_{-1}) and the rate of elimination (k_2) from the carbonium ion intermediate. Both paths go via transition states, the first of which corresponds to the binding of one of the water molecules solvating the carbonium ion, and the second to a structure such as (**5.9**), involving the breaking of a C—H bond. Boyd *et al.* (1960) have estimated that the slower rate of dehydration of *t*-butanol compared with the rate of exchange corresponds to a difference in activation energies

$$E_a^{\text{elim}} - E_a^{\text{exch}} = 13 \cdot 0 \pm 2 \cdot 5 \text{ kJ mol}^{-1}$$

whereas for *t*-amyl alcohol this is only $4 \cdot 2 \pm 2 \cdot 5$ kJ mol^{-1}.

Expressions for k_{dehydr} and k_{exch} clearly show that experimental entropies of activation for dehydration and exchange are complex quantities, whose interpretation must be taken with reserve, until more is known about individual terms constituting them.

This analysis shows that most of the facts on the rates of the alcohol/olefin interconversions can be accounted for satisfactorily in terms of the above three-step mechanism. Certain questions, however, remain unanswered. The fact that dehydration, oxygen exchange, racemization and tritium exchange follow the acidity function H_0 up to quite high acid concentrations suggests that the preprotonation reaction is far from reaching completion, and implies much more negative pK_{BH^+} values for alcohols than most determinations to date suggest (see Chapter 3). This was pointed out by both Noyce and Lane (1962) and Gold and Satchell (1963). Substituent effects are only incompletely clarified so far. The expressions for the rates of exchange and dehydration are both complex, involving a number of constants, which would in general depend on the structure of the substrate in different ways. While the higher basicities of *tert*-alcohols and the greater stability of *tert*-carbonium ions would both tend to enhance the rates of substitution (racemization and exchange) for these alcohols compared with secondary alcohols, the rates of dehydration also strongly depend on the rate of the elimination step. A comparison of the rate of racemization of *sec*-butanol ($\log k/s^{-1} = -10\cdot 0$) and the rate of ^{18}O-exchange of *tert*-butanol ($\log k/s^{-1} = -5\cdot 31$), both in 1 M acid, illustrates the first point (Noyce and Lane, 1962). For secondary carbonium ions elimination is known to be favoured over substitution (Manassen and Klein, 1960). Substituent effects on the rates of dehydration of ring-substituted 2-phenyl-2-propanols are correlated satisfactorily with σ^+ substituent constants with $\rho = -3\cdot 05$, as compared with $\rho = -3\cdot 21$ in the hydration of the corresponding propenes (Deno *et al.*, 1965), showing the dominant effect of the carbonium ion formation upon the reaction rate. The dehydration of β-phenyl-β-hydroxypropionic acids is even more sensitive to substituent effects with $\rho = -4\cdot 6$ (Noyce *et al.*, 1962b).

A reaction closely similar to the addition of water on to double bonds is the addition of alcohols to give ethers, whereas ether formation from alcohols is a substitution on a saturated carbon, like oxygen exchange or racemization in alcohols. Such reactions (in ethanol containing 15 per cent sulphuric acid) involving 2-phenylpropenes on the one hand, and 2-phenyl-2-propanol on the other, have recently been studied by Deno *et al.* (1965). The formation of the ether from the alcohol is 105 times faster than its formation from the olefin, as would be expected in view of the analogy of the mechanism with the three-step mechanism discussed above.

A reaction analogous in a different sense to the dehydration of alcohols is the acid catalysed cleavage of xanthyl-alcohols, in which the xanthyl substituent is carried on a carbon atom adjacent to the carbon carrying the OH group, e.g. XCH_2CH_2OH, where X = xanthyl. Whether the alcohol is primary, secondary or tertiary, the xanthyl cation is ejected from the protonated alcohol (in 10–50 per cent sulphuric acid at 25°C) according to the equation

$$XCH_2CH_2OH + H^+ = X^+ + CH_2{=}CH_2 + H_2O \tag{5.57}$$

with the formation of the olefin (Deno and Sacher, 1965). The ease with which these reactions occur is ascribed to the delocalization of the positive charge into the xanthyl group, which forms an extremely stable xanthyl cation. The triphenylmethyl cation is not ejected under similar conditions. Deno and Sacher (1965) suggest that the transition state for this reaction be pictured as $X \cdots CH_2 {=\!=} CH_2 \cdots OH_2^+$, with the positive charge delocalized throughout the system, implying a simultaneous breaking of X—C and C—O bonds. Support for this view comes from the fact that these reactions follow the H_R acidity function.

5.4.2.2. *The Hydration of* α,β-*Unsaturated Aldehydes and Ketones*

The conversion of α,β-unsaturated aldehydes and ketones into aldols and ketols is a well-known acid-catalysed reaction (Wurtz, 1884), which occurs in dilute and moderately concentrated acid and leads to an equilibrium.

Winstein and Lucas (1937) pointed out that in the series of similar compounds, 2-butene, crotyl alcohol, crotonaldehyde, and crotonic acid, only crotonaldehyde hydrates at a measurable rate in dilute aqueous acid at room temperature. The conjugation between the carbonyl group and the α,β-double bond is clearly of fundamental importance and probably leads to a mechanism of hydration different from that of other olefinic compounds. The reaction was studied for several aliphatic compounds by Lucas and his collaborators (1937–44) and their results are collected in Table 5.8. Acid catalysis was well established, but later extension of the study of the hydration of mesityl oxide and crotonaldehyde into more concentrated solutions showed that these reactions do not follow the H_0 acidity function, but give curved plots of slopes less than unity throughout (Bell et al., 1962). In terms of current views on acidity functions (Chapter 2) this merely means that the rates follow an acidity function which decreases less steeply with acid concentration than does H_0. The rates are higher in sulphuric acid solutions than in perchloric, hydrochloric and nitric acids of the same acidity. Bell et al. (1962) conclude that this is due to general acid catalysis by the hydrogen sulphate ion and support this view by the finding of a large hydrogen isotope effect ($k_H/k_D = 3\cdot4$–4). All these facts can be accounted for in terms of a mechanism proposed by Noyce and Reed (1958). The enhanced basicity of α,β-unsaturated carbonyl compounds has been discussed in Chapter 3. There is hence a rapid preprotonation step:

$$R_2C{=}CH{-}\underset{\underset{R}{|}}{C}{=}O + H_3O^+ \rightleftharpoons H_2O + \left[R_2C{=}CH{-}\underset{\underset{R}{|}}{\overset{+}{C}}{=}OH \leftrightarrow R_2\overset{+}{C}{-}CH{=}\underset{\underset{R}{|}}{C}{-}OH \right]$$

(5.11)

The mesomeric cation **(5.11)** then rapidly reacts with water giving an unsaturated dialcohol **(5.12)**:

$$R_2\overset{+}{C}{-}CH{=}\underset{\underset{R}{|}}{C}{-}OH + H_2O \xrightarrow{\text{fast}} R_2\underset{\underset{OH}{|}}{C}{-}CH{=}\underset{\underset{R}{|}}{C}{-}OH + H^+$$

(5.12)

This is followed by a slow proton transfer to the unsaturated central carbon atom, i.e.

$$HO.CR_2{-}CH{=}CR{-}OH + \text{Acid} \xrightarrow{\text{slow}} R_2\underset{\underset{OH}{|}}{C}{-}CH_2{-}\underset{\underset{R}{|}}{C}{=}OH^+ + \text{Base}$$

(5.13)

leading to the formation of the conjugate acid of the aldol or ketol **(5.13)**. This step is in fact an acid-catalysed keto-enol isomerization, and such reactions are well known to involve general acid catalysis and to show large

TABLE 5.8

Rate constants and equilibrium constants for the hydration of some α,β-unsaturated carbonyl compounds

Compound	Acid	Temperature (°C)	k_{hydr} (s^{-1})	k_{dehydr} (s^{-1})	$K = \dfrac{[\text{alcohol}]}{[\text{olefin}]}$	Reference[†]
(1) *trans*-Croton-aldehyde	1·905 M HNO_3	25	$1·25 \times 10^{-5}$	$1·58 \times 10^{-5}$	0·80	a
	1·806 M HNO_3	35	$3·17 \times 10^{-4}$	$5·16 \times 10^{-4}$	0·61	a
(2) β,β-Dimethyl-acraldehyde	1·04 M HNO_3	25	$3·75 \times 10^{-6}$	$9·61 \times 10^{-6}$	0·39	b
	1·04 M HNO_3	35	$1·042 \times 10^{-5}$	$3·13 \times 10^{-5}$	0·33	b
(3) Mesityl oxide	1 M $HClO_4$	25	$7·83 \times 10^{-5}$	$5·36 \times 10^{-6}$	$14·5 \pm 1$	c
(4) Crotonic acid	1 M $HClO_4$	90	$2·1 \times 10^{-6}$	$4·05 \times 10^{-7}$	5·0	d

† *a*. Winstein and Lucas (1937); *b*. Lucas *et al.* (1944); *c*. Pressman *et al.* (1942); *d*. Pressman and Lucas (1939).

hydrogen isotope effects. The final step is the fast loss of the proton to regenerate the catalyst:

$$\underset{R_2C-CH_2-C=OH^+}{\overset{OH\quad\ R}{|\qquad\quad|}} + \text{Base} \rightleftharpoons \underset{R_2C-CH_2-C=O}{\overset{OH\quad\ R}{|\qquad\quad|}} + \text{Acid}$$

Noyce and Reed (1958) suggested this mechanism on the basis of a study of the dehydration of 4-phenyl-4-hydroxy-2-butanone and its p-nitro and p-methoxy derivatives. The rate of dehydration of this latter compound only was found to follow the H_0 acidity function (slope 1·07), the other two showing a smaller dependence on acidity like the aliphatic systems discussed above. Noyce and Reed (1958) believed therefore that the p-methoxy derivative dehydrated by a carbonium ion mechanism, such as has been discussed for other alcohols in Section 5.4.2.1, but possibly too much significance is attached to this difference in the H_0 dependence of reaction rates.

The rate of dehydration of β-phenyl-β-hydroxypropiophenone is also determined by the rate of enolization and is somewhat less than the rate of dehydration of 4-phenyl-4-hydroxy-2-butanone, analogously to the slower enolization of acetophenone as compared with acetone (Noyce et al., 1959).

5.4.2.3. *Hydration of Triple Bonds*

The hydration of triple bonds is considerably facilitated by the presence of groups which can stabilize carbonium ions by delocalization of the positive charge. Thus 1-alkynyl ethers (**5·14**) and 1-alkynyl thioethers (**5.15**) (where R = alkyl or H, and R' = alkyl)

$$R-C\equiv C-OR' \qquad R-C\equiv C-SR'$$
$$(5.14) \qquad\qquad (5.15)$$

hydrate in aqueous buffer solutions and in very dilute acid to esters and thioesters (Drenth and Hogeveen, 1960; Hogeveen and Drenth, 1963a). It was possible to demonstrate both general acid catalysis (Drenth and Hogeveen, 1960; Hekkert and Drenth, 1961) and a solvent isotope effect $k_H/k_D = 1\cdot 7$ (Stamhuis and Drenth, 1963), consistent with a slow proton transfer to form carbonium ions (**5.16**) and (**5.17**). When R = H, the hydrogen-deuterium

$$\underset{R}{\overset{H}{\diagdown}}C\!\!=\!\!\overset{+}{C}\!-\!OR' \qquad \underset{R}{\overset{H}{\diagdown}}C\!\!=\!\!\overset{+}{C}\!-\!SR'$$
$$(5.16) \qquad\qquad (5.17)$$

exchange of this acetylenic hydrogen is also observed (Hogeveen and Drenth, 1963b). Ethers hydrate more rapidly than thioethers, owing to the greater capacity of oxygen as compared with sulphur to coordinate with the carbonium ion centre.

Phenyl groups are less able to stabilize a carbonium ion and phenyl acetylenes hydrate only in the moderately concentrated sulphuric acid range (20–70 per cent). The reaction was studied for compounds (**5.18**) (R = H, D,

Ph—C≡C—R   X—C₆H₄—C≡C—COOH

(**5.18**)   (**5.19**)

and CH_3) and for *p*-substituted phenylpropiolic acids (**5.19**) by Noyce *et al.* (1965). Large solvent isotope effects were found, $k_H/k_D = 2.98$ (R = H) and 2·25 (R = CH_3) for (**5.18**), and 3·70–4·07 for (**5.19**), and good correlations of rates with the acidity function H_0. Substituent effects of the *p*-substituent are correlated by the σ^+ constants, with $\rho = -4·79$, indicating the pronounced carbonium ion character of the benzylic carbon in the transition state.

All these experimental findings support the following carbonium ion mechanism:

$$\text{Ar—C≡C—COOH} + H_3O^+ \xrightarrow{\text{slow}} \text{Ar—}\overset{+}{\text{C}}\text{=CHCOOH} + H_2O$$

$$\text{Ar—}\overset{+}{\text{C}}\text{=CHCOOH} + H_2O \xrightarrow{\text{fast}} \text{Ar—}\underset{\underset{\overset{+}{O}H_2}{|}}{\text{C}}\text{=CHCOOH}$$

$$\text{Ar—}\underset{\underset{\overset{+}{O}H_2}{|}}{\text{C}}\text{=CHCOOH} + H_2O \underset{\text{fast}}{\rightleftarrows} \text{Ar—}\underset{\underset{OH}{|}}{\text{C}}\text{=CHCOOH} + H_3O^+$$

Finally, a keto–enol transformation into benzoylacetic acid takes place, a reaction which is itself acid–base catalysed. The large negative ρ value shows that the first proton transfer is rate determining (Noyce *et al.*, 1965).

A comparison of the rate of hydration of phenylacetylene with that of styrene shows that phenylacetylene is hydrated 2·3 times more rapidly. The difference is still greater with phenylpropiolic acid as compared with *cis*-cinnamic acid, where the acetylenic compound is hydrated 19 times more rapidly. This means that the vinylic cations are more readily formed in moderately concentrated acid solutions than the corresponding saturated carbonium ions (Noyce *et al.*, 1965).

5.4.2.4. *Cyclodehydrations*

A number of reactions, often termed cyclodehydrations or intramolecular condensations, are known to occur in sulphuric acid solutions and other strongly acid media. Such reactions are favoured by the low water activity in the concentrated acids and often go to completion. As their mechanism has become better understood over the past few decades, it has become clear that they are in fact intramolecular electrophilic substitutions by carbonium, oxocarbonium or carboxonium ion centres as electrophiles.

One of the best known examples of such reactions is the cyclodehydration of o-benzoylbenzoic acid to anthraquinone by concentrated sulphuric acid (Gleason and Dougherty, 1929). It was noted early by Hammett and Deyrup (1932) that this reaction follows the H_0 acidity function, but Long and Paul (1957) subsequently pointed out that the correlation is not very good, and that the reaction should follow the H_R (J_0) acidity function if it occurs by the mechanism suggested by Newman (1942):

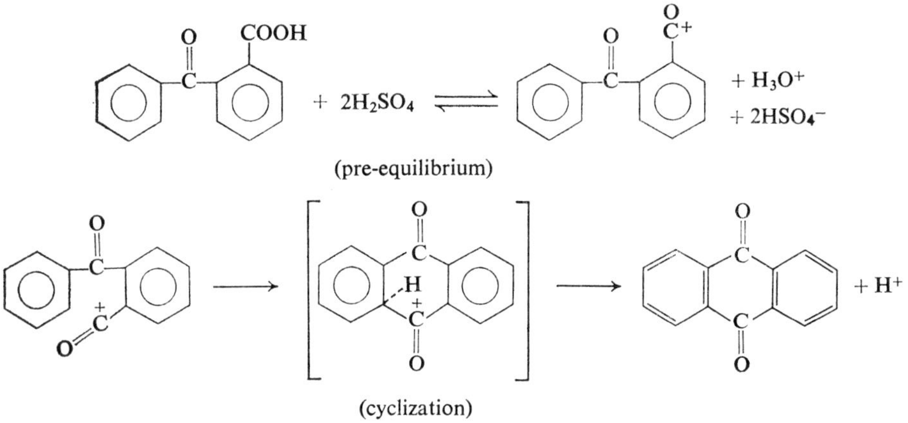

(pre-equilibrium)

(cyclization)

The existence of the pre-equilibrium was deduced from cryoscopic measurements in pure sulphuric acid ($i = 4$) (Newman et al., 1945). Cyclization takes place at higher temperature in the concentrated sulphuric acid region and reaches a limiting rate in 100 per cent sulphuric acid (Deane, 1937), in accordance with the cryoscopic finding that the acid is fully converted into the acylium ion at this acidity. More recently the reaction was re-examined by Vinnik et al. (1959a), who showed that there are in fact two pre-equilibria, the simple protonation of the acid corresponding to a pK_{BH^+} of −8·6 (see Chapter 4), and the complex ionization of the conjugate acid to the oxocarbonium ion at higher acidity. Their results have confirmed that cyclization is the rate-determining step. As the rate reaches a limiting value in the 100 per cent sulphuric acid region, the activation energy for the rate-determining process could be obtained ($E_a = 102·5$ kJ mol^{-1}). Several substituted o-benzoylbenzoic acids were also studied (Vinnik et al., 1962; Vinnik et al., 1963; Ryabova and Vinnik, 1963). Methyl substitution in the ring undergoing electrophilic attack lowers the activation energy, as expected for the above mechanism, whereas NH_3^+ substitution in the other ring has very little effect.

If the reacting species, the oxocarbonium ion, is fully formed in 100 per cent sulphuric acid, it is difficult to explain why the rate of this reaction should show a further steady increase in oleum solutions, such as was found by Deane and Huffman (1943) and by Luder and Zuffanti (1944) at concentrations above 14 per cent "free" sulphur trioxide, and right up to 65 per cent oleum. Luder and Zuffanti (1944) ascribe this to catalysis by sulphur trioxide, but the rôle of this catalyst is not clear in the context of the above mechanism. It seems possible that the increase in rate in oleum solutions is due to a medium effect.

A closely similar reaction, the cyclodehydration of 2,2'-diphenic acid to give fluorenone-4-carboxylic acid (Graebe and Aubin, 1888) also conforms to the above mechanism (March and Henshall, 1962). The rates follow the H_0 acidity function, with slopes in the range 1–1·50, depending on the substituents. The 2,2'-diphenic acid is protonated in >50 per cent sulphuric acid and the protonation is complete in 99 per cent acid, the slope of the linear portion of the $\log I$ vs. $-H_0$ plot being only 0·57. The reason for this is not clear. The fraction of the protonated acid which is converted to the oxocarbonium ion according to the equation

$$\text{(diphenic acid with COOH and } C(OH)_2^+ \text{)} \rightleftharpoons \text{(diphenic acid with COOH and } CO^+ \text{)} + H_2O \quad (5.58)$$

is small in this acidity range and its concentration should therefore be proportional to that of the protonated acid. If the degree of protonation is α and the oxocarbonium ion is the reacting species, a plot of $\log k$ vs. $\log(\alpha/a_{H_2O})$ should be linear. This was in fact found to be so (March and Henshall, 1962). The concentration dependence of the observed activation energy for 2,2'-diphenic acid shows a steady decrease with a limiting value in the region of full protonation, and a further sharp drop between 99·5 and 100 per cent acid, the concentration range in which reaction (5·58) goes to completion. Thus the heat of protonation ($\Delta H_1 = 21$ kJ mol^{-1}) and the heat of complex ionization ($\Delta H_2 = 29$ kJ mol^{-1}) could both be deduced in addition to the true energy of activation ($E_a = 119$ kJ mol^{-1}). Several 5,5'-substituted acids were also studied (March and Henshall, 1962). Electron-releasing substituents accelerate, while electron-attracting substituents retard the reaction. The 5-substituent is *para* to the oxocarbonium group, whereas the 5'-substituent becomes *ortho* to the site of cyclization on the second ring, and would therefore be expected to be predominantly responsible for the observed substituent effect on rate. Since the electronic effects of *ortho* substituents are closely similar to those of *para* substituents, and steric effects in this reaction must be small, $\log k$

values were plotted against σ_p^+ constants of the substituents in Fig. 5.4. With the exception of the chloro substituent, a good correlation is obtained with a large negative slope (−5·5), which is characteristic of electrophilic substitution reactions, although it reflects here the effect of substituents in both rings.

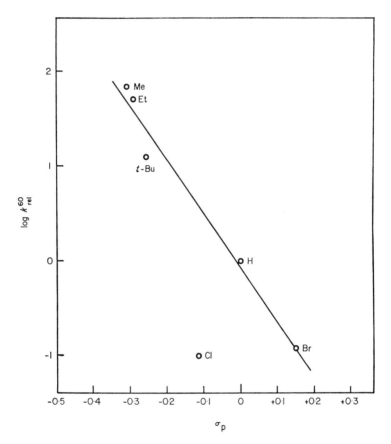

Fig. 5.4. The correlation of the rates of cyclization of 5,5′-substituted-2,2′diphenic acids with σ_p^+ constants of the substituents.

A question of interest in any electrophilic substitution is whether electrophilic attack results in the formation of a σ complex, or in a simultaneous formation of the new bond and the breaking of the C—H bond in the rate-determining step. This question may be answered by isotopic substitution of hydrogen by deuterium (see Section 5.4.6). An attempt to apply this method to reactions of this kind was made by Denney and Klemchuk (1958). When

2-deuterio-2′-carboxybiphenyl is cyclized to fluorenone, both 4-deuterio-fluorenone and fluorenone are formed:

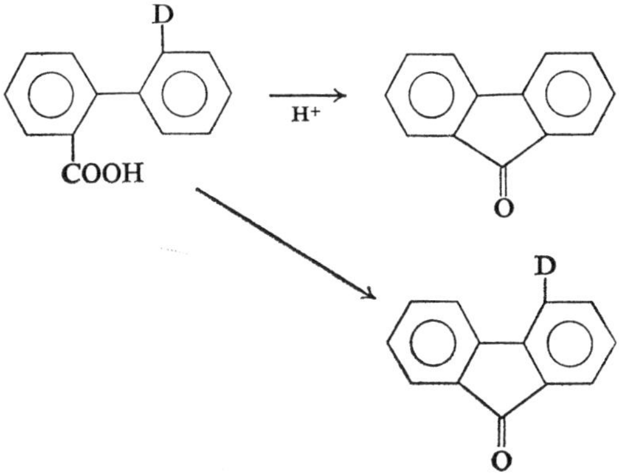

An analysis of the reaction product showed that the ratio of reaction rates for the hydrogen and the deuterium displacement is $k_H/k_D = 1\cdot31 \pm 0\cdot03$ (Denney and Klemchuk, 1958). There is thus a definite hydrogen isotope effect in this reaction, which shows that hydrogen is more readily displaced than deuterium, and that this occurs in the rate-determining step. This means that σ complexes are not formed in these systems, presumably because their formation is strongly disfavoured by geometrical restrictions (the planarity of transition states and products).

A further example of a cyclodehydration involving intramolecular attack on the benzene nucleus, but by a carboxonium ion, is the cyclodehydration of 4-anilino-pent-3-en-2-ones (anils) (Bonner et al., 1955; Bonner and Barnard, 1958). In both concentrated sulphuric and perchloric acid solutions the rates follow the H_0 acidity function, with slopes close to unity. The nitrogen being fully protonated in these media, the acid catalysis must be ascribed to the partial protonation of the carbonyl group, and the rates should be correlated by the H_+ acidity function. Since H_0 and H_+ are parallel functions of acidity (see Chapter 2), a correlation with H_0 is also a correlation with H_+. The precise formulation of the mechanism is complicated by the possibility of tautomerism in the reacting cation (**5.20**), (**5.21**) and (**5.22**), but there is no doubt that

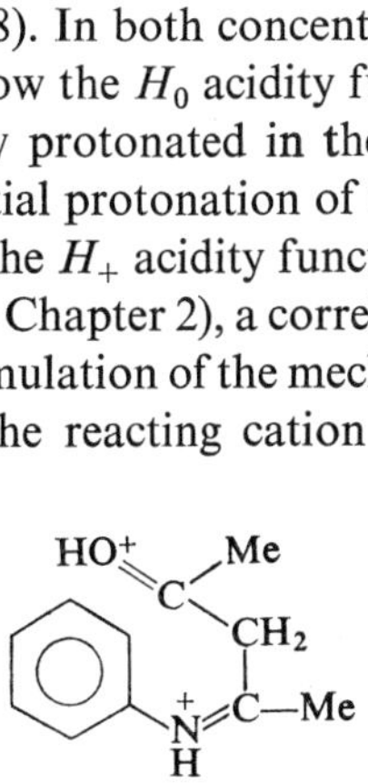

(**5.20**) 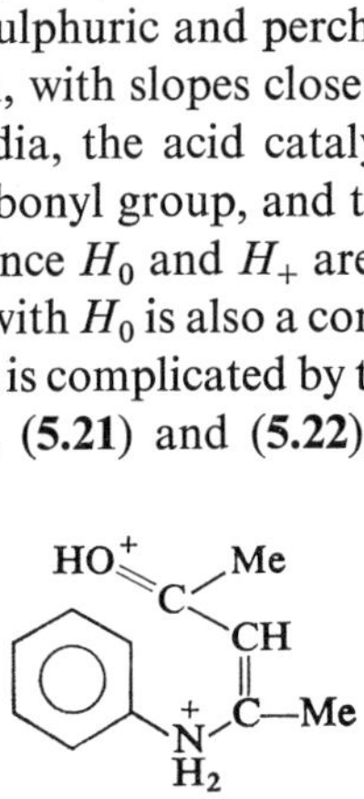 (**5.21**) 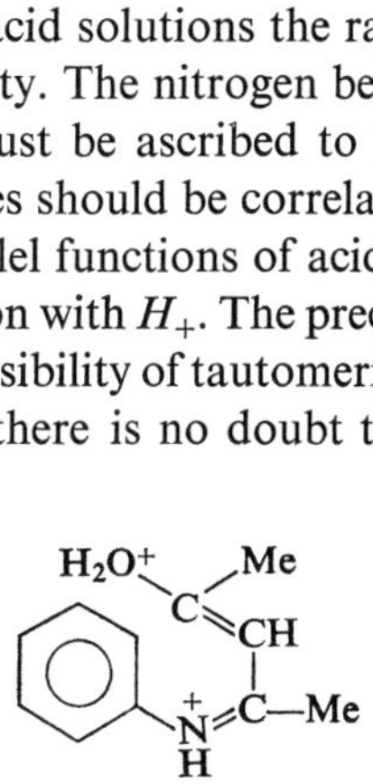 (**5.22**)

intramolecular electrophilic attack is the rate-determining step, because methyl substituents in the ring strongly accelerate the reaction. The same is true of the halogens in *meta*-halogenoanils, which are *para* to the site of electrophilic attack. All halogens activate the nucleus in this reaction in exactly the same way as in aromatic halogenation and nitration, i.e. in the order $F > I > Cl > Br > H$ (Bonner and Barnard, 1958). The rate-determining step must here be followed by the rapid elimination of the elements of water, to give the final product of the reaction, quinoline.

Few cyclodehydrations in aliphatic systems have been studied systematically. Certain geometric requirements must be satisfied in such systems in order to favour reaction. This is the case in 3,3-dimethyl-5-ketohexanoic acid, which undergoes dehydration to dimedone in aqueous sulphuric acid:

$$\begin{array}{c} CH_3 \\ \diagdown \overset{3}{C} \diagup \overset{\overset{4}{CH_2}-\overset{5}{C}\overset{6}{OCH_3}}{} \\ CH_3 \diagup \phantom{\overset{3}{C}} \diagdown \underset{\overset{2}{CH_2}-\overset{1}{COOH}}{} \end{array} \longrightarrow \begin{array}{c} CH_3 \diagdown \diagup CH_2-CO \diagdown \\ C CH_2 \\ CH_3 \diagup \diagdown CH_2-CO \diagup \end{array}$$

Here the two *gem*-methyl groups on C3 constrain the molecule in such a way as to bring the two reacting ends of the molecule close together. The reaction was studied for hexanoic acid and several 6-alkyl derivatives in sulphuric acid solutions over a wide range of acid concentration and temperature (Silbermann

TABLE 5.9

Relative rates and activation parameters in ring closures of 3,3-dimethyl-6-alkyl-5-keto-hexanoic acids (Silbermann and Henshall, 1957)

Alkyl	k/k_H (100°C, 61% H_2SO_4)	E_a (kJ mol^{-1}) (over a concentration range)	$\Delta S^{\ddagger}$ (J K^{-1} mol^{-1}) (100°C, 61% H_2SO_4)
H	1	106·2–98·1	−69·5
CH_3	38	93·0–83·8	−86·2
C_2H_5	14·5	95·5–84·2	−86·2
n-C_3H_7	18·7	89·6–83·0	−90·8
iso-C_3H_7	1	72·5	−148·6

and Henshall, 1957). The correlation of reaction rates with h_0 is good for the 6-ethyl compound and not so good for other derivatives. The observed energies of activation are markedly dependent on the acidity of the solvent, decreasing as the acid concentration increases (Table 5.9). These facts are interpreted in terms of a preprotonation equilibrium of the enolized ketone (enolization occurs in much less acid solutions):

$$\begin{array}{c} CH_3 \\ CH_3 \end{array}\!\!>\!\!C\!\!<\!\!\begin{array}{c} CH_2-C \\ CH_2-COOH \end{array}\!\!\begin{array}{c} OH \\ CHR \end{array} \underset{-H^+}{\overset{H^+}{\rightleftharpoons}} \begin{array}{c} CH_3 \\ CH_3 \end{array}\!\!>\!\!C\!\!<\!\!\begin{array}{c} CH_2-\overset{+}{C} \\ CH_2-C(OH)_2 \end{array}\!\!\begin{array}{c} OH \\ CHR \end{array}$$

The rate-determining step is then the electrophilic attack of the carboxonium ion centre on the double bond of the enol. Strong support for this suggestion comes from the effect of the 6-alkyl substituents on the rate (Table 5.9). The electron-releasing alkyl groups make the $\bar{u}$-electrons more available for donation to the electrophilic centre. The electron release follows the hyperconjugative order (Me > Et). The large negative entropies of activation reflect a considerable loss of rotational freedom in the transition state, and for the isopropyl group in particular an additional large steric effect. The loss of the elements of water occurs either simultaneously with ring closure, or after it, or even before, through the pre-equilibrium formation of small concentrations of oxocarbonium ions.

One of the most interesting examples of cyclodehydration reactions, that of 2-phenyltriarylcarbinols to 9,9-diarylfluorenes, was studied in an unconventional medium, 80 per cent aqueous acetic acid—4 per cent sulphuric acid (Hart and Sedor, 1967). The rate-determining step for highly negatively substituted compounds is here the carbonium ion formation, whereas electron releasing substituents favour the carbonium ion formation and the electrophilic attack becomes rate determining.

5.4.2.5. *Condensations*

Concentrated sulphuric acid is a well-known condensing agent and innumerable references to a variety of condensation reactions in this medium are encountered in the literature. The high acidity coupled with low water activity are obviously the factors favouring these reactions. Few mechanistic studies are, however, available.

A well-known reaction of this kind is the condensation of acetone to mesityl oxide, phorone and mesitylene, which occurs rapidly in concentrated sulphuric acid. In very dilute solution in pure sulphuric acid acetone is simply protonated, but at higher concentrations condensation takes place and the solutions turn yellow and brown. A spectrophotometric examination of these changes led to the conclusion that the reactions were complicated (Nagakura *et al.*, 1957). Indirect inferences about the mechanism of this reaction may be drawn from a study of an analogous reaction between benzaldehyde and acetophenone (Noyce and Pryor, 1955).

The reaction between benzaldehyde and acetophenone in acetic acid containing small concentrations of water and sulphuric acid is first-order in both reactants. Acid catalysis is demonstrated by the fact that the reaction follows the H_0 acidity function. A mechanism involving a rate-determining reaction between the enol of acetophenone and protonated benzaldehyde has been suggested to account for these facts (Noyce and Pryor, 1955):

$$C_6H_5-\underset{\underset{O}{\|}}{C}-CH_3 \rightleftharpoons C_6H_5-\underset{\underset{OH}{|}}{C}=CH_2$$
(enolization)

$$C_6H_5-\underset{\underset{O}{\|}}{C}-H + H^+ \rightleftharpoons C_6H_5-\underset{\underset{^+OH}{\|}}{C}-H$$
(preprotonation)

$$C_6H_5-\underset{\underset{OH}{|}}{C}=CH_2 + {}^+\underset{\underset{H}{|}}{C}-C_6H_5 \xrightarrow{\text{slow}} C_6H_5-\underset{\underset{^+OH}{\|}}{C}-CH_2-\underset{\underset{OH}{|}}{C}H-C_6H_5$$
(5.23)

The hydroxyketone obtained (5.23) which contains a secondary alcohol group undergoes dehydration by the mechanism discussed in Section 5.4.2.2 to give *trans*-benzalacetophenone. This last step in the reaction is also slow, but since the intermediate hydroxyketones are seldom isolated it is probably not rate determining.

5.4.3. ISOMERIZATIONS AND REARRANGEMENTS

The terms isomerizations and rearrangements are sometimes used interchangeably, but more commonly isomerizations are taken to involve relatively minor structural changes, such as transfers of protons from one position in the molecule to another, whereas rearrangements are taken to be reactions in which the basic skeleton of the molecule undergoes a fundamental change and reaction products are usually compounds of a completely different type. Structural changes of hydrocarbon molecules, which consist of fission and reformation of C—C bonds, may be termed either isomerizations or rearrangements.

Many such reactions are acid-catalysed, and some of the best known rearrangements of organic chemistry are catalysed by moderately concentrated or concentrated sulphuric acid in particular. This catalytic action of the strongly acid media is due to their capacity to produce electron-deficient sites in the molecules which act as electrophiles intramolecularly. The mechanisms of some such reactions have been extensively studied and are reasonably well understood. These will be discussed in detail. A variety of other rearrangements of solutes in moderately concentrated or concentrated sulphuric acid are known, but are less well understood. High temperatures and prolonged reaction times often lead to mixtures of products which can sometimes be accounted for by mechanisms or combinations of mechanisms similar to those discussed in this section.

Some other rearrangements which take place in rather dilute acid (such as the benzidine rearrangement) or are not very general reactions (such as the Jacobsen reaction) will not be discussed. The nitramine rearrangement, which involves the migration of a nitro group, will be treated in the section on nitration (Section 5.4.6).

5.4.3.1. *Keto-enol Isomerizations*

Keto-enol isomerizations were described in Section 5.4.2 as rapid processes in moderately concentrated acid solutions. They are well known to be both acid and base catalysed reactions, and to show general acid/base catalysis (see Bell, 1941, 1959). The reaction in either direction leads to an equilibrium between the two forms:

$$\begin{array}{cc} \text{H} \quad \text{O} & \text{OH} \\ | \quad || & | \\ -\text{C}-\text{C}- \rightleftharpoons -\text{C}=\text{C}- \\ | \quad \quad & | \\ (\text{K}) & (\text{E}) \end{array}$$

The thermodynamic equilibrium constant is of course independent of the concentration of the catalyst, but in moderately concentrated acid solutions the concentration ratio, c_E/c_K, will remain constant only if the activity coefficient ratio f_E/f_K does so. This appears to be the case for acetyl-acetone over the concentration range 0–6·5 M sulphuric acid (Schwarzenbach and Wittwer, 1947), but not for cyclohexa-1,2-dione (Bakule and Long, 1963).

The first study of enolization in moderately concentrated acid is due to Zucker and Hammett (1939). The mechanism of this reaction, when catalysed by hydronium ions, was formulated as follows (Pedersen, 1934):

$$\begin{array}{c} \text{O} \quad\quad\quad\quad\quad\quad\quad\quad {}^+\text{OH} \\ || \quad\quad\quad\quad\quad\quad\quad\quad || \\ \text{RR'HC}-\text{C}-\text{R}'' + \text{H}_3\text{O}^+ \rightleftharpoons \text{RR'HC}-\text{C}-\text{R}'' + \text{H}_2\text{O} \end{array}$$
(pre-equilibrium)

$$\begin{array}{c} {}^+\text{OH} \quad\quad\quad\quad\quad\quad\quad\quad \text{OH} \\ || \quad\quad\quad\quad\quad\quad\quad\quad | \\ \text{RR'HC}-\text{C}-\text{R}'' + \text{H}_2\text{O} \xrightarrow{\text{slow}} \text{RR'C}=\text{C}-\text{R}'' + \text{H}_3\text{O}^+ \end{array}$$

The slow step is the transfer of a proton from the α-carbon atom to a water molecule as a nucleophile. The mechanism would thus be described as A-2. In the reverse reaction, the ketonization of an enol, the same step must be rate determining, and since it involves the attack of a proton on the $\bar{u}$-electrons of a double bond it may be described as A-S_E2 (Long and Paul, 1957). Zucker and Hammett (1939) found that the enolization of acetophenone does not follow the H_0 acidity function, but follows more closely the acid concentration (see Chapter 2).

Enolization reactions are usually followed by bromination or iodination of the double bond, but if the proton involved is carried on an asymmetric carbon atom, they may also be followed by the rate of racemization (Ingold and Wilson, 1934). In moderately concentrated acid solutions fast reaction techniques must be used, e.g. the thermal maximum method (Archer and Bell, 1959).

The existence of a preprotonation in acid solution was inferred from the solvent isotope effect on the rate of bromination of acetone ($k_H/k_D = 0.48$) (Reitz, 1936), long before information on its basicity became available (see Chapter 3). The rates of enolization of acetone increase with the acidity of the solution, but the plot of $\log k$ vs. $-H_0$ shows a downward curvature (Archer and Bell, 1959). If at higher acid concentration the preprotonation approaches completion, the rate of enolization should reach a limiting value. The curvature of the $\log k$ vs. $-H_0$ plot for the enolization of acetone was ascribed to this effect (Archer and Bell, 1959), and a limiting rate was observed for acetaldehyde at $-H_0 > 3$ (McTigue and Gruen, 1963), but the pK_{BH^+} values deduced from the H_0 dependence of rate imply much greater basicity of these compounds than found by direct determination (see Chapter 3). It thus appears that this levelling off in the rate-acidity profile is due either to a difference in the solvation of the conjugate acid compared with Hammett indicators, or to decreasing availability of water molecules which take part in the rate-determining step, or to both these effects.

Most A-2 reactions show a rate maximum in the moderately concentrated acid region, with decreasing rates at higher acid concentrations (see Section 5.4.1). Decreasing rates of enolization in concentrated sulphuric acid have been reported for D-α-phenylisocaprophenone (Swain and Rosenberg, 1961). The rate in 85 per cent acid is ten times larger than the rate in 94 per cent acid. The substrate is largely or fully protonated at these acidities (p$K_{BH^+} = -7.39$). The results can be accounted for by assuming that both water and HSO_4^- ions act as nucleophiles in the rate-determining step, water being about 100 times more effective (Swain and Rosenberg, 1961). The rate of this reaction was followed by measuring the rate of racemization. Enolization is the rate-determining reaction for racemization, since the reverse reaction, the ketonization, is much faster in the strongly acidic medium.

A more complex case of keto-enol isomerism is that of cyclohexa-1,2-dione (Bakule and Long, 1963). One of the carbonyl groups is rapidly hydrated in aqueous acid and the ketone is slowly enolized as follows:

5. REACTION MECHANISMS IN SULPHURIC ACID SOLUTIONS

$$\text{(K)} \rightleftharpoons \text{(E)} + H_2O \qquad (5.59)$$

The concentration ratio, c_K/c_E, is 2 in dilute acid and decreases to 0·3 in 6 M perchloric or sulphuric acid. This arises from a seven-fold increase in the f_K/f_E ratio over the same concentration range (i.e. the ketone is salted out by the acid much more than the enol). Both the enolization of the ketone and the ketonization of the enol follow the H_0 acidity function up to 7 M acid with slopes less than unity. The activation parameters, especially the large negative entropy of ketonization (-109 J K^{-1} mol^{-1}), strongly support the presence of a water molecule in the transition state for ketonization (Bakule and Long, 1963).

5.4.3.2. *Cis-trans Isomerizations*

Many *cis-trans* isomerizations of ethylenic compounds are acid-catalysed. The mechanisms of some such reactions, which occur in moderately concentrated acids and in concentrated sulphuric acid, have been substantially elucidated in recent years by Noyce and his school (1959–1963). An unsuspected variety of mechanisms have been shown to obtain, the isomerizations of α,β-unsaturated ketones contrasting with those of α,β-unsaturated carboxylic acids. The fundamental reason for this is the existence of resonance (**5·24**) in the

$$\underset{\beta\quad\alpha}{-\text{CH}=\text{CH}-\overset{\text{O}}{\overset{\|}{\text{C}}}-} \rightleftharpoons \underset{\beta\quad\alpha}{-\overset{+}{\text{CH}}-\text{CH}=\overset{\text{O}^-}{\underset{|}{\text{C}}}-}$$

(5.24)

molecules of the ketones as compared with carboxylic acids, in which the strong resonance within the carboxyl group hinders the involvement of the carbonyl group in the resonance with the α,β-ethylenic bond. Owing to resonance (**5·24**) the basicity of α,β-unsaturated ketones is enhanced (see Chapter 3). They are rapidly protonated in acid solution to give cations (**5.25**),

$$-\text{CH}=\text{CH}-\overset{^+\text{OH}}{\overset{\|}{\text{C}}}- \leftrightarrow -\overset{+}{\text{CH}}-\text{CH}=\overset{\text{OH}}{\underset{|}{\text{C}}}-$$

(5.25)

which are themselves resonance stabilized. Owing to this the positive charge is partly transferred to the β-carbon atom, which hence becomes susceptible to nucleophilic attack by water molecules in aqueous acids:

$$-\overset{+}{C}H-CH=\underset{|}{\overset{OH}{C}}- + H_2O \xrightarrow{\text{slow}} -CH-\underset{|}{\overset{+OH_2}{C}}H=\underset{|}{\overset{OH}{C}}-$$

(5.26)

Studies of the isomerization of *cis*-benzalacetophenone (*cis*-chalcone) into the *trans* isomer (Noyce et al., 1959; Noyce and Jørgenson, 1963) indicate that this nucleophilic attack is the rate-determining step for the parent compound and some negatively substituted derivatives, because the rate-acidity profiles show maxima (at 70–80 per cent sulphuric acid), which are characteristic of the A-2 mechanism. In the hydrated β-hydroxy enol (5.26) there is free rotation about the C_α—C_β bond, and in the subsequent loss of a water molecule the less sterically crowded *trans* isomer of the ketone results. In concentrated sulphuric acid the rate of isomerization of these ketones decreases sharply, owing to the rapid decline in the activity of water. For *cis*-benzalacetophenone the rate reaches a minimum in 90 per cent sulphuric acid, which is followed by a slight increase at still higher acid concentrations. Noyce and Jørgenson (1963) ascribe this to a change in mechanism, brought about by the low activity of water, which is the most effective nucleophile in the medium. As the availability of water decreases, the lifetime of the resonance stabilized carbonium ion increases, so that slow rotation about the partially double C_α—C_β bond becomes the effective unimolecular mechanism of isomerization. The rate of this reaction should remain independent of acid concentration if the substrate is fully protonated. As $pK_{BH^+} = -5.8$, this is so in >90 per cent, and therefore the slight increase in rate at >90 per cent is probably due to a medium effect. The introduction of electron-releasing substituents into the benzene ring strongly enhances the rate of isomerization of *cis*-chalcones by the carbonium ion mechanism, owing to the stabilization of the carbonium ion by charge delocalization, which also greatly reduces the double bond character of the C_α—C_β bond. Noyce and Jørgenson (1963) estimate a ρ constant of -7. Unlike this high sensitivity to substituent effects of the carbonium ion mechanism of isomerization, the effect of substituents on the isomerization by the A-2 mechanism is small.

The solvent isotope effect on the isomerization of *cis*-chalcone (in 5 per cent aqueous dioxane—3 M sulphuric acid) shows that the reaction is faster in the deuterated solvent ($k_H/k_D = 0.52$) (Noyce et al., 1961), consistent with an A-2 mechanism. Also no incorporation of deuterium in the *trans*-chalcone occurs, which proves that no C—H bonds are formed or broken in the reaction, in support of the proposed hydration-dehydration mechanism of isomerization.

In contrast to the behaviour of α,β-unsaturated ketones, the *cis-trans* isomerization of cinnamic acid follows an addition–elimination sequence, in which the first step is the normal protonation of the ethylenic bond, followed by the hydration of the resulting carbonium ion (Noyce et al., 1962a). The mechanism of such reactions was discussed in Section 5.4.2. The solvent isotope effect found in this reaction ($k_H/k_D = 2\cdot15$–$2\cdot46$ in 64–71 per cent acid) conclusively proves that protonation of the double bond of cis-cinnamic acid is the slow step in the reaction (Noyce et al., 1962b). This step also leads to the incorporation of deuterium into the *trans* acid. Substituent effects on the isomerization of cis-cinnamic acid are correlated by the σ^+ constants with a ρ value of $-4\cdot3$, characteristic of reactions involving carbonium ion formation (Noyce and Avarbock, 1962).

5.4.3.3. *Rearrangements of Carbonium Ions*

Rearrangements of carbonium ions are not restricted to strongly acid media. The same carbonium ions, which are obtained in strongly acid media by complex ionization of alcohols, are obtained in other media by loss of some other electronegative groups (e.g. halogens) or by deamination reactions, and rearrange in the same way. Some rearrangements involving dialcohols, aldehydes and ketones are, however, primarily observable in strongly acid media and therefore emphasis in this section will be placed on a discussion of such reactions. A fuller discussion of carbonium ion rearrangements has recently been published (Bethell and Gold, 1967).

Rearrangements of carbonium ions are usually so directed as to give products with increased conjugation or hyperconjugation or with lesser steric strain. The carbonium ion mechanism of isomerization of some cis-chalcones discussed in the preceding section may be regarded as an example of the latter kind. Very good examples of rearrangements leading to more stable products owing to increased conjugation or hyperconjugation are allylic rearrangements. These may be represented by the equation

$$\text{R—CH—CH=CH}_2 \rightleftharpoons \text{R—CH=CH—CH}_2\text{X} \qquad (5.60)$$
$$\quad\;\;|$$
$$\;\;\text{X}$$

where X may be Cl, Br, OH_2^+ or other electronegative group, and R may be hydrogen, an alkyl or an aryl group. The rearrangement occurs through an intermediate mesomeric allylic cation (**5.27**). If R = H the rearrangement

$$\left[\begin{array}{c} \text{H} \;\; \text{H} \\ \;\;| \;\;\;\; | \\ \text{R—C}\text{---}\text{C}\text{---}\text{CH}_2 \end{array} \right]^+$$

(**5.27**)

is observable only by isotopic labelling of one of the end carbon atoms and is very slow. If R is a group which can conjugate or hyperconjugate with the allylic cation and with the double bond in the reaction product, the reaction is much faster.

The intermediate formation of carbonium ions in acid-catalysed allylic rearrangements of alcohols has been demonstrated by studying concurrent rearrangement and ^{18}O-exchange (Bunton et al., 1958a). The rearrangement reaction (5.61) in acid solution is accompanied by ^{18}O exchange at both

$$\text{Ph—CH(OH)—CH=CH}_2 \rightleftharpoons \text{Ph—CH=CH—CH}_2\text{OH} \tag{5.61}$$

positions 1 and 3, and also by racemization of the alcohol (Goering and Dilgren, 1960). The oxygen exchange at position 3 is faster than that at position 1, as would be expected from the greater stability of the primary alcohol, and is also faster than the rearrangement reaction. All these reactions follow the H_0 acidity function (slope $\log k$ vs. $-H_0 = 1 \cdot 3$), consistent with a preprotonation step. The slow step is the formation of the carbonium ion, which is partitioned in the ratio 60:40 between the reverse reaction and the rearrangement.

When groups which stabilize the allylic system in allylic alcohols are absent, a completely different rearrangement takes place. Thus 2-methylallyl alcohol rearranges to isobutyraldehyde in 12 per cent sulphuric acid (Hearne et al., 1941):

$$\text{CH}_2\text{=C(CH}_3\text{)—CH}_2\text{OH} \rightleftharpoons (\text{CH}_3)_2\text{CH—CHO} \tag{5.62}$$

Presumably the same rearrangement would occur with allyl alcohol (Currell and Fry, 1956). There are two protonation sites in such a molecule: a primary alcohol group, whose loss would lead to an allylic cation (**5.28**), and a double bond, whose protonation leads to the formation of a tertiary carbonium ion (**5.29**):

$$\left[\text{CH}_2\text{---C(CH}_3\text{)---CH}_2\right]^+ \qquad \text{CH}_3\text{—}\overset{+}{\text{C}}(\text{CH}_3)\text{—CH}_2\text{OH}$$

(**5.28**)   (**5.29**)

The course of the reaction shows that the protonation of the double bond is faster than the loss of water from the conjugate acid of the alcohol and that the more stable intermediate (**5.29**) is obtained (Currell and Fry, 1956). This rearranges by a 1,2-shift of the hydride ion, i.e.

$$\text{CH}_3\text{—}\overset{+}{\text{C}}(\text{CH}_3)(\text{H})\text{—CHOH} \longrightarrow \left[\text{CH}_3\text{—C(CH}_3)(\text{H})\text{—CH=}\overset{+}{\text{O}}\text{H} \leftrightarrow \text{CH}_3\text{—C(CH}_3)(\text{H})\text{—CH}\overset{+}{\text{—}}\text{OH}\right] \tag{5.30}$$

to give the protonated aldehyde. The direction of this rearrangement implies that the carboxonium ion, in which the charge is delocalized between the carbon and the oxygen atoms (5.30), and which probably hydrogen-bonds with the solvent, is the more stable cation. The final step in the reaction is the fast deprotonation of the aldehyde (aldehydes are protonated only at higher acidities, see Chapter 3). The protonation of the double bond is probably the rate-determining step in this reaction, although the rearrangement could also be slow. By varying the conditions isobutylene glycol can also be formed by nucleophilic attack of water on the carbonium ion, as well as isobutylene glycol-isobutyracetal (Hearne et al., 1941).

The well-known pinacol-pinacolone rearrangements are reactions of the same type, except that the carbonium ions are obtained by heterolysis of protonated dialcohols. The reaction takes its name from the symmetrical dialcohol pinacol (5.31), which rearranges in acid solution to pinacolone (5.32), according to the equation

$$\underset{(5.31)}{\underset{\overset{|}{\text{OH}}\ \overset{|}{\text{OH}}}{\text{CH}_3-\overset{\overset{\text{CH}_3}{|}}{\text{C}}-\overset{\overset{\text{CH}_3}{|}}{\text{C}}-\text{CH}_3}} \rightleftharpoons \underset{(5.32)}{\underset{\overset{|}{\text{CH}_3}}{\text{CH}_3-\overset{\overset{\text{CH}_3}{|}}{\text{C}}-\overset{\overset{\text{O}}{\|}}{\text{C}}-\text{CH}_3}} + \text{H}_2\text{O} \quad (5.63)$$

The reaction takes place in dilute and moderately concentrated acid at higher temperature (73°C). A preprotonation step was demonstrated by Bunton et al. (1956) both by a correlation of rate with h_0 and by a solvent isotope effect ($k_H/k_D = 0.45$ in 1·86 N sulphuric acid). This is followed by a carbonium ion formation step:

$$\underset{\overset{|}{_+\text{OH}_2}\ \overset{|}{\text{OH}}}{\text{CH}_3\text{C}-\overset{\overset{\text{CH}_3}{|}}{\underset{|}{\text{C}}}-\text{CH}_3} \rightarrow \underset{\overset{|}{\text{OH}}}{\text{CH}_3-\overset{\overset{\text{CH}_3}{|}}{\underset{+}{\text{C}}}-\overset{\overset{\text{CH}_3}{|}}{\text{C}}-\text{CH}_3} + \text{H}_2\text{O} \quad (5.33)$$

This is reversible, as shown by oxygen exchange (Bunton et al., 1958b). As with other alcohols (Section 5.4.2.1), this is a slower reaction than preprotonation. Apart from recombination, the carbonium ion (5.33) undergoes a rearrangement:

$$\text{CH}_3-\underset{\underset{\text{CH}_3}{\diagdown}}{\overset{\overset{\text{CH}_3}{|}}{\underset{+}{\text{C}}}}-\overset{\overset{\text{CH}_3}{|}}{\text{C}}-\text{OH} \longrightarrow \left[\text{CH}_3-\overset{\overset{\text{CH}_3}{|}}{\underset{\underset{\text{CH}_3}{|}}{\text{C}}}-\overset{\overset{\text{CH}_3}{|}}{\underset{\underset{\text{OH}}{|}}{\text{C}^+}} \longleftrightarrow \text{CH}_3-\overset{\overset{\text{CH}_3}{|}}{\underset{\underset{\text{CH}_3}{|}}{\text{C}}}-\overset{\overset{\text{CH}_3}{|}}{\underset{\underset{_+\text{OH}}{\|}}{\text{C}}} \right]$$

(5.34)

This gives the protonated ketone (**5.34**), which deprotonates rapidly in dilute acid. The migration of the methyl group is represented here as an irreversible process, which it is in dilute acid, but in concentrated acid this step has also been shown to be reversible by isotopic labelling (^{14}C) of one of the methyl groups in the ketone (Rothrock and Fry, 1958). After a sufficiently long time in concentrated sulphuric acid, the isotopic label becomes equally distributed among the four methyl groups. It follows that the carbonium ion originally formed may undergo either recombination with water or rearrangement, both reactions being reversible, the latter at least in concentrated sulphuric acid. Bunton *et al.* (1958b) estimated that recombination is about 1·5 times faster than rearrangement, and consequently the rate of formation of the carbonium ion is 2·5 times the rate of rearrangement. Pinacol and pinacolone are also obtained from tetramethyl epoxide by acid-catalysed ring opening, which is in fact a faster method of obtaining the carbonium ion (**5.33**) (Pocker, 1959).

With unsymmetrically substituted glycols two questions arise: which of the two hydroxyl groups is lost, and which group migrates onto the carbonium centre. The answer to the first question follows from the general principle that the most stable carbonium ion should be obtained, i.e. the carbon carrying substituents which are most able to delocalize the positive charge will be the one to lose its hydroxyl group preferentially. For example, the tertiary-secondary glycol (**5.35**) rearranges to give the aldehyde (**5.36**) (Tiffeneau and Dorlencourt, 1909):

$$\begin{array}{c} \text{Ph} \quad \text{Ph} \\ | \quad\quad | \\ \text{H--C------C--Me} \\ | \quad\quad | \\ \text{OH} \quad \text{OH} \\ (\textbf{5.35}) \end{array} \rightleftharpoons \begin{array}{c} \text{H} \quad \text{Ph} \\ | \quad | \\ \text{O=C--C--Me} + \text{H}_2\text{O} \\ \quad\quad | \\ \quad\quad \text{Ph} \\ (\textbf{5.36}) \end{array} \qquad (5.64)$$

showing that the intermediate is the more stable tertiary cation (**5.37**), rather than the secondary cation (**5.38**). The reaction product shows moreover that

$$\begin{array}{c} \text{Ph} \quad \text{Ph} \\ | \quad | \\ \text{H--C--C--Me} \\ | \quad + \\ \text{OH} \\ (\textbf{5.37}) \end{array} \qquad \begin{array}{c} \text{Ph} \quad \text{Ph} \\ | \quad | \\ \text{H--C--C--Me} \\ + \quad | \\ \quad \text{OH} \\ (\textbf{5.38}) \end{array}$$

of the two groups that can migrate on to the carbonium centre, it is the aryl group that does so. In this type of reaction aryl groups migrate more readily than alkyl groups, which in turn migrate more readily than hydrogen. This is the order in which these groups are able to delocalize positive charge, and their "migratory aptitudes" are believed to be related to this ability. The effects of substituents in the phenyl group on its "migratory aptitude" certainly

5. REACTION MECHANISMS IN SULPHURIC ACID SOLUTIONS 241

supports this view, since there is a correlation with σ^+ constants ($\rho \simeq -3$) (Heck and Winstein, 1957).

This simple picture does not even extend to all tertiary–secondary glycols, and in most tertiary–primary and secondary–primary glycols, it is the 1,2-hydride shift that occurs in preference to the migration of other groups, so that ketones result, rather than aldehydes. For example, 1-methyl-2-phenyl-glycol (**5.39**) gives benzylmethyl ketone (**5.40**) upon a final loss of a proton in dilute acid (Tiffeneau, 1907).

$$\underset{\underset{\text{OH OH}}{|\quad|}}{\overset{\overset{\text{H H}}{|\quad|}}{\text{Me—C—C—Ph}}} \underset{-H_2O}{\overset{H^+}{\rightleftharpoons}} \underset{\underset{\text{OH}}{|}}{\overset{\overset{\text{H H}}{|\ \searrow|}}{\text{Me—C—C—Ph}}}$$

(**5.39**)

$$\left[\underset{\text{Me—C—CH}_2\text{Ph}}{\overset{^+\text{OH}}{\overset{\parallel}{}}} \longleftrightarrow \underset{\text{Me—C—CH}_2\text{Ph}}{\overset{\text{OH}}{\overset{|}{\underset{+}{}}}} \right]$$

(**5.40**)

In concentrated sulphuric acid, where aldehydes and ketones are fully protonated, rearrangements of aldehydes into ketones occur, e.g.

$$\text{Me}_2\text{PhC—CHO} \rightleftharpoons \text{MePhCH—COMe} \quad (5.65)$$
$$(5.41)$$

(Orékhov and Tiffeneau, 1926). These reactions are in fact double migrations and again do not follow the order of migratory aptitudes of groups. In the present example, the first migration in the carboxonium ion to give (**5.42**) is

$$\underset{\underset{\text{Ph}}{|}}{\overset{\overset{\text{Me OH}}{|\ \searrow\ |}}{\text{Me—C—C—H}}} \longrightarrow \underset{\underset{\text{Ph H}}{|\quad|}}{\overset{\overset{\text{OH}}{|}}{\text{Me—C—C—Me}}}$$

(**5.42**)

followed by a hydride shift, resulting in the ketone (**5.41**) in concentrated acid. Protonated ketones are more stable than protonated aldehydes, because alkyl and aryl groups stabilize the carboxonium ion charge by hyperconjugation or by conjugation, and this may account for their formation in concentrated sulphuric acid. However, why a methyl group is preferred on the carboxonium centre to a phenyl group in the example (5.65) is not immediately clear. It may be pointed out that the same carbonium ion (**5.42**) would be obtained in the pinacolic rearrangement of the glycol (**5.43**) in dilute sulphuric acid, but would rearrange with methyl shift to give the aldehyde, PhMe$_2$C—CHO.

$$\underset{(\mathbf{5.43})}{\underset{\overset{|}{\text{OH}}\quad\overset{|}{\text{OH}}}{\text{Ph—}\overset{\overset{\text{Me}}{|}}{\text{C}}\text{—}\overset{\overset{\text{Me}}{|}}{\text{C}}\text{—H}}} \xrightarrow[-\text{H}_2\text{O}]{+\text{H}^+} \underset{(\mathbf{5.42})}{\underset{\overset{|}{+}\quad\overset{|}{\text{OH}}}{\text{Ph—}\overset{\overset{\text{Me}}{|}}{\text{C}}\text{—}\overset{\overset{\text{Me}}{|}}{\text{C}}\text{—H}}}$$

More recent work on these rearrangements, mainly by Collins and his collaborators (1953–59), has thrown some light on these rather confusing facts. The reversibility of the carbonium ion formation was confirmed by a demonstration that *threo-* and *erythro-*1,2-diphenyl-1-*p*-tolyl-ethylene glycols are interconverted in dilute acid (3·13 N aqueous ethanolic sulphuric acid) more rapidly than they rearrange (Collins *et al.*, 1959). The rearrangement of triarylethylene glycols (**5.44**) and (**5.45**) was shown to give an aldehyde (**5.46**) and one or two ketones (**5.47**)—when more than one kind of aryl group is

$$\underset{(\mathbf{5.44})}{\underset{\overset{|}{\text{OH}}\quad\overset{|}{\text{OH}}}{\text{Ph}_2\text{C}\text{——}\text{CHPh}}} \begin{array}{c}\nearrow\;\text{Ph}_3\text{C—CHO}\quad(\mathbf{5.46})\\ \\ \searrow\;\text{Ph}_2\text{CH—COPh}\quad(\mathbf{5.47})\end{array}$$

$$\underset{\substack{(\mathbf{5.45})\\(\text{An}=p\text{-anisyl})}}{\underset{\overset{|}{\text{OH}}\quad\overset{|}{\text{OH}}}{\text{PhAnC}\text{——}\text{CHPh}}} \begin{array}{c}\nearrow\;\text{Ph}_2\text{AnC—CHO}\quad(\mathbf{5.46})\\ \\ \searrow\;\begin{cases}\text{Ph}_2\text{CH—COAn}&(\mathbf{5.47a})\\ \text{PhAnCH—COPh}&(\mathbf{5.47b})\end{cases}\end{array}$$

present—depending on conditions. In the rearrangement of triphenylethylene glycol (**5.44**) in concentrated sulphuric acid the ketone (**5.47**) is the exclusive product, but in 31·5 per cent sulphuric acid and other less acidic solvents (formic acid and oxalic acid hydrate) both ketone (**5.47**) and aldehyde (**5.46**) are obtained in varying proportions, which means that no single factor determines the rearrangement. Finally, the hydride transfer was shown to be intramolecular by deuterium isotope labelling, and it was established at the same time that hydrogen migrates faster than deuterium (Collins *et al.*, 1959).

In order to explain deviations from the expected migratory sequence of groups it was suggested that, apart from electronic factors, non-bonded interactions between groups in the transition states for rearrangement (**5.48**) and (**5.49**) must also be taken into account, and that therefore configuration and conformation of the starting material are also important (Raaen and Collins, 1958; Collins *et al.*, 1959). Non-bonded compressions in the carbonium ion and in the product probably also play a part. The relative

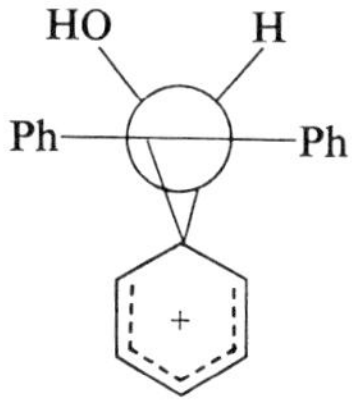

(5.48)

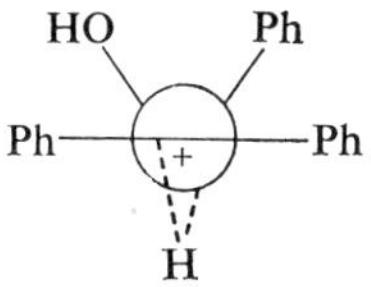

(5.49)

migration sequence methyl 1, ethyl 17, *t*-butyl > 4000, found by Stiles and Mayer (1959), is explainable in these terms.

In view of the influence of all these other factors on the direction of groups in migrations, the intrinsic (electronic) migratory aptitudes of groups often become obscured, and the weighting of the relative importance of the various factors in rearrangements of carbonium ions still remains a difficult problem.

The complexity and unpredictability of these reactions is perhaps best illustrated by the difference in the rearrangement behaviour of *cis*- and *trans*-1,2-dimethylcyclohexane-1,2-diol and *cis*- and *trans*-1,2-dimethylcyclopentane-1,2-diol. In aqueous perchloric acid up to 4 M both *cis* and *trans* derivatives of cyclohexane interconvert and rearrange via a carbonium ion intermediate, the rearrangement products being 2,2-dimethylcyclohexanone and 1-acetyl-methylcyclopentane (Bunton and Carr, 1963a). The *cis* and *trans* derivatives of cyclopentane, however, differ considerably in their rearrangement behaviour (Bunton and Carr, 1963b). The *cis*-diol does not exchange its oxygen atoms with the solvent, but rearranges to the *trans*-diol, 2,2-dimethylcyclopentanone and a tar. The *trans*-diol undergoes extensive oxygen exchange with the solvent and gives tar as the major product. The *cis*-diol rearranges more readily probably because the methyl group can migrate to the forming carbonium ion centre, without hindrance from the leaving water molecule.

Even some derivatives of the stable cyclic allylic cations undergo rearrangements, apparently due mainly to steric repulsions. For example, the 1,3,4,4,5,5-

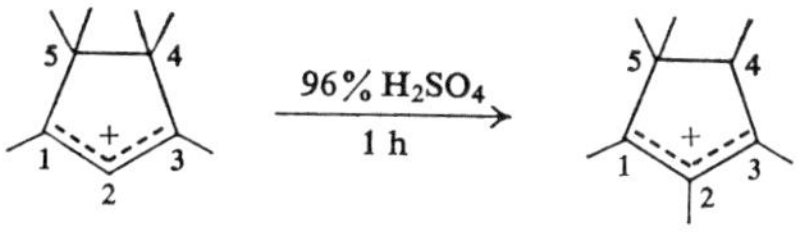

hexamethylcyclopentenyl cation rearranges fairly rapidly to the 1,2,3,4,5,5-hexamethylcyclopentenyl cation (Deno *et al.*, 1963b). Interesting observations on the mobility of protons attached to the ring have also been made by following the hydrogen-deuterium exchange by n.m.r. Significantly, only hydrogens α to the allylic system undergo exchange, but not the hydrogen at C-2 (Deno *et al.*, 1963b).

The considerable stability of the cycloalkenyl cations is probably largely responsible for the disproportionations and rearrangements of some tertiary cycloalkyl cations in sulphuric acid that have been recently demonstrated by n.m.r. (Deno and Pittman, 1964). Tertiary alicyclic cations, obtained from tertiary alcohols (or alkenes), may undergo rearrangement if present in high dilution (10^{-5} M), as, for example,

At higher concentrations intermolecular hydride transfers occur, which lead to the formation of alkenyl cations and mixtures of saturated hydrocarbons.

The rapid polymerization of t-butyl cations in concentrated sulphuric acid was mentioned in Chapter 4. Some light has recently been thrown on the complex processes involved by a study of n.m.r. spectra of the solutions of t-butanol, isobutylene, and trimethylpentenes, which all yield in the first instance tertiary carbonium ions in 96 per cent sulphuric acid (Deno et al., 1964a). The results may be summarized as follows:

$$\left.\begin{array}{l} t\text{-BuOH} \\ (CH_3)_2C{=}CH_2 \\ \text{trimethylpentenes} \end{array}\right\} \xrightarrow{96\% \ H_2SO_4} \begin{array}{l} 50\% \text{ alkanes} \\ (\text{mainly } C_4\text{—}C_{18}) \\ 50\% \text{ cyclopentenyl cations} \\ (\text{mainly } C_{10}\text{—}C_{18}) \end{array}$$

and it can be seen that overall the reactions are disproportionations, coupled with a multitude of alkylations and rearrangements. The greater stability of cyclopentenyl cations compared with other alkenyl cations is reflected in their dominance amongst the products.

Concentrated sulphuric acid is often used as a rearrangement agent for saturated hydrocarbons and their halogenated products. The hydrocarbons are vigorously stirred with sulphuric acid. The presence of a tertiary carbon atom is essential for reaction and three kinds of rearrangement are observed (Roebuck and Evering, 1953): rapid *cis-trans* isomerizations of methyl-substituted cyclohexanes, shifts of a methyl group along a hydrocarbon chain, and relatively slow shifts of a methyl group around a hydrocarbon ring. Racemization of optically active hydrocarbons and isotopic exchange with the solvent also occur (Burwell et al., 1954). The mechanism of these reactions probably involves initiation by oxidation of an alkane at the tertiary position to form a carbonium ion. This rearranges before abstracting hydride ions from other molecules of alkanes and generating more carbonium ions. The rearrangements of saturated hydrocarbons in concentrated sulphuric acid are thus chain reactions (Burwell et al., 1954), which sometimes lead to equilibria (Maury et al., 1954).

Finally some recent observations on the behaviour of cyclopropane in concentrated sulphuric acid deserve mention. Baird and Aboderin (1963, 1964) have found that cyclopropane exchanges hydrogen with 7·5 M deuterosulphuric acid and simultaneously undergoes solvolysis to 1-propanol (and its hydrogen sulphate), with deuterium entering all three positions of the carbon skeleton. The distribution of deuterium in the aliphatic chain after reaction in 98 per cent acid is C-1 28 per cent, C-2 28 per cent, and C-3 43 per cent (Deno et al., 1968). These results are interpreted in terms of the formation of protonated cyclopropane cations, which represent an equilibrium between forms (5.50) and (5.51). The initially formed 1-propyl hydrogen sulphate rearranges to the

$$
\underset{(5.50)}{\overset{D\cdots CH_2}{\underset{CH_2-CH_2}{\diagup\diagdown}}} \quad \rightleftharpoons \quad \underset{(5.51)}{\overset{CH_2D}{\underset{CH_2\cdots CH_2}{\diagup\diagdown}}}
$$

2-propyl hydrogen sulphate in 60–98 per cent sulphuric acid, because the secondary carbonium ion is more stable than the primary. The stability of the cyclopropyl cation is estimated to be intermediate between the 1-propyl and the 2-propyl cations (Deno et al., 1968).

5.4.3.4. *The Beckmann Rearrangement*

The Beckmann rearrangement of ketoximes is a reaction of the same general type as the pinacolic rearrangements discussed in the preceding section, except that the electron deficient centre is nitrogen, i.e.

$$\underset{R'}{\overset{R}{\diagup}}C=N\overset{}{\diagdown}_{OH} \longrightarrow \underset{R'}{\overset{O}{\diagdown}}C-N\overset{R}{\underset{H}{\diagup}}$$

the products being *N*-substituted amides. The reaction is acid catalysed under a variety of conditions, the most common reaction media being concentrated sulphuric acid and ether solutions of phosphorus pentachloride. Esters of the oximes also undergo the same type of rearrangement. The reaction has been known to be intramolecular since the turn of the century (Lobry de Bruyn and Sluiter, 1904) and many other features have long been firmly established, e.g. that the group R substitutes on the nitrogen always on the side opposite the leaving OH group and that it does so without racemization, i.e. without parting with its bonding electrons. The history and the present position on the Beckmann rearrangement have recently been reviewed in detail by Smith (1963).

Hammett and Deyrup (1932) showed that in concentrated sulphuric acid the rearrangement of acetophenone oxime follows the H_0 acidity function. Isotopic studies (^{18}O-labelling) demonstrated that the carbonyl oxygen which appears in the reaction product arises from the solvent and is not transferred intramolecularly (Brodskii and Micluchin, 1941). Information on substituent effects has been more fragmentary until recently, except that in the rearrangements of aliphatic ketoximes it has been clearly established that it is always the larger alkyl group that migrates preferentially, presumably because it determines in the first place the predominance of the *anti* configuration in the oxime itself (McLaren and Schachat, 1949). The conversion of the *syn* and *anti* forms is acid catalysed. Alicyclic oximes of necessity yield lactams.

On the basis of these facts the two most likely possibilities for the slow stage in the reaction have commonly been taken to be either the removal of water from the N—OH$_2^+$ group, as suggested by Stieglitz and Leech (1914), or the actual transfer of the organic radical to an electron deficient nitrogen obtained by loss of OH (Pearson and Cole, 1955). Also the possibility that both steps occur simultaneously in a concerted mechanism has been considered (McNulty and Pearson, 1959). The three suggested transition states for the case of acetophenone oxime may be represented as follows:

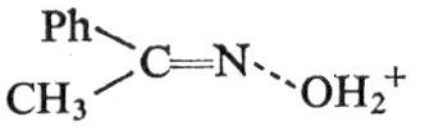

(5.52) 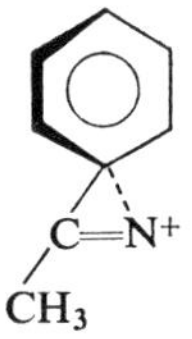 (5.53) 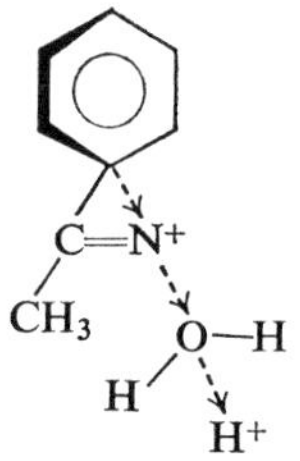 (5.54)

In the latter two, the migrating group is often assumed to acquire considerable carbonium ion character, as in (5.55). The possibility that a sulphate ester

(5.55)

is formed as an intermediate in rearrangements in sulphuric acid seems unlikely, because the sulphate ester undergoes the rearrangement much faster than the oxime itself (Pearson and Ball, 1949), so that the slow step in the reaction could only be its formation.

More recent work on the Beckmann rearrangement and on the basicities of oximes (see Chapter 3) has thrown doubt on all the formulations of transition states given above. First of all, oximes are much more basic than ketones and, since acetophenone is fully protonated in about 85 per cent acid, oximes would certainly be expected to be so. Therefore, as McNulty and Pearson (1959) have pointed out, the dependence of the rate of rearrangement on acidity in concentrated sulphuric acid should be ascribed to an interaction of the protonated oxime with another proton, which they suggest occurs in the transition state (**5.54**). The first protonation of the oxime, however, leads to an N-protonated species (see Chapter 3). The possibility of a second protonation on the oxygen now arises, according to the equation

$$>C=\overset{+}{N}<\overset{H}{OH} + H^+ \rightleftharpoons >C=\overset{+}{N}<\overset{H}{OH_2^+} \quad (5.66)$$

and this could occur either in a pre-equilibrium or in a rate-determining step (the transition state (**5.54**) implies a slow concerted second protonation). All experience with oxygen protonation would favour the first assumption. A second protonation would be expected to follow the H_+ acidity function, but as H_0 and H_+ are parallel functions of acidity (see Chapter 2), the experimentally observed correlations with H_0 (Hammett and Deyrup, 1932; McNulty and Pearson, 1959: Vinnik and Zarakhani, 1963) are consistent with a second preprotonation. The resulting species is now clearly unstable, the N—O bond being strained by the proximity of two positive charges. The cation could thus break up, either rapidly in a pre-equilibrium step (5.67),

$$>C=\overset{+}{N}<\overset{H}{\overset{+}{OH_2}} \rightleftharpoons >C=\overset{+}{N}: + H_3O^+ \quad (5.67)$$

as assumed by Vinnik and Zarakhani (1963), or in a slow step, possibly rate-determining for the rearrangement. If this were a pre-equilibrium step the reaction would be expected to follow the H_R acidity function, and if the rearrangement is carried out in an ^{18}O-enriched solvent, the unrearranged oxime would be expected to contain the ^{18}O label. While the first criterion is uncertain in view of the multiplicity of acidity functions, the second criterion has recently been applied to the rearrangement of acetophenone oxime and no ^{18}O exchange was found (Gregory et al., 1968). This rules out a pre-equilibrium loss of water from the nitrogen, and suggests that the breaking of the N—O bond may be rate determining.

Corroborative evidence for this comes from an examination of substituent effects on the rate of rearrangement of acetophenone oximes. These are correlated by the σ constants much better than by σ^+, with a ρ value of $-1\cdot 5$ (Gregory et al., 1968). A correlation with σ^+ would be expected for either transition states (**5.53**) or (**5.54**), in view of the importance of the positive charge on the ring, as in (**5.55**). Hence neither of these is consistent with the substituent effects on rate. If the breaking of the N—O bond is rate determining, the substituent effects should be two-fold: on the position of the preprotonation equilibrium and on the rate of the bond-fission itself. Substituent effects on the first protonation of oximes should be governed by the σ^+ constants, as are those of ketones, but for the second protonation which occurs on the oxygen, no conjugation between *para* substituents and the protonation centre is possible. Hence a correlation with the σ constants would be expected. Any charge delocalization that transfers the positive charge from the nitrogen into the ring should retard the fission of the N—O bond, so that the rate of fission of the bond should be dependent on the σ^+ constants. The observed dependence of $\log k$ values on the σ constants of the substituents therefore probably means that the dominant effect of the substituents is on the position of the second pre-equilibrium.

Unsymmetrically *m*- and *p*-substituted benzophenone oximes give equal amounts of each possible benzanilide (Bachmann and Barton, 1938), which means that substituents have no effect on the probability of migration. This is determined by the position of the *syn-anti* equilibrium, both forms in this case being present in equal amounts.

All *ortho* substituents accelerate the rearrangement of aryl ketoximes more than the *para* substituents (Pearson and Cole, 1955), including the nitro group. This is probably due to steric inhibition of the resonance stabilization of the oximes (Smith, 1963).

Propiophenone oxime rearranges faster than acetophenone oxime, i.e. the order is Et > Me (McNulty and Pearson, 1959). In terms of the discussed mechanism, this is consistent with an enhanced basicity of the substrate, coupled with a probable steric effect on the leaving water molecule.

All this evidence is consistent with the slow fission of the N—O bond as rate determining in a Beckmann rearrangement, the loss of the proton from the nitrogen being synchronous with it. The resulting species containing the electron deficient nitrogen is unstable and rearranges according to

$$\begin{array}{c} Ph \\ CH_3 \end{array}\!\!\!\!C\!\!=\!\!\overset{+}{N}\!:\quad\longrightarrow\quad \begin{array}{c} \\ CH_3 \end{array}\!\!\!\!\overset{+}{C}\!\!=\!\!N\!\!\begin{array}{c} Ph \\ \; \end{array}$$

in a fast reaction, which may be regarded as an electrophilic attack by the positive nitrogen on the σ electrons of the phenyl-carbon bond. Since it is always the larger group that migrates, the leaving water molecule probably shields the nitrogen until rearrangement is complete. The final step in the reaction is the hydration of the carbonium centre.

An alternative mechanism of Beckmann rearrangement has recently been discovered by Conley and Lange (1963), and its occurrence in polyphosphoric and sulphuric acid media also demonstrated (Hill et al., 1965). This mechanism involves the fragmentation of the unstable electron deficient species according to the equation

$$\underset{R'}{\overset{R}{>}}C\overset{+}{=}\overset{..}{N}: \longrightarrow R^+ + R'-C\equiv N \qquad (5.67)$$

to give a carbonium ion and a nitrile. This "Beckmann fission" takes place when R is a group which can furnish a carbonium ion of considerable stability. In a subsequent step the carbonium ion attacks the nitrogen in the nitrile. This mechanism is clearly intermolecular.

5.4.3.5. *The Schmidt Rearrangements*

Another rearrangement in which electron-deficient nitrogen is the cause of the reaction, and which is therefore related to the Beckmann rearrangement is the acid-catalysed rearrangement of organic azides. The reaction is known as the Schmidt rearrangement, although rearrangements of alkyl and aralkyl azides were discovered by Curtius and Darapsky (1901) long before the work of Schmidt (1924). In synthetic organic chemistry acid-catalysed rearrangements of azides normally take place as part of the so-called Schmidt reaction, in which a number of oxygen bases (alcohols, aldehydes, ketones, carboxylic acids) or olefins dissolved in concentrated sulphuric acid are made to react with sodium azide in chloroform solution, so that azide formation is the first step in the reaction. The formation of the azide and its rearrangement are two mechanistically distinct steps, and rearrangements of azides, separately prepared, can be studied. The intermediacy of azide formation can be demonstrated in some instances by the isolation of azides alongside the rearrangement products, e.g. in the Schmidt reaction of 9-*tert*-butylfluorenol (Arcus and Lucken, 1955; Coombs, 1958a and 1958b). Only in the Schmidt reactions of aldehydes and ketones the intermediates are not known as separately obtainable compounds. Only the rearrangment part of the Schmidt reactions will be discussed here, whereas further reference to azide formation will be made in Section 5.4.7. A very detailed discussion of Schmidt rearrangements has recently been written by Smith (1963).

In general an organic azide may be represented in the form **(5.56)** where

$$\underset{Y}{\overset{X}{\underset{|}{R-C-N_3}}} \qquad \underset{R'}{\overset{R}{>}}C=N-N_2^+ \qquad R-C\overset{O}{\underset{N_3}{<}}$$

$$\textbf{(5.56)} \qquad \textbf{(5.57)} \qquad \textbf{(5.58)}$$

R = alkyl, aryl or hydrogen, and X and Y = alkyl, aryl or OH groups. If one of the groups is OH, then in strongly acid solution (5.56) presumably changes to the hypothetical intermediate (5.57), an iminodiazonium ion or alkylidene azide. Carboxylic acid azides are represented by (5.58). Azide groups are linear, with the organic group attached to the terminal nitrogen at an angle; this is 120° in methyl azide according to Livingston and Rao (1960). In strong acid solution the azide group is protonated most probably on the α-nitrogen, as it is easiest to envisage the loss of nitrogen from such a species according to the reaction

$$\begin{array}{c} X \\ | \\ Y-C-N-N\equiv N \\ | \\ R \end{array} \overset{H}{\underset{+}{}} \rightarrow \begin{array}{c} X \\ | \\ Y-C-\overset{..}{N}H \\ | \quad + \\ R \end{array} + N_2$$

This leaves an electron-deficient nitrogen centre, which induces the migration of one of the aryl or alkyl groups or hydrogen:

$$\begin{array}{c} X \\ | \\ Y-C-\overset{..}{N}H \\ | \quad + \\ R \end{array} \longrightarrow \begin{array}{c} X \\ | \\ Y-C-NHR \\ + \end{array} \qquad (5.59)$$

The hydrolysis of (5.59) yields an amine and a variety of products, depending on what X and Y are. For example, benzyl azide rearranges in warm 1:1 sulphuric acid or in concentrated hydrochloric acid by both phenyl and hydrogen migration, the first resulting in aniline and formaldehyde, and the second in benzaldehyde and ammonia (Curtius and Darapsky, 1901). Alkylidene azides yield amides, i.e.

$$\underset{R'}{\overset{R}{\diagdown}}C=N-\overset{+}{N_2} \xrightarrow{-N_2} \underset{R'}{\overset{R}{\diagdown}}C=\overset{..}{\underset{+}{N}}$$

$$\underset{}{\overset{O}{\underset{\|}{R-C-NHR'}}} \xleftarrow{H_2O} R-\overset{+}{C}=N-R'$$

except when one of the R groups yields a very stable carbonium ion, in which case the loss of nitrogen leads directly to a nitrile:

$$\underset{R'}{\overset{R}{\diagdown}}C=\overset{..}{\underset{+}{N}} \longrightarrow R'-C\equiv N + R^+$$

Nitriles are commonly obtained from aldehydes by loss of the CH proton from the intermediate.

Carboxylic acid azides give amines having one less carbon atom, by the following mechanism (Newman and Gildenhorn, 1948):

$$R-C\underset{\underset{H}{N-N_2^+}}{\overset{O}{\diagup}} \xrightarrow{-N_2} R\underset{+}{\overset{O}{\diagup}}C-\ddot{N}H \longrightarrow \underset{+}{O=C}-\overset{H}{\underset{\cdot\cdot}{N}}-R$$

$$\downarrow + H_2O$$

$$RNH_2 + CO_2$$

All these rearrangements take place with complete retention of configuration (Campbell and Kenyon, 1946) and hence the migrating group does not part with its electrons in the reaction. Aryl azides do not rearrange in acid solution, but decompose to give aniline derivatives (Smith and Brown, 1951).

The rates of the Schmidt rearrangements of benzhydryl azides and 1,1-diarylethyl azides in sulphuric acid solution in glacial acetic acid show good correlations with h_0 (Gudmundsen and McEwen, 1957). This is consistent with preprotonation as the first step in the reaction. Hydrazoic acid itself has been shown to be very weakly basic (see Chapter 3) and so are probably organic azides. Electron-releasing substituents accelerate the reactions and $\log k$ values are correlated quite satisfactorily with σ substituent constants ($\rho = -2.26$). This is consistent with a substituent effect on the basicity of the substrate and on the breaking of the $N-N_2$ bond. Any assistance of nitrogen loss by the migrating group through a bridged transition state such as (**5.60**)

(**5.60**)

would require a correlation with σ^+ constants of the substituents, and since a correlation with σ constants is found, this does not seem to occur. Substituent effects are thus consistent with the loss of nitrogen as the rate-determining step.

A study of the Schmidt reaction of a number of unsymmetrically *m*- and *p*-substituted benzhydrols has shown that both benzaldehydes and substituted benzaldehydes are obtained, the product ratio varying with the substituent in a way which demonstrates that electron-releasing substituents promote the migration of the ring, whereas electron-withdrawing substituents retard it (Tietz and McEwen, 1955). In fact an excellent correlation of the logarithm of the product ratio (benzaldehyde/substituted benzaldehyde) with σ constants was obtained ($\rho = -2\cdot03$), and a similar result was obtained for products from the Schmidt reaction of 1,1-diphenylethylene ($\rho = -2\cdot11$) (Tietz and McEwen, 1955). Since a transition state of type (5.60) would require a correlation with σ^+ constants, the precise significance of these results is not clear, but they do indicate that migratory aptitudes of groups in the Schmidt rearrangement of arylalkyl azides are similar to those in the pinacolic rearrangements.

In alicyclic and some other cyclic systems the Schmidt rearrangement, like the Beckmann rearrangement, leads to ring enlargements. So, for example, the acid-catalysed rearrangement of 9-azidofluorene yields phenantridine:

$$\text{(5.68)}$$

The rates of this reaction in acetic acid/sulphuric acid are also proportional to h_0 (Arcus and Evans, 1958). The unimolecular rearrangement of the conjugate acid is thus the rate-determining step, the loss of nitrogen probably preceding the bond rearrangement. The reaction is general for other azidofluorenes. When one of the rings is substituted it migrates preferentially if the substituent is electron releasing, and is preferentially retained on the carbon if the substituent is electron attracting (Arcus and Lucken, 1955). This behaviour is thus closely similar to that found in non-cyclic systems.

In the Schmidt reaction of aldehydes and ketones, the rearranged products are closely similar to those in the Beckmann rearrangement of the corresponding oximes. It appears thus that geometrical isomerism of the intermediate iminodiazonium ion determines the products of migration. So unsymmetrically p-substituted benzophenones subjected to the Schmidt reaction give isomeric benzanilides in the ratio 1:1, regardless of the nature of the substituent (Smith and Horwitz, 1950). Also, products obtained from alkyl aryl ketones in the Schmidt reaction are closely similar to those in the Beckmann rearrangement of the corresponding oximes (Smith and Antoniades, 1960). All this means that it is the loss of nitrogen that is rate determining in these reactions, the remaining fragment being then identical with that in the Beckmann rearrangement.

In the acid-catalysed rearrangement and decomposition of acid azides the rearranged intermediate, $\overset{+}{\text{RNHCO}}$, may alternatively be obtained by the protonation of an isocyanate. It has proved possible in fact to isolate an isocyanate from a Schmidt reaction of 4-phenanthrene-carboxylic acid in a mixture of trifluoroacetic acid and trifluoroacetic anhydride, i.e. under conditions where further hydrolysis could not occur (Rutherford and Newman, 1957).

Substituent effects in the Schmidt reaction of substituted benzoic acids in concentrated sulphuric acid are correlated satisfactorily with σ constants of the substituents with $\rho = -2\cdot 14$ (Briggs and Littleton, 1943). A correlation with σ constants is consistent with a substituent effect on the preprotonation equilibrium and loss of nitrogen, and suggests that the loss of nitrogen is the slow stage in this reaction, as in the rearrangements of other azides.

5.4.3.6. Diphenyltriketone-benzoin Rearrangement

The rearrangement of diphenyltriketone (**5.61**) in 1:1 sulphuric acid into benzoin (**5.62**) and benzyl (**5.63**) involves steps similar to a pinacolic rearrangement combined with a decarboxylation reaction (Schonberg and Azzam, 1958). The details of the mechanism have not been firmly established, but the following scheme seems possible:

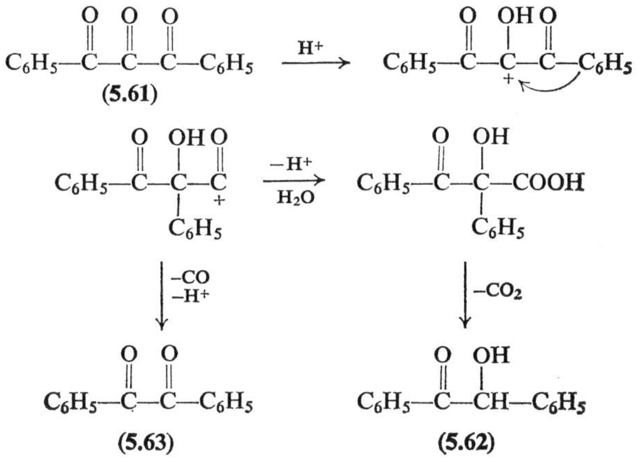

$$C_6H_5-\underset{(5.61)}{\overset{O}{\underset{\|}{C}}-\overset{O}{\underset{\|}{C}}-\overset{O}{\underset{\|}{C}}-C_6H_5} \xrightarrow{H^+} C_6H_5-\overset{O}{\underset{\|}{C}}-\overset{OH}{\underset{|}{C}}-\overset{O}{\underset{\|}{\overset{+}{C}}}-C_6H_5$$

$$C_6H_5-\overset{O}{\underset{\|}{C}}-\overset{OH}{\underset{|}{C}}-\underset{\underset{C_6H_5}{|}}{\overset{O}{\underset{\|}{\overset{+}{C}}}} \xrightarrow[H_2O]{-H^+} C_6H_5-\overset{O}{\underset{\|}{C}}-\underset{\underset{C_6H_5}{|}}{\overset{OH}{\underset{|}{C}}}-COOH$$

$$\downarrow \begin{array}{c}-CO\\-H^+\end{array} \qquad\qquad \downarrow -CO_2$$

$$\underset{(5.63)}{C_6H_5-\overset{O}{\underset{\|}{C}}-\overset{O}{\underset{\|}{C}}-C_6H_5} \qquad \underset{(5.62)}{C_6H_5-\overset{O}{\underset{\|}{C}}-\overset{OH}{\underset{|}{CH}}-C_6H_5}$$

The rearranged carbonium ion intermediate is partitioned between two parallel reaction paths leading to products (**5.62**) and (**5.63**). The decarbonylation and decarboxylation reactions involved in these transformations will be discussed in greater detail in Section 5.4.4.

5.4.4. DECARBONYLATIONS AND DECARBOXYLATIONS

It has been known since the last century that many substances are decomposed by concentrated sulphuric acid with the evolution of gases, mostly carbon monoxide, carbon dioxide, and sulphur dioxide (the latter usually at higher temperature). So, for example, formic acid and oxalic acid were known to decompose quantitatively to give carbon monoxide, and a 1:1 mixture of carbon monoxide and carbon dioxide respectively. The decomposition of phenylacetic acids was under investigation early in the century (Bistrzycki and Reintke, 1905; Bistrzycki and Siemiradzki, 1906), when the first thorough kinetic study of the decomposition of oxalic acid in concentrated sulphuric acid also appeared (Bredig and Lichty, 1906; Lichty, 1907). The kinetics of the decarbonylation of several other acids (formic, malic, citric and triphenylacetic) were studied in the 1923-34 period. The rôle of sulphuric acid in these reactions was often thought to be that of a dehydrating agent until Hammett (1935) showed that the rates of all these reactions are determined by the H_0 acidity function of the strongly acid solvent, and that the acid therefore acts as a catalyst. The same conclusion about the decomposition of formic acid was reached earlier from purely chemical evidence (Senderens, 1927).

The formally similar decarbonylations of aromatic aldehydes, discovered by Bistrzycki and Fellmann (1910), and Bistrzycki and Ryncki (1912), might have been assumed to involve the intermediate formation of formic acid, but have been shown in the more recent past to be electrophilic substitutions by the hydrogen ion. The closely similar decarboxylations of organic acids are reactions of the same kind, only more complicated. These reactions could therefore equally well have been discussed in Section 5.4.5.

5.4.4.1. Decarbonylations of Organic Acids

The acids which undergo decarbonylation in concentrated sulphuric acid are of a few specific types. Apart from formic and oxalic acid, these are triphenyl- and diphenylacetic acid, and some β-hydroxy and β-keto acids. Gillespie and Leisten (1954a) pointed out that the common feature of all these acids is that they give relatively stable carbonium ions upon decarbonylation. However, benzoylformic acid gives benzoic acid (Elliott and Hammick, 1951), since the benzoyl cation is unstable in concentrated sulphuric acid (see Chapter 4). The carbon monoxide evolved in this decarbonylation definitely arises from the carboxyl group (Banholzer and Schmidt, 1956).

On the basis of his H_0 correlations Hammett (1940) proposed the following mechanism for these decarbonylations:

$$R-COOH + H_2SO_4 \rightleftharpoons R-C\begin{matrix}O\\OH_2^+\end{matrix} + HSO_4^-$$

$$R-C\begin{matrix}O\\OH_2^+\end{matrix} \longrightarrow RCO^+ + H_2O$$

$$RCO^+ \longrightarrow R^+ + CO$$

(where R = H for formic acid). A summary of the H_0 correlations for all decarbonylations so far studied is given in Table 5.10. The slopes are usually interpreted as indicating whether the monoprotonated or the diprotonated acids are the unstable species (Hammett, 1940; Elliott and Hammick, 1951). Integral slopes of one or two mean that one or two degrees of protonation

TABLE 5.10

Summary of the rate-acidity function correlations for decarbonylation reactions

Acid	Temperature (°C)	Concentration range (weight % H_2SO_4)	H_0 range	Slope (log k vs. $-H_0$)	Reference†
(1) Formic	15–45	85–98·9	−8·3 to −10·6	1·17	a, b, c
(2) Oxalic	25	90–96	−9·0 to −9·9	~1·2	d
		98–100	−10·3 to −12	2·0	
(3) Malic	20–50	96·5–99·2	−10 to −10·66	0·9	e, f
(4) Citric	15, 25, 35	95–98·8	−9·7 to −10·5	2	g
(5) Benzoylformic	18·1–32·8	94–98	−9·6 to −10·3	2·1	h
	15	85·5–99·9	−8 to −9·4	1·8	i
	25		−7·9 to −9·8	1·75	
(6) Triphenylacetic	12, 22	93–97	−9·4 to −10	2·6	j

† a. Schierz (1923); b. Schierz and Ward (1928); c. DeRight (1933); d. Lichty (1907); e. Whitford (1925); f. De Right (1934); g. Wiig (1930a); h. Elliott and Hammick (1951); i. Vinnik et al. (1959b); j. Dittmar (1929).

are very incomplete over the acid concentration range in which decarbonylation rates are measured. A unit slope, however, does not rule out the possibility that one stage of protonation may be complete, and that a second protonation forms the unstable species, a correlation with H_0 being then in fact a correlation with the H_+ function. Likewise a slope of two does not rule out the possibility that a first protonation is complete, and a second and a third are incomplete and increase with increasing acidity. What are the real protonation equilibria in concentrated sulphuric acid of the acids in Table 5.10 cannot be decided without a knowledge of their pK_{BH^+} values. Few such values are known (Table 5.11). The only acid for which this is known with accuracy, from an extrapolation of conductance measurements to zero time, is oxalic acid (Liler, 1963). The values for benzoylformic acid (Vinnik et al., 1959b) and diphenylacetic acid (Vinnik and Ryabova, 1962) have been obtained spectrophotometrically and also involve extrapolations in obtaining extinction coefficients of fully protonated acids. For other acids in Table 5.11 estimates of pK_{BH^+} values for the first protonation have been made using their pK_a values in aqueous solution and the correlation of Fig. 3.11, p. 130. The estimate for oxalic acid is in very good agreement with the experimentally obtained value. This agreement gives some weight to the other estimates. Comparing the estimated pK_{BH^+} values with the acidity ranges over which kinetics were studied (Table 5.10), it can be seen that oxalic acid is incompletely monoprotonated in pure sulphuric

TABLE 5.11

Measured and estimated pK_{BH^+} values of acids which undergo decarbonylation in concentrated sulphuric acid

Acid	pK_1 (in aqueous soln.)†	pK_2	pK_{BH^+} (estimate)‡	pK_{BH^+}	Reference§
(1) Formic	3·75		−8·7		
(2) Oxalic	1·27	4·28	−11·8 (stat. corr.)	−11·64 (stat. corr.)	a
(3) Malic	~3·7 (estimate)	5·14	~−8·7		
(4) Citric	3·128	4·761	−9·1 (stat. corr.)		
(5) Benzoylformic	—		—	−9·5	b
(6) Triphenylacetic	3·96		−8·4		
(7) Diphenylacetic	3·94		−8·4	−9·57	c

† All pK_a values are from Kortüm et al. (1961).
‡ Estimated from the linear correlation of pK_{BH^+} with pK_a (Fig. 3.11, p. 130) for the first protonation.
§ a. Liler (1963); b. Vinnik et al. (1959b); c. Vinnik and Ryabova (1962).

acid, whereas diprotonation is immeasurably small (Liler, 1963). Both extents of protonation increase with increasing acidity and hence a slope of log$k$ vs. $-H_0$ of two is obtained. The H_0 correlation also indicates an incomplete double preprotonation for benzoylformic acid (Elliott and Hammick, 1951). Vinnik et al. (1959b) report a limiting rate of decomposition in 100 per cent acid, which is consistent with their estimate of the second $pK_{BH_2^{++}}$ value of −10·9 for this acid. Malic acid is probably fully monoprotonated and the second protonation leads to decomposition. Citric acid also seems to be extensively monoprotonated, so that the second and third protonations lead to decomposition—it may be recalled that dicarboxylic acids are extensively diprotonated in concentrated sulphuric acid (see Chapter 3).

Apart from the question of how many protons are added to the acid molecules in the pre-equilibrium step, the structure of the protonated carboxyl group is especially important for the mechanism. Hammett (1940) believed that the proton was added on the hydroxyl oxygen of the carboxyl group. Evidence presented in Chapter 3 shows, on the contrary, that acids are carbonyl protonated. The hydroxyl protonated species could, however, be present in undetectably small concentrations, sufficient to lead to reaction, but this would imply a high instability of such species. Thus the probable unstable species in the decarbonylation of acids 1–5 from Table 5.10 are **(5.64)**–**(5.68)**. The fact

5. REACTION MECHANISMS IN SULPHURIC ACID SOLUTIONS

$$\underset{(5.64)}{\text{H}-\text{C}{\overset{\displaystyle O}{\underset{\displaystyle OH_2^+}{\diagup\!\!\!\diagdown}}}} \qquad \underset{(5.65)}{\overset{\displaystyle HO}{\underset{\displaystyle HO}{}}\!\!{\overset{+}{\text{C}}}-\text{C}{\overset{\displaystyle O}{\underset{\displaystyle OH_2^+}{\diagup\!\!\!\diagdown}}}}$$

Structures (5.64)–(5.68) shown.

that the presence of a hydroxyl group in (5.66) and (5.67) is essential for the stabilization of the resulting carboxonium ions (5.69) and (5.70) suggests that

$$\begin{array}{c} \text{H}-\overset{+}{\text{C}}=\text{OH} \\ | \\ \text{CH}_2 \\ | \\ \text{C(OH)}_2^+ \\ (5.69) \end{array} \qquad \begin{array}{c} \text{CH}_2-\text{C(OH)}_2^+ \\ | \\ \overset{+}{\text{C}}=\text{OH} \\ | \\ \text{CH}_2-\text{C(OH)}_2^+ \\ (5.70) \end{array}$$

they remain unchanged apart from possibly competing for protons with the carboxyl groups.

According to the Hammett mechanism, either complex ionization or decomposition of the oxocarbonium ion may be rate determining. If complex ionization were very fast, correlations of reaction rates with the H_R acidity function would be expected. This has in fact been found only for triphenylacetic acid (Deno and Taft, 1954). Good correlations with H_0 for other acids in Table 5.10 are a strong indication that this is not so in these instances. Studies of these reactions in ^{18}O-labelled sulphuric acid confirm this conclusion: no ^{18}O is exchanged between the protonated acids 1–5 and the solvent under the conditions of decomposition, since the carbon monoxide evolved contains no ^{18}O, but some enrichment in ^{18}O was found for triphenylacetic acid (Ropp, 1958, 1960). This means that complex ionization is rate determining for acids 1–5, or that both steps may be synchronous. A further confirmation of this is obtained from kinetic isotope effects.

Unimolecular decarbonylation reactions from Table 5.10 have provided particularly suitable examples for the study of kinetic isotope effects, mainly because they yield quantitatively gaseous reaction products which can be subjected directly to mass-spectrometric analysis (^{13}C) or radioactivity measurements (^{14}C). These reactions have therefore figured prominently amongst examples used to test the predictions of the theory of absolute reaction rates regarding the effect of atomic mass upon the rates of bond fission. The decomposition of ^{14}C formic acid is one of the early examples of this: the reaction of the ^{12}C acid is 11·11 per cent faster at 273 K and 8·59 per cent faster at 298 K than that of the ^{14}C acid (Ropp et al., 1951). These values are in good agreement with an estimate of the carbon mass effect on the C—O bond frequency (Eyring and Cagle, 1952) and suggest that the breaking of the C—O bond is rate determining. If this is so, the deuterium kinetic isotope effect on the rate of decomposition of formic acid should only be secondary. According to Ropp (1960) formic-d acid decomposes much more slowly than formic acid ($k_H/k_D = 1\cdot72$ at $0\cdot15°C$ and $1\cdot49$ at $25\cdot2°C$). Hence the effect may not be secondary, but a concerted breaking of both the C—O and C—H bonds may be taking place in the rate-determining step. The ^{18}O kinetic isotope effect would be extremely valuable in deciding this issue, but this has not yet been studied. The temperature dependence of the ^{13}C kinetic isotope effect in this reaction is also consistent with the view that the C—O stretching vibration and the related bending vibrations are lost in the transition state (Bigeleisen et al., 1962).

The decomposition of oxalic acid offers the possibility of studying the ^{13}C and ^{14}C isotope effects intramolecularly (Fry and Calvin, 1952). The decomposition of the isotopically labelled acid results in isotope fractionation:

$$\begin{array}{c} \overset{*}{\text{C}}\text{OOH} \\ | \\ \text{COOH} \end{array} \begin{array}{c} \overset{k}{\nearrow} \\ \\ \underset{k'}{\searrow} \end{array} \begin{array}{c} \overset{*}{\text{C}}\text{O}_2 + \text{CO} + \text{H}_2\text{O} \\ \\ \text{CO}_2 + \overset{*}{\text{C}}\text{O} + \text{H}_2\text{O} \end{array} \qquad (\text{C}^* = {}^{13}\text{C or }{}^{14}\text{C})$$

The heavier isotope is found preferentially in the carbon dioxide fragment ($k/k' = 1\cdot032$ for ^{13}C and $1\cdot067$ for ^{14}C), which means that ^{12}C—O bonds in the protonated carboxyl group break more readily than bonds to the heavier carbon isotopes. This result is again consistent with the fission of the C—O bond in the rate-determining step, but does not rule out the synchronous breaking of the C—O and C—C bonds. Similar observations were made on ^{14}C-labelled benzoylformic acid (Ropp, 1960).

In summary, isotope effects on rate are consistent either with a rate-determining C—O bond breaking, or with a concerted C—O and C—R bond breaking. The concerted mechanism seems more attractive, because the stability of the carbonium ion R^+ would have a more direct influence upon the rate of these reactions if such a mechanism obtained.

The partial enrichment in ^{18}O of the carbon monoxide evolved in the decarbonylation of triphenylacetic acid in $95\cdot4$ per cent sulphuric acid enriched in ^{18}O (Deno and Taft, 1954), which suggests a complex ionization pre-equilibrium for this acid, requires the breaking of the C—C bond to be rate determining. No ^{14}C isotope effect in this reaction has, however, been found (Ropp, 1960); therefore, the mechanism of this reaction cannot be regarded as established.

The foregoing discussion shows that there are a number of questions concerning the mechanism of decarbonylation reactions that remain unanswered. Even the supposition that the structure of the protonated carboxyl group in the unstable species is

$$-C\overset{O}{\underset{OH_2^+}{\diagdown}}$$

may be questioned, since no other evidence for the existence of such a species is available (see Chapter 3). The alternative possibility that carbonyl protonated acids decompose according to

$$R-C\overset{OH}{\underset{OH}{\diagdown}}_+ \longrightarrow R^+ + CO + H_2O$$

if R^+ is a carbonium ion of greater stability than the carboxonium ion, is not ruled out by any of the kinetic isotope effects, except that it would require the deuterium kinetic isotope effect in the decomposition of formic acid to be regarded as primary. Its considerable magnitude would not be inconsistent with this. In the absence of information on the ^{18}O kinetic isotope effect this alternative cannot be ruled out.

Some observations on the rate-acidity profiles in decarbonylation reactions also require further study. For example, there is an unexplained rate maximum in the decarbonylation of citric acid at >98·8 per cent sulphuric acid (Wiig, 1930a), and a similar rate maximum in the decomposition of oxalic acid in oleum solutions at 14 per cent sulphur trioxide (Wiig, 1930b).

Few other decarbonylation reactions appear to be free of side reactions. Diphenylacetic acid, for example, which yields only 40 per cent of the theoretical yield of carbon monoxide, undergoes sulphonation in concentrated sulphuric acid as a parallel reaction (Vinnik and Ryabova, 1962). The reactions of benzilic acid in concentrated sulphuric acid, which lead to carbon monoxide evolution and to intensely coloured solutions (von Dobeneck and Kiefer, 1965), are believed to occur via the initial formation of a carbonium ion (5.71)

$$\underset{Ph}{\overset{Ph}{\diagdown}}C\underset{OH}{\overset{COOH}{\diagup}} + H^+ \longrightarrow \underset{Ph}{\overset{Ph}{\diagdown}}C_+\overset{COOH}{\diagup} + H_2O$$

(5.71)

which undergoes decarbonylation and is also converted into a number of other condensed and complicated carbonium ions, some of which are responsible for the colouration of the solutions.

5.4.4.2. Decarbonylations of Aromatic Aldehydes

Our present understanding of these reactions is due entirely to the work of Schubert and his collaborators (1954–58). They have shown that formic acid is not an intermediate in these reactions, because formic acid decarbonylates faster than aromatic aldehydes under the same conditions (Schubert and Zahler, 1954). Aromatic aldehydes decompose in >60 per cent sulphuric acid at higher temperatures (80–100°C) giving an almost quantitative yield of carbon monoxide. Electron-releasing substituents are essential to achieve reaction under these conditions and 2,4,6-trimethyl-, 2,4,6-triethyl- and 2,4,6-triiso-propylbenzaldehyde were studied using both spectrophotometric and gas volumetric methods (Schubert and Zahler, 1954). The fact that electron-releasing substituents favour the reaction in which apparently a positive fragment ($\overset{+}{\text{CHO}}$) is lost, indicates that the reaction is not a simple breakdown of the aldehyde or the protonated aldehyde, but that it probably involves proton-addition at a site under the direct influence of the electron-releasing substituents, i.e. at the ring itself. The rate-acidity profiles for these reactions show maxima at acid concentrations of 70 to 90 per cent, the shapes of the curves being different depending on substituents. The rapid oxygen protonation of these aldehydes can be studied spectrophotometrically and their pK_{BH^+} values have been determined (Schubert and Zahler, 1954). With a half-protonation for mesitaldehyde at 64 per cent acid ($pK_{BH^+} = -4\cdot6$), any mechanism involving O-protonated aldehydes would lead to a constant rate at higher acidities where protonation is complete. While this is almost true for mesitaldehyde (rate almost independent of acidity between 75 and 95 per cent acid, but decreasing towards 100 per cent acid), it is certainly not true for 2,4,6-triethyl and 2,4,6-triisopropyl derivatives, which show pronounced rate maxima at 85–90 per cent and at 80 per cent acid respectively. No simple correlation of rate with acidity function obtains for these reactions. To account for these facts Schubert and Zahler (1954) proposed the following mechanism:

$$\text{ArCHO} + \text{H}^+ \underset{k_{-1}}{\overset{k_1}{\rightleftarrows}} \text{ArCH}\overset{+}{\text{OH}}$$

(preprotonation)

$$\text{ArCHO} + \text{HA}_i \underset{k_{-2}}{\overset{k_2}{\rightleftarrows}} \overset{+}{\text{Ar}}\!\!\begin{array}{c}\diagup\text{H}\\\diagdown\text{C--H}\\\|\\\text{O}\end{array} + \text{A}_i$$

(ring protonation)

$$\overset{+}{\text{Ar}}\!\!\begin{array}{c}\diagup\text{H}\\\diagdown\text{C--H}\\\|\\\text{O}\end{array} + \text{A}_i \xrightarrow{k_3} \text{ArH} + \text{CO} + \text{HA}_i$$

(decomposition)

In this mechanism HA_i are acids (H_3O^+ and H_2SO_4) and A_i conjugate bases (H_2O and HSO_4^-) present in the solvent mixtures. This mechanism shows the reaction as an electrophilic displacement of the fragment HCO^+ and implies general acid catalysis in ring protonation and general base catalysis in its reverse and in the decomposition. This last step accounts for the decrease in rate observed at >90 per cent sulphuric acid, where the concentrations of both H_2O and HSO_4^- decrease sharply. The rôle of oxygen protonation is simply in controlling the concentration of the free base, which is the subject of electrophilic attack. Some information on the relative rates of the two reactions of the ring-protonated intermediate was obtained from deuterium isotope effects (Schubert and Burkett, 1956). A comparison of the rates of decarbonylation of mesitaldehyde and mesitaldehyde-d_1 shows that the protio compound decarbonylates faster over the acid concentration range 60–100 per cent ($k_{ArHCO}/k_{ArDCO} = 1\cdot8-2\cdot8$). This means that the removal of the formyl proton is at least partly rate controlling.

Solvent isotope effects on the decarbonylation of mesitaldehyde proved more difficult to interpret, but are essentially consistent with the above mechanism (Schubert and Burkett, 1956). The same is true of solvent isotope effects in the decarbonylation of 2,4,6-triisopropyl benzaldehyde, in which the presence of the bulky isopropyl groups hinders the approach of bases for removal of the ring proton, and hence favours the removal of the formyl proton (Schubert and Myhre, 1958).

The mechanism proposed by Schubert and Zahler (1954) accounts satisfactorily for most experimental observations on the rates of decarbonylation of aldehydes. Some details, however, remain incompletely clarified. For example, it is not clear why some rate-acidity profiles show an extensive rate plateau (mesitaldehyde), whereas others show pronounced rate maxima (2,4,6-triethyl- and 2,4,6-triisopropylbenzaldehyde), although the structures of the substrates are closely similar.

It is interesting that 2,4,6-trimethoxybenzaldehyde, which is a relatively strong oxygen base ($pK_{BH^+} = -2\cdot1$) does not undergo decarbonylation at all, but breaks up quantitatively in moderately concentrated mineral acids into 1,3,5-trimethoxybenzene and formic acid (Burkett *et al.*, 1959). The reason for this complete change of reactivity of the aldehyde with substitution of still more electron-releasing groups remains unexplained. A mechanism similar to the above was suggested for this reaction, except that the third step was assumed to be a nucleophilic attack by water molecules on the carbonyl carbon. A formally similar acid-catalysed reaction, the deacylation of aromatic ketones with strong sulphuric acid to give acetic acid and a hydrocarbon, follows the H_0 acidity function, and is believed to involve the decomposition of the conjugate acid of the ketone (Schubert and Latourette, 1952).

5.4.4.3. *Decarboxylations of Aromatic Carboxylic Acids*

The decarboxylations of aromatic carboxylic acids in strongly acid solutions are reactions closely similar to the decarbonylations of aromatic aldehydes.

The reaction occurs principally with aromatic systems which afford considerable carbonium ion stabilization. The best known example is that of mesitoic acid (Schubert, 1949), which decomposes in 81·8–100 per cent sulphuric acid at 50–90°C into carbon dioxide and mesitylene (which is sulphonated in >88 per cent acid). The rate of this reaction follows the H_0 acidity function up to −7 (Long and Varker, 1957), passes through a maximum at 80–85 per cent acid, and falls to zero in 100 per cent sulphuric acid (Schubert, 1949). There is some evidence of catalysis by sulphuric acid molecules.

The decarboxylations of 2,4,6-trialkylbenzoic acids in concentrated sulphuric acid are complicated by the formation of oxocarbonium ions (Schubert et al., 1954). The conversion of mesitoic acid to the oxocarbonium ion is negligible in <90 per cent sulphuric acid, however, and its pK_{BH^+} could be determined as −7·3. The maximum in the rate of decarboxylation at 80–85 per cent acid can thus be interpreted in the same way as the maxima in the rates of decarbonylation of aromatic aldehydes (see Section 5.4.4.2). The activating effects for decarboxylation of the methyl, ethyl and isopropyl groups in the *para* position are approximately the same, but the rates of decarboxylation of the 2,4,6-trialkylbenzoic acids fall in the order: triisopropyl ≫ triethyl > trimethyl. This may be explained in terms of a steric effect (Schubert et al., 1954).

The decarboxylation of 2,4,6-trimethoxybenzoic acid is complicated by the formation of oxocarbonium ions at much lower acidities (the conversion to RCO^+ is complete in 64 per cent perchloric acid) and the rate-determining step has not been identified with certainty (Schubert et al., 1955).

A recent example of a decarboxylation which is complicated in a different way is that of azulene-1-carboxylic acid (Longridge and Long, 1968b). The mode of ionization of this acid in moderately concentrated mineral acids could not be unambiguously established (Longridge and Long, 1968a). The acid undergoes decarboxylation in 0·01–6 M perchloric or sulphuric acid. In very dilute acid (<0·03 M) the reaction is acid-catalysed, but a gradual increase in rate in the range 0·5–6 M perchloric or sulphuric acid is ascribed to a positive salt effect. Beyond 6 M acid a rapid decrease in rate is observed due to equilibrium protonation of the ring at the 3-carbon. However, this should be in equilibrium with the 1-carbon protonation, and the concentrations of both species should increase with increasing acid concentration. In spite of this the rate of decarboxylation decreases rapidly (Long, 1968).

5.4.5. ELECTROPHILIC AROMATIC SUBSTITUTIONS BY THE HYDROGEN ION

Strongly acid media have been described in Chapters 3 and 4 as media giving rise to positively charged species, either by protonation or by complex ionization of solutes. Such species can act as electrophiles and effect electrophilic substitution reactions. The most important electrophiles in the medium are, however, the hydrogen ions themselves, but undissociated acid molecules may also play this part.

Electrophilic substitutions by the hydrogen ion in strongly acid media are often treated as acid-catalysed reactions, although, strictly speaking, hydrogen ions are consumed in the reaction, according to the equation

$$ArX + H^+ = ArH + X^+ \tag{5.69}$$

except when X^+ reacts further to regenerate a hydrogen ion. An electrophilic substitution by the hydrogen ion which may legitimately be regarded as acid-catalysed, is the aromatic hydrogen exchange reaction in which one isotope replaces another without altering the acidity of the medium. The theoretical significance of the hydrogen exchange reaction is considerable, since it represents the simplest type of electrophilic substitution.

General and mechanistic aspects of hydrogen exchange have been reviewed recently by Gold (1964) and by Willi (1965). The subject is under intensive study by several groups of investigators and the past few years have brought important new developments.

Within the past decade electrophilic displacements of other groups by the hydrogen ion have also received considerable attention. A number of such reactions have been reviewed by Willi (1965). A few reactions of this type have already been discussed in Section 5.4.4. Some further examples will be discussed in the present Section.

5.4.5.1. *Hydrogen Isotope Exchange*

Studies of hydrogen isotope exchange began very soon after the discovery of heavy hydrogen in 1933. It was Ingold *et al.* (1934) who first recognized that aromatic hydrogen isotope exchange was an electrophilic displacement reaction, with characteristics closely similar to those of other electrophilic substitution reactions, e.g. regarding the directive effects of substituents. Ingold *et al.* (1936b) obtained the following sequence for the activating effect of substituents on this reaction: $O^- > NMe_2 > OMe > H > SO_3H$. *Ortho* and *para* positions were generally activated much more strongly than the *meta* position. The reactivity of various ring positions in this reaction, as in other electrophilic aromatic substitutions, is usually expressed in terms of the ratio of rates of exchange at a particular position relative to that of *one* position in benzene.

In deuteration reactions deuterium uptake may in general take place at more than one position (*ortho* and *para* positions especially showing similar reactivity), and therefore information on the reactivity of particular ring positions in exchange is more conveniently and more accurately obtainable from studies of the loss of heavy hydrogen isotopes from specifically labelled compounds to a light reaction medium. This approach has been favoured in more recent hydrogen exchange studies. Loss of deuterium is usually followed

TABLE 5.12

Partial rate factors for exchange in benzene derivatives (PhX)

X	Solvent system	Isotope exchanged	Partial rate factor *p*-	*o*-	*m*-	Reference†
OCH_3	8·53 M H_2SO_4	D	55,000	23,000	—	*a*
OCH_3	10·43 M $HClO_4$	D	—	—	0·25	*a*
F	$CF_3CO_2H—H_2SO_4—H_2O$ 82·36:14·47:3·17	T	1·57	—	—	*b*
Cl	$CF_3CO_2H—H_2SO_4—H_2O$ 82·36:14·47:3·17	T	0·119	—	0·0017	*b*
Br	$CF_3CO_2H—H_2SO_4—H_2O$ 82·36:14·47:3·17	T	0·068	—	—	*b*
I	$CF_3CO_2H—H_2SO_4—H_2O$ 82·36:14·47:3·17	T	0·081	—	—	*b*
CH_3	71·34% H_2SO_4	T	250	250	5·0	*c*
t-C_4H_9	71·34% H_2SO_4	T	180	170	—	*c*
C_6H_5	$CF_3CO_2H—H_2SO_4—H_2O$ 95·31:2·21:2·48	T	143	133	—	*d*

† *a*. Satchell (1956); *b*. Eaborn and Taylor (1959); *c*. Eaborn and Taylor (1961a); *d*. Eaborn and Taylor (1961b).

by infrared spectra or by mass spectrometry, and loss of tritium by radioactivity measurements.

A considerable body of information is available on the reactivity of various ring positions in substituted benzenes, obtained in a variety of acidic solvents (Tables 5.12 and 5.13). The relative reactivities are not strictly medium independent. For example, a comparison of the reactivity of the *p*-methyl with the *p*-*t*-butyl derivative in Table 5.12 shows the latter to be less reactive in the sulphuric acid medium, but the reverse is true for media trifluoroacetic

TABLE 5.13

Partial rate factors for exchange in toluene at 25°C

Weight % H_2SO_4	Isotope exchanged	Partial rate factors *p*-	*o*-	*m*-	Reference†
68	D	83	83	1·9	*a*
71·34	T	250	250	5·0	*b*
73·24	T	254	—	4·99	*c*
75·30	T	243	—	5·02	*c*
77·67	T	—	—	5·42	*c*
79·80	T	—	—	5·72	*c*
81·14	T	—	—	5·70	*c*

† *a*. Gold and Satchell (1956); *b*. Eaborn and Taylor (1961a); *c*. Eaborn and Taylor (1960).

acid/water and perchloric acid/trifluoroacetic acid/water (Eaborn and Taylor, 1961a). Acidic media containing trifluoroacetic acid as a component have the advantage over aqueous sulphuric acid and perchloric acid of offering enhanced solubility of organic substrates, but their acidities are less readily measurable and less well defined owing to lower dielectric constants. Concentrated sulphuric acid may be unsuitable owing to sulphonation as a side reaction. Mixtures of perchloric acid, trifluoroacetic acid and water are preferable if sulphonation is liable to cause serious difficulty. Such media have been used in studies of hydrogen exchange of polycyclic hydrocarbons, which have shown that naphthalene is strongly activated in this reaction compared with benzene (see Gold, 1964) and phenanthrene and anthracene are even more so, e.g. the 9-position of anthracene by a factor of 3×10^7 (Bancroft et al., 1964). All this information is consistent with the view that delocalization of the partial positive charge imparted to the system in the electrophilic attack is extremely important in determining reaction rates. Since *ortho* and *para* positions show closely similar reactivities in most systems (see Table 5.12), steric effects on hydrogen exchange are small; in fact, for toluene in sulphuric acid solution they are nil (Table 5.13).

The attack of an electrophile on the aromatic system may be imagined to occur in a single-step bimolecular displacement reaction, i.e.

$$\text{ArH} + \text{X}^+ \rightarrow \text{ArX} + \text{H}^+$$

or via the formation of an intermediate in a two-step reaction:

$$\text{ArH} + \text{X}^+ \longrightarrow \left[\text{Ar}\begin{array}{c}\diagup \text{H}\\ \diagdown \text{X}\end{array}\right]^+ \longrightarrow \text{ArX} + \text{H}^+$$

In aromatic hydrogen exchange the electrophilic reaction is the transfer of a proton to a ring carbon. Consistent with experience that such reactions are slow, aromatic hydrogen exchange of the more reactive substrates has been shown to be generally acid-base catalysed in aqueous buffer solutions (for 1,3,5-trimethoxybenzene by Kresge and Chiang, 1959, and 1961b, and for azulene by Colapietro and Long, 1960, and Gruen and Long, 1967). For a variety of substrates in moderately concentrated acid solutions the reaction follows the H_0 acidity function quite well (Table 5.14). This led to suggestions that in concentrated acid solutions the reaction may possibly occur through a preprotonation equilibrium (Gold and Satchell, 1955a, 1956), but since Long and Paul (1957) have shown that slow proton transfer reactions should also follow the acidity function this assumption is unnecessary. The more difficult question of whether acidic species other than the H_3O^+ ion in sulphuric acid solutions also catalyse the reaction was examined by Kresge et al. (1965), who

TABLE 5.14

Slopes of plots of log k_{exch} vs. $-H_0$ for dedeuteration and detritiation of benzene derivatives

Substrate	Acid	Slope	Reference‡
1,3,5-Trimethoxybenzene-2-*t*	0–3 M HClO$_4$	1·07	a
1,3-Dimethoxybenzene-4-*t*	1·1–5·2 M HClO$_4$	1·14	b
p-Cresol-*o*-*d*	1·0–6·2 M H$_2$SO$_4$	1·08	c
1,2,3-Trimethoxybenzene-4,6-*t*	3·5–5·2 M H$_2$SO$_4$	1·0	d
Anisole-*p*-*d*	3·5–7·6 M H$_2$SO$_4$	1·18	e
Anisole-*o*-*d*	5·2–7·9 M H$_2$SO$_4$	1·18	e
p-Cl-phenol-*o*-*d*	5·4–9·9 M H$_2$SO$_4$	1·02	c
p-NO$_2$-phenol-*o*-*d*	10·1–16·0 M H$_2$SO$_4$	~0·8†	c
Toluene-*p*-*d*	7·7–12·3 M H$_2$SO$_4$	1·46	f
Toluene-*p*-*t*	11·0–12·8 M H$_2$SO$_4$	1·60	g
Toluene-*m*-*t*	12·1–14·3 M H$_2$SO$_4$	Not const.	g
Benzene-*d*	10·9–14·9 M H$_2$SO$_4$ 14·2–15·9 M H$_2$SO$_4$	1·1 → 1·8	h i
Benzene-*t*	12·1–15·2 M H$_2$SO$_4$ 14·2–15·9 M H$_2$SO$_4$	1·3 → 1·8	h i

† Probably affected by hydrogen bonding of the NO$_2$ group to the solvent. In >80 per cent acid protonation occurs.

‡ *a*. Kresge and Chiang (1961); *b*. Kresge *et al.* (1967); *c*. Gold and Satchell (1955b); *d*. Satchell (1959); *e*. Satchell (1956); *f*. Gold and Satchell (1956); *g*. Eaborn and Taylor (1960); *h*. Gold and Satchell (1955b); *i*. Olsson and Russell (1969).

compared the rates of detritiation of 1,3-dimethoxybenzene-4-*t* in sulphuric acid and perchloric acid solutions of the same acidity and found that the rates in sulphuric acid solutions were 2·5–3 times higher. In the same solutions the equilibrium protonation of azulenes is greater in perchloric acid than in sulphuric acid and, assuming that this also holds for 1,3-dimethoxybenzene, the rates of exchange in sulphuric acid were estimated to be as much as 5·5 times higher than in perchloric acid. The excess rate could be ascribed to catalysis by HSO$_4^-$ ions and H$_2$SO$_4$ molecules. In this connection it is interesting to note that log k vs. $-H_0$ slopes in the more concentrated sulphuric acid tend to be greater than unity (Table 5.14) and that for benzene in particular there is a distinct increase in slope in >80 per cent sulphuric acid (Fig. 5.5), i.e. in the region in which the concentration of sulphuric acid molecules rises significantly (see Fig. 1.1, p. 19). Therefore these increases in slope may be attributed to catalysis by sulphuric acid molecules.

In view of the firmly established general acid catalysis in aromatic hydrogen exchange, the reaction should be formulated as involving a generalized proton donor (HA) as follows:

$$\text{ArL} + \text{HA} = \text{ArH} + \text{LA} \tag{5.70}$$

where L stands for D or T. The fact that the conjugate base A is also involved in any transition state that may be formulated probably accounts for differences in partial rate factors for exchange in various media and in particular

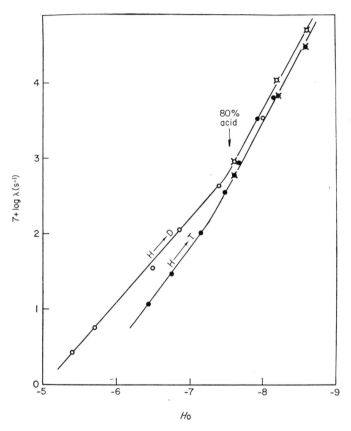

FIG. 5.5. Correlations of dedeuteration and detritiation rates of benzene with the H_0 acidity function. (At <80 per cent acid data of Gold and Satchell, 1955b; at >80 per cent acid data of Olsson and Russell, 1969.)

for the smaller reactivities of *ortho* positions as compared with *para* in some media (see Table 5.12).

Whether reaction (5.70) occurs as a one-step displacement with a possible transition state (**5.71**), or as a two-step reaction via the formation of an intermediate (**5.72**) is the next fundamental question. The analogous question

$$\text{Ar} \overset{\text{H}\cdots}{\underset{\text{L}\cdots}{\diagdown}} \text{A} \qquad \left[\text{Ar} \overset{\text{H}}{\underset{\text{L}}{\diagdown}} \right]^+$$

(5.71)   (5.72)

in some other electrophilic substitutions has been answered in favour of the formation of an intermediate by the use of hydrogen isotopes (see Section 5.4.6). The use of isotopes seemed difficult in a reaction which is itself an isotope exchange (Gold, 1964). However, detailed analyses of primary isotope effects (T vs. D exchange) and solvent isotope effects (D_3O^+ vs. H_3O^+ as electrophiles) in hydrogen exchange have proved extremely valuable tools in deciding the issue.

If an intermediate is formed in a hydrogen exchange reaction, then, owing to its high symmetry, it would be expected to lose either of the hydrogen isotopes at closely comparable rates. The effect of mass on the rate of breaking a bond involving hydrogen is predictable from the vibrational frequency of the bond. Assuming that all differences except those in the zero-point vibrational energy are insignificant, that isotopic substitution affects only the vibrations of the bond which holds the isotope, and that this may be treated as a harmonic oscillator, Swain *et al.* (1958) derived the following relationship between the relative rates of breaking the bond of hydrogen isotopes to a heavier fragment:

$$\frac{k_H}{k_T} = \left(\frac{k_H}{k_D}\right)^{1.442} \tag{5.71}$$

In an intermediate of type (**5.72**) the mass of the aromatic fragment is so large that bond breaking to a hydrogen isotope would be expected to depend only upon its own mass, the mass of the other isotope present (H, D or T) introducing only a minor secondary isotope effect.

Knowledge of the relative rates of the breakdown of the intermediate can be used in conjunction with the steady state treatment of the two-step mechanism for the reaction to predict the primary and the solvent isotope effects on deuterium and tritium exchange. If the reaction mechanism is represented by the sequence

$$\text{ArL} + \text{HA} \underset{k_2 H}{\overset{k_1 H}{\rightleftarrows}} \left[\text{Ar} \overset{\text{H}}{\underset{\text{L}}{\diagdown}} \right]^+ + \text{A}^- \xrightarrow{k_2 L} \text{ArH} + \text{LA}$$

(L = D or T, A^- = any conjugate base) in which the second step is made irreversible by the low concentration of L^+ in the reaction medium, then the observed rate constant for exchange, obtained from the steady state treatment, is given by

$$k_{\text{exch}}^{L} = \frac{k_1^H k_2^L}{k_2^L + k_2^H} = \frac{k_1^H}{1 + k_2^H/k_2^L} \quad (5.72)$$

A comparison of dedeuteration and detritiation rates for a given aromatic system can now provide evidence for a two-step mechanism or against it, since the denominator in this equation is calculable from observed rates of exchange, assuming that the protonation rate constant k_1^H is independent of the isotope present in the aromatic molecule. On this assumption the relationship

$$\frac{k_{\text{exch}}^D}{k_{\text{exch}}^T} = \frac{1 + k_2^H/k_2^T}{1 + k_2^H/k_2^D} = \frac{1 + (k_2^H/k_2^D)^{1.442}}{1 + k_2^H/k_2^D} \quad (5.73)$$

should hold.

Indirect indications that intermediates may be formed in hydrogen exchange reactions come from demonstrations that stable carbonium ions of the type ArH_2^+ are formed in certain aromatic systems (see Chapter 3). Comparisons of dedeuteration and detritiation rates of substrates under the same conditions support this, since they conform with the predictions of equation (5.73), from which the ratios k_2^H/k_2^D have been calculated for several substrates (Table 5.15). All the values are within the limits expected for a primary hydrogen

TABLE 5.15

Relative rates of hydrogen vs. deuterium loss from the intermediate
(Longridge and Long, 1967)

Compound	Log (relative exchange rate)	ΔpK	k_2^H/k_2^D
C_6H_6	0	—	3·4
$C_6H_5CH_3$-m-d	0·8	—	3·4
$C_6H_5CH_3$-o-d	2·6	—	4·6
$C_6H_5CH_3$-p-d	2·6	—	5·5
$C_6H_5OCH_3$-o-d	4·3	—	7·2
$C_6H_5OCH_3$-p-d	4·8	—	6·7
			6·9†
1,3,5-$C_6H_3(OCH_3)_3$-2-d	10·0	−3·5	6·7
			9·0‡
Azulene-1-d	11·5	0	9·2
Guaiazulene-2-sulphonate-3-d	12·2	1·1	7·4
4,6,8-Trimethylazulene-1-d	12·8	2·3	9·6
Guaiazulene-3-d	13·0	3·3	6·0

† Value reported by Russell (1969).
‡ Value reported by Kresge and Chiang (1967).

isotope effect (see Bell, 1959), and provide strong evidence for a two-step mechanism. Strictly the figures should be corrected for a secondary isotope effect (i.e. in order to refer to the loss of D and H from the same residue), but the magnitude of this correction is not known with certainty. The observed relative rates of loss of deuterium and tritium might be expected to be independent of medium, but do in fact depend on acid concentration; for example, Russell (1969) found for the *para* position of anisole $k_{\text{exch}}^{\text{T}}/k_{\text{exch}}^{\text{D}} = 0.460$ in 65 per cent sulphuric acid and 0.509 in 57 per cent acid. This means that Swain's formula (5.71) may be less reliable than is normally assumed.

Solvent isotope effect on detritiation reactions also lends strong support to the two-step mechanism. For a one-step detritiation, the reaction should be faster in ordinary water (with H_3O^+ as electrophile) than in heavy water (with D_3O^+ as electrophile). Measurements have shown that the reverse is the case: $k_{H_2O}/k_{D_2O} = 0.625$ for p-cresol-o-t (Gold et al., 1959 and 1960) and $k_{H_2O}/k_{D_2O} = 0.591$ for 1,3,5-trimethoxybenzene-2-t (Kresge and Chiang, 1962 and 1967). This can readily be accounted for in terms of the two-step mechanism, for which the relative rates are given by

$$\left(\frac{k_{H_2O}}{k_{D_2O}}\right)_{\text{ArT}} = \frac{k_1^H}{k_1^D} \cdot \frac{1 + (k_2^D/k_2^T)}{1 + (k_2^H/k_2^T)} \qquad (5.74)$$

and, since Swain's formula gives $k^D/k^T = (k_H/k_D)^{0.442}$, this becomes

$$\left(\frac{k_{H_2O}}{k_{D_2O}}\right)_{\text{ArT}} = \frac{k_1^H}{k_1^D} \cdot \frac{1 + (k_2^H/k_2^D)^{0.442}}{1 + (k_2^H/k_2^D)^{1.442}} \qquad (5.75)$$

For 1,3,5-trimethoxybenzene Table 5.15 gives k_2^H/k_2^D as 6·7 or 9·0, and it can thus be seen that the second ratio will be $\ll 1$ and will outweigh the first ratio ($k_1^H/k_1^D > 1$), hence producing the reversal of the solvent isotope effect observed. Similarly, the deuteration rate of 1,3,5-trimethoxybenzene is higher than the dedeuteration rate of the labelled substrate in the light medium, a result also accountable in terms of the formation of an intermediate (Kresge and Chiang, 1967). Secondary isotope effects in this reaction have also been analysed and shown to amount to 10–15 per cent (Kresge and Chiang, 1967). They are believed to be due to differences in bond hydridization for various isotopes.

A direct proof of the two-step mechanism for azulene as substrate has been provided by Challis and Long (1965a and 1965b) and by Longridge and Long (1967), who have measured directly the fast protonation and deprotonation rates of azulene in 1·5–4·0 M perchloric acid by a stop-flow method. The reaction is reversible, so that $k = k_f + k_r$ is obtained from the kinetic measurements, but this can be combined with the known equilibrium constant to yield the rate constants for the forward and the reverse reaction. The experimentally obtained protonation rate constant agrees closely with k_1^H calculated from the rate constant for exchange (Schulze and Long, 1964b). Kinetic studies thus corroborate the direct observation of the intermediate by n.m.r. spectroscopy (see Chapter 3).

Since the denominator in the expression for the rate constant for exchange (5.72) is a quantity of the order of magnitude of 5–10 for all the substrates (see Table 5.15), it is clear that vast differences in the reactivity of various substrates are primarily due to differences in the rates of protonation. Therefore structural effects on the rates of exchange (Tables 5.12 and 5.15) reflect almost exclusively the effects of substituents on the protonation rates, which have not been measured directly for any other substrate apart from azulene. Relative exchange rates from Table 5.15 give rise to a Hammett plot vs. σ^+ from which a ρ value of -6.1, applicable to the protonation rates (i.e. corrected for differences in k_2^H/k_2^D) may be obtained. Several ρ values for hydrogen exchange reactions collected in Table 5.16 show that the introduction of mesomerically electron-donating substituents into the benzene ring (—OH, —NR$_2$) reduces the ρ value, and the more effective the donor group the greater becomes the reduction in the value of ρ. Clearly, the electron demand of the proton is here met to a considerable extent by the electron-donor substituent present, so that the sensitivity of the reaction rate to other substituents is reduced.

TABLE 5.16

Some ρ values in the hydrogen exchange of benzene derivatives

Compound	Isotope exchanged	Number of substituents	ρ	Reference†
X-benzenes	T	15	-8.2	a
X-benzenes	D	4	-5.6	b
p-X-phenols	o-D	3	-4.7	c
p-X-anilines	o-H	4	-3.51	d
p-X-N,N-dimethylanilines	o-H	11	-3.42	e

† a. Stock and Brown (1963), p. 96; b. Based on data in Table 5.15; c. Based on data of Gold and Satchell (1955a) and referring to 55 per cent acid; d. Bean and Katritzky (1968); e. Ling and Kendall (1967).

The relative rates of the breakdown of the intermediate (Table 5.15) reveal a trend of considerable theoretical interest, namely that there is a maximum in the ratio k_2^H/k_2^D when the pK difference between the catalysing acid medium and the substrate falls to zero. This implies complete symmetry of the transition state with respect to the position of the proton. A maximum in the primary isotope effect has been predicted theoretically for such a transition state (Westheimer, 1961; Bigeleisen, 1964; Melander, 1960; O'Ferrall and Kouba, 1967).

Information on activation parameters for hydrogen exchange and for the protonation and deprotonation reactions is at present limited to the azulene system, for which thermodynamic parameters for the protonation equilibrium are also available (Challis and Long, 1965a and 1965b). On the basis of these data a free energy profile has been constructed for a two-stage A-S_E2 mechanism for azulene, as shown in Fig. 5.6.

An unexpected observation in the protonation–deprotonation studies on azulene is that deprotonation rates are also acid dependent—Challis and Long (1965a, 1965b) found $k_f = 1 \cdot 52 h_0^{1 \cdot 26}$ and $k_r = 110 h_0^{-0 \cdot 68}$. There is no simple way of accounting for this anticatalysis. The entropy of activation for deprotona-

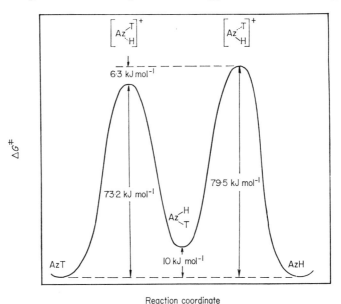

Reaction coordinate

FIG. 5.6. The free energy diagram for hydrogen–tritium exchange in azulene (after Challis and Long, 1965a).

tion, $\Delta S_r^{\ddagger} = -46$ J K^{-1} mol^{-1}, suggests a net binding of two water molecules in forming the transition state, which may be rationalized by assuming that two water molecules presumably solvate the azulenium ion, whereas the proton lost to the solvent becomes solvated by four water molecules. Also unusual is the large difference in the slopes of the plots of log [AzH$_2^+$]/[AzH] vs. $-H_0$ (1·94) and log k_f vs. $-H_0$ (Challis and Long, 1965a and 1965b). This suggests considerable differences in the activity coefficient behaviour of the azulenium ion and the transition state for its formation.

Correlations of exchange rates with acidity functions may be said to be still one of the most confused aspects of the subject. Correlations with H_0 have been used almost exclusively and are mostly well obeyed (Table 5.14), although for carbonium ion formation H_R' should be the relevant function. The matter is further complicated, especially in sulphuric acid solutions, by the probable participation of other acidic species, apart from the H_3O^+ ion, in the catalysis. Knowledge of the activity coefficients of substrates is limited and that of protonated substrates strictly unobtainable. Ideas on solvation are often used to rationalize observations, but only in a qualitative way.

5.4.5.2. Protodeboronation Reactions

Protodeboronation of areneboronic acids is a reaction which shows many characteristics similar to those of hydrogen exchange and of typical electrophilic aromatic displacements in general. In the protodeboronation of *p*-methoxybenzeneboronic acid in moderately concentrated sulphuric, phosphoric and perchloric acid the rates are correlated by the acidity function H_0, but fall in the order $H_3PO_4 \gg H_2SO_4 > HClO_4$, which suggests general acid catalysis by species other than the hydronium ion. For the more reactive 2,6-dimethoxybenzeneboronic acid general acid catalysis has been demonstrated in aqueous phosphate buffer solutions (Kuivila and Nahabedian, 1961a). The slopes of the H_0 correlations, while generally tending to be smaller than in the hydrogen exchange reaction, show distinct increases in the region about 70 per cent sulphuric acid (Nahabedian and Kuivila, 1961), probably owing to increasing contribution by undissociated sulphuric acid molecules in the proton transfers (cf. Fig. 5.5). All these facts indicate a slow proton transfer in the rate-determining step of the reaction. This conclusion is corroborated by solvent isotope effects (Kuivila and Nahabedian, 1961b; Nahabedian and Kuivila, 1961), which show that the reactions are faster in a light hydrogen medium than in a deuterated medium by factors between 1·6 and 3·7, depending on acidity and the nature of the substrate.

By analogy with hydrogen exchange and other electrophilic substitutions, it has been suggested (Kuivila and Nahabedian, 1961b) that the reaction takes place via the formation of an intermediate:

$$\text{C}_6\text{H}_5\text{B(OH)}_2 + \text{HA} \underset{k_{-1}}{\overset{k_1}{\rightleftharpoons}} [\text{C}_6\text{H}_6\text{B(OH)}_2]^+ + \text{A}^-$$

$$[\text{C}_6\text{H}_6\text{B(OH)}_2]^+ + \text{H}_2\text{O} \xrightarrow{k_2} \text{C}_6\text{H}_6 + \text{B(OH)}_3 + \text{H}^+$$

However, independent evidence for its formation has not been provided.

Deboronation reactions of substituted benzeneboronic acids are faster than dedeuteration reactions by factors of 15 to 66, depending on the substrate and acidity (Nahabedian and Kuivila, 1961) and therefore the reaction could conceivably become a single-step displacement, at least for the activated substrates. The rates of reaction for variously substituted benzeneboronic acids (in 74·5 per cent sulphuric acid) are, however, satisfactorily correlated by the σ^+ constants, which makes any change of mechanism with substitution unlikely. The high negative ρ value of $-5\cdot 2$ is wholly consistent with a transition state in which a substantial positive charge is imparted to the benzene ring. No independent information on the rate of the proton transfer step is available but, as in hydrogen exchange reactions, substituent effects on the observed rates should primarily arise from the protonation step.

Activation parameters obtained for variously substituted benzeneboronic acids show that differences in activation energy are largely responsible for differences in reactivity, the variations in entropies of activation being relatively small (Nahabedian and Kuivila, 1961). The entropies of activation are highly negative (mostly between $-64\cdot 9$ and $-82\cdot 8$ J K^{-1} mol^{-1} in moderately concentrated sulphuric acid solution), but a smaller value ($-50\cdot 2$ J K^{-1} mol^{-1}) is found for p-methoxybenzeneboronic acid in 30 per cent sulphuric acid. In an attempt to explain these differences Nahabedian and Kuivila (1961) suggest that for the less reactive substrates the boron atom is more electrophilic and may become coordinated to the bisulphate ion, either in a pre-equilibrium step or simultaneously with the proton transfer. However, as mentioned above, the good correlation with σ^+ makes changes of mechanism with substitution unlikely, but coordination of the boron atom to water molecules in the medium probably occurs for all the substrates prior to any reaction. It may be that water molecules become more tightly bound and that hydrogen bonding of the OH groups to the solvent is also strengthened in the transition state, all of which would give rise to negative entropies of activation, but any attempt to account for minor differences in these quantities is bound to be highly qualitative in the present state of our knowledge.

5.4.5.3. *Displacements of Other Groups by the Hydrogen Ion*

A number of other groups are displaced from aromatic systems by the hydrogen ion, particularly organometallic groups containing elements of Group IV. For example, a protodesilylation reaction is, in fact, a solvolytic cleavage of a C—Si bond (Eaborn, 1953):

$$ArSiR_3 + H_3O^+ = ArH + R_3SiOH + H^+ \tag{5.76}$$

where R = alkyl. Although such reactions were first reported in concentrated sulphuric acid (e.g. Kipping, 1907), extreme conditions of acidity are quite unnecessary. Eaborn and Pande (1960) have estimated the ease of displacement of such groups from the benzene system relative to hydrogen as follows:

X =	H	SiEt$_3$	GeEt$_3$	SnEt$_3$	PbEt$_3$
Relative reactivity	1	10^4	10^6	10^{10}	10^{13}

It can be seen that all these organometallic groups are displaced much more readily than the boronic acid group (see Section 5.4.5.2), and substituent effects have therefore been studied in dilute acid solutions (aqueous perchloric acid mixtures with ethanol and methanol) by Eaborn and collaborators (1956–61) and in sulphuric acid/acetic acid/water mixtures by Deans and Eaborn (1959). The reactions are of considerable interest for comparison with hydrogen exchange (Section 5.4.5.1) and deboronation (Section 5.4.5.2), especially in regard to substituent effects on reactivity, and for this reason the results of Hammett correlations are summarized in Table 5.17.

TABLE 5.17

ρ values for the displacement of some organometallic groups from X—C$_6$H$_4$MR$_3$ systems by the hydrogen ion

Reaction	Group displaced	ρ value (plot vs. σ^+)	ρ value† (plot vs. $\sigma + r(\sigma^+ - \sigma)$)	r	References§
Desilylation	—Si(CH$_3$)$_3$	−4·6	−5·0	0·7	a
Degermylation	—Ge(C$_2$H$_5$)$_3$	−3·9	−4·6	0·65	b
Destannylation	—Sn(C$_6$H$_{11}$)$_3$	~−2·8‡	−3·8	0·4	c
Deplumbylation	—Pb(C$_6$H$_{11}$)$_3$	~−1·8‡	−2·5	0·4	d

† Equation proposed by Yukawa and Tsuno (1959).
‡ Very poor correlation.
§ *a*. Based on data of Eaborn (1956); *b*. Deans and Eaborn (1959); *c*. Eaborn and Waters (1961); *d*. Eaborn and Pande (1961).

Good correlations with σ^+ constants, typical of electrophilic aromatic substitutions, are found in desilylation and degermylation reactions, as in dedeuteration and deboronation, and indicate that formation of intermediates of the type

$$\underset{\text{Ar}}{\overset{+}{\diagdown}}\genfrac{}{}{0pt}{}{H}{MR_3}$$

takes place in these reactions. The correlations become progressively worse, however, the more electropositive the element involved, so that destannylation and deplumbylation reactions are better correlated by the Yukawa and Tsuno's equation, which is applicable to systems where resonance interactions of the substituents with the reaction centre are not operative to the full extent (Yukawa and Tsuno, 1959). This suggests that the positive charge of the proton in destannylation and deplumbylation reactions is not so much delocalized into the benzene ring in the transition state, but is rather transferred largely to the metallic atom. This means that the electrophilic displacement here approaches a single-step process, and that intermediates are so unstable as to be treated as transition states. The gradual deterioration of the correlation with σ^+ constants, as measured by the coefficient r, shows that it is impossible to draw a sharp dividing line between the reactions in which unstable intermediates are formed and those in which only transition states occur. Destannylation and deplumbylation reactions may be said to approach the latter situation closely. The trend is clearly due to the decreasing strength of the C—M bond, but the formation of complexes $ArMR_3\cdots OH_2$ (where M = Sn or Pb) would also facilitate the electrophilic attack of the proton on the benzene ring (Eaborn and Pande, 1960). The decreasing ρ values with increasing electropositive character of the group being displaced mean that greater fractions of the positive charge reside on it, and smaller fractions on the ring, so that the sensitivity of the reaction to electronic effects of the substituents becomes progressively smaller in the series Si > Ge > Sn > Pb (Eaborn, 1956). The higher ρ values for dedeuteration (−5·6) and deboronation reactions (−5·2) may also be said to fit within this series.

5.4.6. ELECTROPHILIC AROMATIC SUBSTITUTIONS BY OTHER ELECTROPHILES

Sulphuric acid/water mixtures are suitable reaction media for some of the best known and most extensively studied electrophilic aromatic substitutions, such as nitration, because the strong acidity of the solutions favours the formation of positively charged species, usually by some complex ionization process. The advantage of sulphuric acid/water mixtures as media for electrophilic substitutions over other solvents is that acidity scales developed for these media (see Chapter 2) facilitate the analysis of kinetic data and provide a basis for quantitative comparisons of the reactivities of a wide range of substrates. Information so obtained is of great importance for the general theory of reactivity of aromatic compounds. Substituent effects are one of the most important aspects of the broader question of reactivity. For aromatic electrophilic substitutions they have been discussed in considerable detail by Stock and Brown (1963). In the present Section use will be made of the methods of treatment and ideas developed by these authors. The brief outline on substituent effects in Chapter 3, however, also provides an essential minimum basis for the understanding of this Section.

Apart from electrophiles that may be generated in sulphuric acid/water mixtures, the medium itself contains hydrogen ions and sulphur trioxide as electrophiles in a number of combined forms. Reactions involving hydrogen ions as electrophiles have been discussed in Section 5.4.5. The reactivity of sulphur trioxide electrophiles will be discussed in the present Section.

5.4.6.1. Nitration

Nitration is one of the most important electrophilic aromatic substitutions. It has played a prominent part in the elucidation of the nature of the interactions of substituents with aromatic systems, as manifested through directive effects. Nitration is an irreversible second-order reaction, as has already been shown by the pioneering work of Martinsen (1905 and 1907). It takes place with a variety of nitrating agents in a variety of reaction media (see, for example, Gillespie and Millen, 1948, and de la Mare and Ridd, 1959). Nitric acid in sulphuric acid/water mixtures has been recognized for a long time as one of the most effective nitrating media, in which the nitrating agent has been well established as the nitronium ion (NO_2^+). In more recent times nitronium tetrafluoroborate, $NO_2^+BF_4^-$, has been shown to be an extremely effective nitrating agent in aprotic solvents such as sulpholane and acetonitrile (Olah et al., 1956; Kuhn and Olah, 1961; Ciaccio and Marcus, 1962).

Evidence for the existence of nitronium ions in solutions of nitric acid and dinitrogen pentoxide in concentrated sulphuric acid has been reviewed in Chapter 4. Nitronium ions are present in spectroscopically detectable amounts only in solutions of nitric acid in >85 per cent sulphuric acid. The conversion of nitric acid to nitronium ions according to the equation

$$HNO_3 + 2H_2SO_4 \rightleftharpoons NO_2^+ + H_3O^+ + 2HSO_4^- \quad (5.77)$$

is complete in about 90 per cent sulphuric acid. The rates of nitration of aromatic substrates in sulphuric acid/water mixtures are extremely sensitive to the concentration of sulphuric acid and have been shown to parallel the ionization of triarylcarbinols in sulphuric acid/water mixtures (Westheimer and Kharasch, 1946; Bennett et al., 1947), i.e. to follow the H_R acidity function (Deno and Stein, 1956). This represents kinetic evidence for nitronium ion as the nitrating agent, since the rates of nitration as a second-order reaction are given by

$$\text{Rate} = k_{\text{obs}}[\text{ArH}][\text{HNO}_3] \quad (5.78)$$

and can only depend on acidity if the concentration of one or the other effective reactant depends on acidity. This is the case with the nitronium ion, owing to the existence of the equilibrium (5.77). The slopes of the plots of $\log k$ vs. H_R are, however, not always close to unity. Moodie et al. (1964) found that plots of $\log k$ vs. $-(H_R + \log a_{H_2O})$ are straight lines of near unit slope for a wide range of substrates, but slopes of plots vs. $-H_R$ are usually higher (e.g. 1·6 for m-nitrotoluene (Tillett, 1962)). This is not unexpected because the activity coefficient behaviour of species involved in equilibrium (5.77) will be different from that of triarylcarbinol indicators. In the nitronium ion the positive charge is relatively localized and accessible for solvation, as compared with the highly delocalized positive charge in the triarylcarbonium ions. Also, the formation of nitronium hydrogen sulphate (Deno et al., 1961) would affect the slopes. The slopes themselves are thus of no particular significance, but the linear and parallel plots of $\log k$ vs. $-H_R$ provide a convenient basis for comparisons of the reactivity of various substrates. This is one of the important advantages which nitrations in sulphuric acid/water mixtures have over nitrations in other media, where relative reactivities can only be established by the method of competitive nitration developed by Ingold and Shaw (1927).

Since nitric acid is fully converted to the nitronium ion in about 90 per cent sulphuric acid, the rates of nitration should become independent of acidity at approximately this acid concentration. This is not what is observed. Rather than a levelling off in the rate, maxima in the rate-acidity profiles are observed for a variety of substrates at about 87–92 per cent sulphuric acid (see Gillespie and Millen, 1948). This has already been established for nitrobenzene in the pioneering work of Martinsen (1905). No entirely satisfactory explanation of these maxima has yet been found, although there is a considerable weight of evidence in favour of their being due essentially to a salt effect. The question will be discussed at the end of this Section.

As a second-order reaction between the aromatic substrate and the nitronium ion, nitration could occur either as a single-step displacement of a proton by the nitronium ion, or via the formation of an intermediate in a two-step process:

$$\text{ArH} + \text{NO}_2^+ \underset{k_{-1}}{\overset{k_1}{\rightleftharpoons}} \text{Ar}\overset{+}{\underset{\text{NO}_2}{\diagdown}}\!\!{}^{\text{H}}$$

$$\text{Ar}\overset{+}{\underset{\text{NO}_2}{\diagdown}}\!\!{}^{\text{H}} + \text{B} \xrightarrow{k_2} \text{ArNO}_2 + \text{BH}^+$$

The second step is the removal of a proton by a base. This latter mechanism has been accepted for nitration since Melander (1950) demonstrated that the rate of displacement of tritium is about the same as the rate of displacement of protium ($k_T/k_H = 0.74$–0.84 was reported as the lower limit for a variety of substrates). This could not be so if the aryl–hydrogen bond were broken in the transition state, but if it is broken in the second, fast step, no isotope effect would be observed. For the two-step mechanism the steady state approximation leads to the following expression for the observed rate constant:

$$k_{\text{obs}} = \frac{k_1 k_2 [\text{B}]}{k_2 [\text{B}] + k_{-1}} \tag{5.79}$$

(For simplicity only one basic species is assumed to be involved). It is clear that only if $k_2[B] \gg k_{-1}$ the observed rate constant will show no hydrogen isotope effect. The much greater rate of proton loss as compared with the reversal of the first step appears thus to obtain in nitrations of benzene and its derivatives in sulphuric acid/water mixtures. There is certainly a sufficiently high concentration of bases in these mixtures for the removal of the proton (water molecules and HSO_4^- ions). The high rate of breakdown of the intermediate and the irreversibility of the reaction imply an asymmetrical free energy versus reaction coordinate diagram, as shown in Fig. 5.7, which may

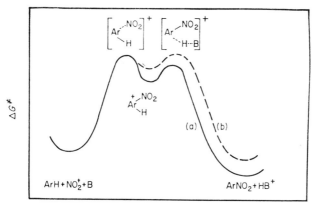

FIG. 5.7. The free energy diagram for the nitration of aromatic compounds: (a) Benzene; (b) Anthracene.

be compared with the highly symmetrical situation in hydrogen exchange (Fig. 5.6). The preliminary formation of $\bar{u}$-complexes between the nitronium ion and the $\bar{u}$-electron cloud prior to the rate-determining formation of the intermediate (the σ-complex) is neither required nor ruled out by the kinetic data, because the formation of such rather unstable complexes would be fast and would merely represent a pre-equilibrium in the reaction.

The high stability of the nitrobenzenes is due to resonance interaction of the nitro group with the aromatic system (**5.73**). This can only take place if

(**5.73**)

the nitro group can become coplanar with the benzene ring. If this is not possible, the nitrocompound may be destabilized and nitration may become reversible. This situation obtains in the nitration of anthracene, where the product, 9-nitroanthracene, is destabilized owing to steric inhibition of resonance and yields nitric acid (81 per cent), anthraquinone (21 per cent) and appreciable amounts of polymer and soluble sulphonic acids upon reaction with concentrated sulphuric acid in trichloroacetic acid at 95°C (Gore, 1957). This means that the free energy versus reaction coordinate diagram for this substrate is less asymmetrical than for nitrobenzene (dashed line in Fig. 5.7). As the rates of the two reactions for the breakdown of the intermediate become more comparable, a hydrogen isotope effect in nitration might be expected.

This point was tested by Cerfontain and Telder (1967a). In order to avoid complications due to hydrogen exchange, anthracene-9-d and naphthalene-1,4-d_2 were nitrated under aprotic conditions (with $NO_2^+BF_4^-$ in sulpholane and acetonitrile). The following isotope effects were found for anthracene-9-d:

$$k_H/k_D = 2\cdot6 \pm 0\cdot3 \ (30°C) \text{ in sulpholane and}$$
$$k_H/k_D = 6\cdot1 \pm 0\cdot6 \ (0°C) \text{ in acetonitrile.}$$

The corresponding figures for naphthalene-1,4-d_2 are close to unity. It follows that $k_2[B] \approx k_{-1}$ for anthracene, whereas for naphthalene $k_2[B] > k_{-1}$, as in the nitrations of benzene and its derivatives in sulphuric acid/water mixtures.

The rate-determining step in these nitrations is therefore the formation of the benzenium ion intermediate. Any substituent effects on the rate and the direction of the nitro group into o-, m- and p-positions occur in this step. The quantities of theoretical interest are the reactivities of individual positions in a substituted benzene relative to that of *one* position in benzene itself. These quantities can be obtained only if the total relative rate (k_{rel}) and the composition of the reaction products (so-called orientation) are known. The fact that two *ortho* and two *meta* positions are present must be taken into account, so that relative rates at o-, m- and p-positions, the so-called partial rate factors, for substituent R, are given by the expressions:

$$\left. \begin{array}{l} f_o^R = \dfrac{\% \ ortho \times k_{rel}}{2 \times \frac{1}{6} \times 100} \\[2mm] f_m^R = \dfrac{\% \ meta \times k_{rel}}{2 \times \frac{1}{6} \times 100} \\[2mm] f_p^R = \dfrac{\% \ para \times k_{rel}}{2 \times \frac{1}{6} \times 100} \end{array} \right\} \quad (5.80)$$

Orientation is not very dependent on the medium, because it is primarily determined by intramolecular electronic effects, but nitration in non-aqueous

TABLE 5.18

Relative rates and partial rate factors for the nitration of some substituted benzenes

Substituent	k_{rel}†	Orientation (%)			Partial rate factors			Reference‡
		o	m	p	o	m	p	
—NMe$_3^+$	1.75×10^{-8}	—	89	11	—	4.2×10^{-8}	1.0×10^{-8}	a
—NHMe$_2^+$		—	78	22	—	1.23×10^{-7}	7.1×10^{-8}	a
—NH$_2$Me$^+$		—	70	30	—	5.7×10^{-7}	4.9×10^{-7}	a
—NH$_3^+$		—	62	38	—	1.62×10^{-6}	1.95×10^{-6}	a
—NO$_2$	6×10^{-8}	6.8	91.8	1.4	1.22×10^{-8}	1.65×10^{-7}	0.5×10^{-8}	b
—CN	3.8×10^{-7}		80.5			9.1×10^{-7}		c
—F	1.5×10^{-1}	12.4	—	87.6	5.6×10^{-2}	—	7.9×10^{-1}	d
—Cl	6.3×10^{-2}	29.6	0.9	69.5	5.6×10^{-2}	1.7×10^{-3}	2.62×10^{-1}	e
—Br	6.3×10^{-2}	36.5	1.2	62.4	6.9×10^{-2}	3.8×10^{-3}	2.36×10^{-1}	e
—I	1.8×10^{-1}	38.3	1.8	59.7	2.06×10^{-1}	9.7×10^{-3}	6.4×10^{-1}	d
—CH$_3$	21–27	58.4	4.4	37.2	37–47	2.8–3	47–62	f
—C(CH$_3$)$_3$	15.7	12.0	8.5	79.5	5.5	4.0	75	f
—Ph	40 ± 2	68	<1	32	41.5	<0.6	38	g
—OCH$_3$	~70	31	2	67	~65	4.2	294	h

† Rate relative to the total rate for benzene taken as unity.

‡ a. Brickman et al. (1965) (referring to nitration in 98 per cent H$_2$SO$_4$ at 25°C); b. Estimate by Tillett (1962); c. Estimate based on rate data of Deno and Stein (1956) and orientation data of Baker et al. (1928); d. Data taken from de la Mare and Ridd (1959), p. 85; e. Relative rate based on data of Deno and Stein (1956) and orientation data as under d, p. 85; f. As under d, p. 88; g. Simamura and Mizuno (1957) referring to nitration by nitric acid in acetic anhydride; h. Rate estimated by Deno and Stein (1956) and orientation data as under d, p. 53 (both refer to nitration in sulphuric acid/water mixtures).

solvents with nitrating agents such as acetylnitrate tends to favour *ortho* substitution owing to the operation of a special mechanism (Knowles and Norman, 1961). Relative rates are more medium dependent, especially when protonation may intervene in acidic media. Protonation of a substituent group also fundamentally alters the orientation.

Relative rate and orientation data given in Table 5.18 have been selected to refer to sulphuric acid/water mixtures as nitrating media wherever possible

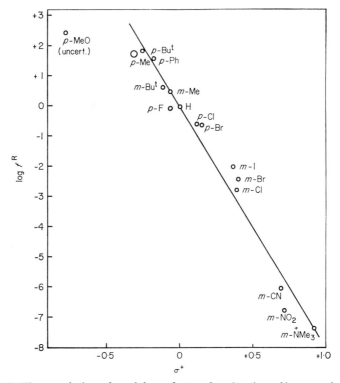

FIG. 5.8. The correlation of partial rate factors for nitration of benzene derivatives with σ^+ constants.

and to substrates which do not undergo protonation in the range of acidities studied. For substitution at the *meta* and the *para* positions the logarithms of the partial rate factors have been plotted *vs.* σ^+ constants in Fig. 5.8. The correlation, which includes estimates for some of the most deactivating substituents, is reasonably satisfactory in view of the fact that relative rate and orientation data from a variety of sources have been used in obtaining partial rate factors. This treatment of aromatic substitution in terms of a linear free energy relationship with σ^+ constants (see Chapter 3) has superseded the earlier approach in terms of a selectivity relationship (see Stock and Brown, 1963). The ρ value of $-7\cdot9$ obtained from Fig. 5.8 is greater than the value of -6 based on a narrower range of substituents (Stock and Brown, 1963). The constants ρ in electrophilic aromatic substitutions are a measure of the selectivity or discriminating power of electrophiles between substitution at *m*- and *p*-positions. They can be correlated with the stability of the electrophiles in a qualitative way, the more stable electrophiles being the more discriminating. In this sense the nitronium ion is less discriminating than, for example, molecular bromine ($\rho = -12\cdot1$ for non-catalytic bromination), but more discriminating than the ethyl cation for which $\rho = -2\cdot4$ for Friedel–Crafts ethylation (Stock and Brown, 1963). It is approximately comparable with the hydrogen ion in hydrogen exchange reactions (see Table 5.16).

One of the puzzling features of substituent effects on nitration is the greater activating effect of the *p-t*-butyl group as compared with methyl, where a hyperconjugative order Me > *t*-Bu might have been expected (Knowles *et al.*, 1960). The same point was examined by Utley and Vaughan (1968) for the nitration of *p*-alkylphenyltrimethylammonium ions, but the analysis is here complicated by steric effects. The effect of other positive poles (PMe_3^+, $AsMe_3^+$ and $SbMe_3^+$) has also been studied recently (Gastaminza *et al.*, 1968). The rates increase with the atomic weight of the atom bearing the formal charge.

As already mentioned, protonation of a substituent group has a profound effect on the reactivity of the benzene nucleus. It leads to overall strong deactivation and, by binding an electron pair involved in conjugation with the aromatic system, it also changes the orientation of the nitro groups in substitution from *ortho-para* to *meta*. When the substrate undergoes a protonic change in acidic media, nitration can occur in principle either through the base or through the conjugate acid, since they are always present in equilibrium. With substrates such as aniline in concentrated sulphuric acid the base is present in negligibly small concentrations, but as it is much more reactive, it is possible that it might contribute at least to some extent to the overall rate of nitration. This question was investigated for the nitration of aniline, *N*-methyl- and *N,N*-dimethylaniline in concentrated sulphuric acid by Brickman and Ridd (1965), Brickman *et al.* (1965) and Hartshorn and Ridd (1968), who concluded that the reaction occurred exclusively through the respective anilinium ions. The rate-acidity profiles for these substrates are closely similar to that for the trimethylphenylammonium ion (Fig. 5.9), but the deactivating effect of the positive poles decreases with the increasing number of protons in the ammonium ion (see Table 5.18). This may be ascribed to the spreading of the positive charge by hydrogen bonding with the solvent (Brickman *et al.*, 1965). The nitration of some other substrates that undergo prototropic change in acid solutions has been studied in recent years from the same point of view (for example, quinoline and isoquinoline by Moodie *et* al., 1963b, cinnoline and 2-methylcinnolinium perchlorate by Gleghorn et al., 1968, and 2-phenylpyridine and its N-oxide by Katritzky and Kingsland, 1968).

The remaining major question in the nitration behaviour of aromatic substrates in concentrated sulphuric acid is the occurrence of rate maxima in the 87–92 per cent sulphuric acid region. If these were associated exclusively with the state of the nitrating agent in solution, they would be expected to occur at precisely the same acid concentration for all substrates. The variation in the position of the maxima for various substrates, which is small but significant, rules out this possibility. The second hypothesis that progressive

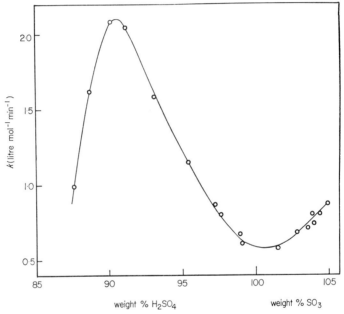

FIG. 5.9. Rate of nitration of trimethylphenylammonium ion in concentrated sulphuric acid (data of Gillespie and Norton, 1953).

protonation of the substrate leads to deactivation and to the consequent fall in rate could account for part of the decrease in rate for nitrobenzene and anthraquinone, but cannot account for the rate maximum in the nitration of trimethylphenylammonium ions (Fig. 5.9) (de la Mare and Ridd, 1959). Bennett et al. (1947) suggested that the falling concentration in this region of the HSO_4^- ions, which are needed for the removal of the proton in the substitution reaction, may account for the fall in rate. This requires the presence of a primary hydrogen isotope effect in this acidity region, which has been shown to be absent (Bonner et al., 1953). Thus, the hypothesis of bisulphate ion catalysis could not be sustained. De la Mare and Ridd (1959) suggested that the variation in rate in this acidity region is due to an environmental effect, i.e. to a variation in the activities of the reactants and the transition state with the change of medium. The activity coefficients of the nitronium ion, the aromatic compound, and the transition state, which resembles the intermediate, may all be very different functions of the concentration of hydronium hydrogen sulphate in sulphuric acid. The fall in rate between 90 per cent and 100 per cent acid may thus be regarded as a salt effect. Consistent with this view, other hydrogen sulphate salts also increase the rates of nitration of benzenesulphonic acid in anhydrous sulphuric acid (Surfleet and Wyatt, 1965), and so does sodium tetrahydrogenosulphatoborate, an observation which rules out a specific catalytic effect of the HSO_4^- ion (Akand and Wyatt, 1967). The rates of nitration in anhydrous sulphuric acid of two substrates of different charge types, the trimethylphenylammonium ion and 1-chloro-4-nitrobenzene, are similarly increased by basic solutes, which makes it unlikely that the Brønsted theory of salt effects is applicable to these solutions in the same sense as it is to dilute aqueous solutions (Bonner and Brown, 1966). Non-electrolytes reduce the rates of nitration of benzenesulphonic acid up to a concentration of 0·25 mol kg^{-1}, in accordance with a dilution effect upon the self-dissociation of the anhydrous acid, but at still higher concentrations the rate increases again (Akand and Wyatt, 1967). Specific effects thus appear to be present also (Bonner and Brown, 1966).

Finally, N-nitration and the related nitramine rearrangement deserve brief mention. Easy N-nitration to form N-nitramines is associated with weak basicity of the N-atom (Chute et al., 1948). The reaction takes place usually under not too strongly acidic conditions, i.e. the free base is necessary for the reaction. The nitrating agents are nitric acid and dinitrogen pentoxide. The rates of N-nitration of the guanidinium ion have been shown to follow the H_R acidity function, which suggests that the effective nitrating agent is the NO_2^+ ion (Williams and Simkins, 1952 and 1953). The guanidinium ion is not protonated further under the conditions of nitration (71·5–83 per cent sulphuric acid), but the reaction is reversible, and the rates of denitration follow the H_0 acidity function with a slope of 1·32 in sulphuric acid (Lockhart, 1966). The denitration of some substituted nitroguanidines shows similar behaviour in aqueous sulphuric, perchloric and methanesulphonic acids (Lockhart, 1966).

The mechanism of N-nitration of primary and secondary aromatic amines, as formulated by Hughes et al. (1958), involves two steps:

$$ArNHMe + NO_2^+ \rightarrow Ar\overset{+}{N}HMeNO_2$$
$$Ar\overset{+}{N}HMeNO_2 + B \rightarrow ArNMeNO_2 + HB^+$$

Either of these could be slow, but since the reaction is reversible, the situation is similar to that in the nitration of anthracene (Fig. 5.7). Slow proton loss in the second step could occur at high acidities, and should lead to a primary isotope effect in the reaction. This has been found in the *N*-nitration of *N*-methyl-2,4,6-trinitroaniline in sulphuric acid ($k_H/k_D = 1 \to 4\cdot 8$ with increasing acidity). The two-step mechanism is thus substantiated (Halevi *et al.*, 1965).

Aryl nitramines undergo rearrangement in moderately or strongly acid solutions in which the nitro group enters almost exclusively at the *ortho* position relative to the amino group, but a small proportion of the *para* product is formed as well. This is in distinct contrast with the products of *C*-nitration under the same conditions, where the *para* isomer predominates (Hughes and Jones, 1950). The acid-catalysed rearrangement of *N*-nitroaniline follows the H_0 acidity function in both aqueous sulphuric and perchloric acid up to 4 M, and the solvent isotope effect, $k_H/k_D = 0\cdot 30$, identifies the acidity dependence of the rate as being due to a preprotonation equilibrium (Banthorpe *et al.*, 1964). Similar behaviour is found in the rearrangement of *N*-nitro-1-naphthylamine (Banthorpe and Thomas, 1965). The rearrangement has been shown to be purely intramolecular by the fact that ^{15}N labelled nitrate in the reaction medium does not enter into the rearrangement product (Brownstein *et al.*, 1958; Banthorpe *et al.*, 1965). Deuteration at the *ortho* position, where the nitro group predominantly enters, does not affect noticeably the rate of rearrangement, but leads to more *para* product (Banthorpe *et al.*, 1964). The question of how the nitro group is transferred intramolecularly from the nitrogen to the *ortho*, and particularly to the *para* position, has caused a good deal of speculation. The rates of rearrangement of *p*-substituted *N*-nitro-*N*-methylanilines are reported to be satisfactorily correlated by the σ^+ constants with $\rho = -3\cdot 9$ (White *et al.*, 1961). This seems to rule out the heterolysis of the N—N bond as the rate-determining step, and hence also a suggestion (Dewar, 1963) that the nitronium ion so obtained forms transiently a $\bar{u}$-complex with the aromatic system before attacking the *ortho* and *para* positions. Banthorpe *et al.* (1964) favour a mechanism previously suggested by Brownstein *et al.* (1958), which involves initial isomerization to an *N*-nitrite, which then forms the *para* product via two cyclic transition states:

It is difficult to justify the initial step of this so-called "cartwheel" mechanism. The matter remains highly controversial. Some evidence is available, however, that denitration accompanies the nitramine rearrangement, since nitration of other sufficiently activated substrates in the same reaction medium has also been observed (Hughes and Jones, 1950).

5.4.6.2. *Nitrosation and Diazotization*

Like nitration, nitrosation is a reaction that takes place in a variety of reaction media in which the nitrosating agents are compounds of the nitrosonium ion with various Brønsted bases, NOX; but in acid solution, unlike nitration which takes place via the nitronium ion, nitrosation does not necessarily involve the nitrosonium ion as the nitrosating agent. At low acidities the reaction is second order in nitrous acid, probably owing to a pre-equilibrium dehydration (5.81) which yields dinitrogen trioxide as the nitrosating agent (Hughes *et al.*,

$$2HNO_2 \rightleftharpoons N_2O_3 + H_2O \tag{5.81}$$

1958). At >1 M acid the reaction order with respect to nitrous acid becomes one (Kalatzis and Ridd, 1966). The state of nitrous acid in moderately concentrated and concentrated sulphuric acid has been discussed in Chapter 4. Two nitrosating species may be present in such media, protonated nitrous acid and the nitrosonium ion (NO^+). Only nitrosations occurring via these nitrosating agents will be discussed in this section. Wider aspects of nitrosation have been discussed by de la Mare and Ridd (1959).

Nitrosating agents may attack aromatic rings, amine or amide nitrogen atoms, or other unsaturated systems, but many of these reactions are complicated by side reactions or by further reactions of the reaction products.

Nitrosation as an electrophilic aromatic substitution is normally limited to the most strongly activated substrates, i.e. aromatic amines and phenols. The well-known nitrosation of *N,N*-dimethylaniline takes place in quite strongly acidic solution (1:1 hydrochloric acid) to yield almost entirely the *p*-nitroso product. The orientation in this substitution shows that it is the amine that undergoes reaction, and not its conjugate acid. The almost exclusive attack on the *p*-position suggests that the nitrosating agent is highly selective and therefore relatively stable in solution. The small proportion of the *ortho* product, which appears to be rapidly oxidized to the *o*-nitro derivative, is probably due to the steric effect of the *N*-methyl groups.

With *N*-alkyl anilines in weakly acidic solutions *N*-nitrosation is the preferred reaction. There is a close similarity between the rates of nitrosation of *N*-methylaniline and the rates of diazotization of aniline in perchloric acid (up to 6·5 M), which shows that the first step in this latter reaction is also *N*-nitrosation (Kalatzis and Ridd, 1966). *N*-nitroso compounds are not very stable at high acidities and *N*-nitrosation becomes reversible. So in concentrated sulphuric acid the more stable p-nitrosoamines are obtained from both secondary and some primary aromatic amines (Blangley, 1938), but deactivation of the ring towards electrophilic attack may change the balance of the reaction in favour of N-nitrosation and diazotization.

The nitrosation of phenol takes place in quite dilute sulphuric acid (~0·5 N) and yields mainly p-nitrosophenol, with only 6 per cent of the *ortho* product (Veibel, 1930). The less reactive substrates (benzene and alkylbenzenes) apparently undergo nitrosation in concentrated sulphuric acid, but the nitroso compounds react further to give diazonium salts (Tedder, 1957). This may be ascribed to the reducing action of the nitrosonium ions (de la Mare and Ridd, 1959).

Few systematic studies of substituent effects on these reactions are available. A study of substituent effects on the rates of diazotization of aniline, which are determined by the rates of N-nitrosation as the first step in the reaction, suggests that the mechanism is a good deal more complicated than might have been expected (de Fabrizio *et al.*, 1966). Substituent effects are inconsistent with the assumption that the free amine is the reactive species, as assumed earlier (Hughes *et al.*, 1958), but can be interpreted in terms of an initial attack of the nitrosating agent on the aromatic ring of the protonated amine, followed by proton loss from the arylammonium ion and migration of the nitrosating agent to the nitrogen.

The acidity dependence of the diazotization of aniline at high acid concentrations, where nitrous acid is largely converted into nitrosonium ions, also appears inconsistent with any mechanism involving a rate-determining nitrosation of the amine (Challis and Ridd, 1960). The plots of $\log k$ *vs.* $-H_0$ are linear with slopes $-2·4$ in sulphuric acid and $-2·1$ in perchloric acid, i.e. much greater than the slope of -1 that would be expected for a reaction of the free amine. Moreover, there is a solvent isotope effect, $k_H/k_D \simeq 10$, which is much less at lower acidities. This suggests that a proton loss occurs in the rate-determining step, whereas the nitrosation of the amine may occur in a pre-equilibrium to give $Ph\overset{+}{N}H_2NO$. A higher rate of reaction in sulphuric acid compared with the rate in perchloric acid (by a factor of 20) can then be ascribed to the catalysis of the proton removal by the hydrogen sulphate ions (Challis and Ridd, 1960). Conflicting conclusions from this study, and the above mentioned study of substituent affects, show that the mechanism of N-nitrosation of amines is still not well understood.

By contrast the rates of diazotization of benzamide in 35–57 per cent sulphuric acid follow the H_R acidity function, thus showing that the nitrosating agent is the NO^+ ion (Ladenheim and Bender, 1960). They are also consistent with the attack of nitrosonium ions upon the unprotonated benzamide as the slow stage of the reaction.

5.4.6.3. *Chlorination, Bromination and Iodination*

The variety of media and the variety of agents that lead to halogenation of aromatic compounds have been described in detail by de la Mare and Ridd (1959). The conditions under which these reactions occur in acid solutions suggest that positive species containing halogens are the active agents. The most extensively studied acid-catalysed halogenation is bromination, which takes place in relatively dilute acid with many substrates. Concentrated sulphuric acid and even oleum are, however, known to be needed for the halogenation of some deactivated aromatic substrates (Derbyshire and Waters, 1950a, 1950b and 1950c; Arotsky *et al.*, 1966). Few systematic studies of halogenation reactions in moderately concentrated and concentrated acid are available.

The most common source of the halogenating agents at lower acidities are the hypohalous acids. Since the reactions are acid catalysed, as was first shown by Shilov and Kaniaev (1939), it is assumed that the hypohalous acids ionize as follows:

$$XOH + H^+ \rightleftharpoons XOH_2^+ \quad (5.82)$$
$$XOH_2^+ \rightleftharpoons X^+ + H_2O \quad (5.83)$$

where $X = Cl$, Br or I. No distinction between the two possible halogenating agents, XOH_2^+ and X^+, has been made so far. The rates of bromination in dilute acid have been shown to follow the H_0 acidity function with a slope of unity at $H_0 > 0$ and of 1·7 at $H_0 < 0$ (Shilov and Kaniaev, 1939; Derbyshire and Waters, 1950b), suggesting that the H_R function would correlate the rates somewhat better. This may mean that Br^+ is the brominating agent. Some support for the view that the chlorinium ion, Cl^+, is the effective reagent has been obtained from the observation that the rate of the acid catalysed chlorination of some very reactive substrates (e.g. anisole) is independent of the concentration of the aromatic compound (de la Mare *et al.*, 1950; de la Mare *et al.*, 1954). This means that the production of the chlorinating agent is rate determining. Since the proton transfer (5.82) is unlikely to be slow, it seems probable that reaction (5.83) becomes rate determining under such conditions.

The chlorination reaction requires much higher acidities for the same rate of reaction than does bromination (Derbyshire and Waters, 1951) and this appears to be due to the smaller basicity of hypochlorous acid compared with hypobromous acid, although it could also be due to the greater reactivity of the positive brominating agent (Derbyshire and Waters, 1951). Thus, both the nature and the relative reactivities of these halogenating agents are still uncertain. What is certain is that positive halogenating agents are more reactive than neutral halogen molecules and hypohalous acids (de la Mare *et al.*, 1954; Derbyshire and Waters, 1950a), so that in general the following order holds: positive halogen > X_2 > HOX.

Chlorinating and brominating agents can also arise in acid solution by the protonation of N-chloro- and N-bromo-amines and amides. For example, the cation R_2NHCl^+ derived from morpholine is a chlorinating agent for phenol (Carr and England, 1958) and N-bromo-succinimide is a brominating agent in the presence of concentrated sulphuric acid (Schmid, 1946). It is conceivable that dehalogenation to produce a positive halogen cation precedes the aromatic halogenation by these halogenating agents (Derbyshire and Waters, 1950a).

Positive halogenating agents can also be produced in aqueous acids by the action of oxidizing agents on the solutions of the halogens. So, for example, a positive brominating agent may be generated in sulphuric acid/water mixtures by the action of potassium bromate on bromine solutions (Krafft, 1875; Derbyshire and Waters, 1950b).

In pure sulphuric acid it has been shown that iodine may exist as I_5^+ or I_3^+ cations (see Chapter 4), depending on the reaction used for the preparation of the solutions. A positive iodinating agent may also be produced in concentrated sulphuric acid by disproportionation of iodine, which is fostered by the presence of silver sulphate. Its exact nature is not known, although it has been assumed to be the iodinium ion, I^+ (Derbyshire and Waters, 1950c). Chlorination and bromination may also be carried out under the same conditions (Kiamud Din and Choudhury, 1963) and have also been assumed to involve Cl^+ and Br^+ ions, respectively.

As in other electrophilic aromatic substitutions, the positive halogenating agents may react with the aromatic systems either in a single-step displacement reaction, or in a two-step mechanism involving the formation of an intermediate. A study of the acid-catalysed bromination of benzene and hexadeuterobenzene has shown that they are brominated at the same rate (de la Mare et al., 1957), which is consistent with a two-step mechanism, in which the second step is fast.

Substituent effects are reasonably well established only in acid-catalysed bromination. The rates are correlated by the σ^+ constants as in other electrophilic aromatic substitutions. The ρ value of $-6\cdot2$ is less than that for nitration, and much less than that for bromination by molecular bromine with $\rho = -12\cdot1$ (Stock and Brown, 1963).

As in other electrophilic substitutions, the state of substrates that undergo prototropic change in acid solution will affect the direction of the halogens to particular sites in the molecule. In concentrated sulphuric acid all common basic substances are fully protonated, and the probability that the conjugate acid will react with the halogenating agent rather than the more reactive base, which, however, is present in negligible concentration, is quite high. It has, for example, been shown that it is the quinolinium ion that undergoes bromination (de la Mare et al., 1960) and chlorination (Kiamud Din and Choudhury, 1963) in concentrated sulphuric acid to give the 5- and the 8-halogenated derivatives. This is in close agreement with nitration in concentrated sulphuric acid, which also yields the 5- and the 8-substituted products (Dewar and Maitlis, 1957).

5.4.6.4. *Amination*

Aromatic aminations have been observed to accompany the Schmidt reaction, when benzene or toluene are used as solvents for hydrazoic acid (Schmidt, 1924). Hydrazoic acid decomposes in sulphuric acid on warming to give hydroxylamine in considerable amount, which is then oxidized by the solvent (sulphur dioxide appears in solution). Hydrazoic acid is fully protonated in concentrated sulphuric acid (see Chapter 3), and its decomposition may be assumed to yield a highly reactive electrophile, presumably NH_2^+ (Hoop and Tedder, 1961). There has been some interest in direct aminations by this electrophile in recent years. Compounds of the hydroxylamine class may also act as aminating agents under conditions of acid catalysis and, apart from sulphuric acid, aluminium chloride may act as catalyst (Kovacic and Bennett, 1961). Competitive amination of benzene and toluene in sulphuric acid solutions shows that the reactive species in sulphuric acid solution is of lower activity, presumably owing to solvation (Kovacic *et al.*, 1964). From the reported selectivity factor a ρ value of $-3\cdot 4$ to $-3\cdot 9$ may be obtained, showing that this electrophile is quite highly reactive.

Under the conditions of these aminations many side reactions may take place, apart from the oxidation of the reagent, such as sulphonations. So, for example, the amination of mesitylene yields mesidine, diaminomesitylene and 3-amino-2,4,6-trimethylbenzenesulphonic acid (Hoop and Tedder, 1961).

It is interesting that the presence of benzenesulphonic acid enhances the yield of aniline or of toluidines, when benzene or toluene are used as solvents for hydrazoic acid in the presence of concentrated sulphuric acid (Sidhu *et al.*, 1966). This effect has been traced to the formation of benzenesulphonyl azide, which is therefore a more effective aminating agent than hydrazoic acid itself.

5.4.6.5. *Sulphonation and Desulphonation*

Sulphonation is an important reaction in concentrated sulphuric acid and oleum media, which is of interest *per se* because of the importance of sulphonated compounds, but which must also be borne in mind as a possible interfering reaction in the study of other reactions in these media. Sulphonation has been shown to be slow at acidities at which hydrogen exchange and nitration occur at measurable rates (Ingold *et al.*, 1936a), but has been found to complicate kinetic and equilibrium studies in many other reactions. As in hydrogen exchange reactions, one of the species in sulphuric acid solution is responsible for the reaction, i.e. sulphur trioxide, possibly as free, or in the form of compounds or ions present in the reaction medium. One of the most difficult problems in formulating the mechanism of sulphonation in sulphuric acid media has been the precise identification of this reactive species.

Extensive monographs have recently appeared on all aspects of sulphonation and desulphonation by Gilbert (1965) and by Cerfontain (1968), the latter devoted to mechanistic aspects in particular. Therefore only a general discussion will be presented here, in which desulphonation will also be treated as the reverse of the sulphonation reaction, although it is in fact an electrophilic substitution by the hydrogen ion. The justification for this is that, by the principle of microscopic reversibility, both reactions must occur via the same mechanism.

Sulphonation is unusual amongst electrophilic aromatic substitutions in being a reversible reaction. The reason for this reversibility was thought to be the negative charge of the sulphonate anion, which would favour the displacement of SO_3 by the hydrogen ion (Baddely et al., 1944), and this view has persisted up to the present time. It is certainly true that the sulphonate anion must be more reactive towards hydrogen ions in electrophilic displacement than the sulphonic acid molecule, but the charge may none the less play only a secondary part in the question of the stability of the reaction products of sulphonation. It could be argued that sulphonic acids and sulphonate anions are less stable than other products of aromatic electrophilic substitutions simply because they are less resonance stabilized. The strength of benzenesulphonic acid, for which a fairly reliable acid dissociation constant, $K_a = 2·79_5 \times 10^{-3}$, has been reported (Kortüm et al., 1961), is a reflection of the interactions between the benzene ring and the sulphonic acid group. The acid is much weaker than sulphuric acid, which indicates an electron-releasing effect of the phenyl group towards the sulphur atom. This may be largely inductive, owing to the partial positive charge on the sulphur, but there is some evidence that it is in part mesomeric, based on the fact that electron-withdrawal by the SO_3^- group is considerably enhanced when interaction with an electron-releasing group across the benzene ring is possible. This is illustrated by the values of substituent constants for the SO_3^- group, which are $\sigma_p^- = +0·49$ as compared with $\sigma_p = +0·37$ (Zollinger and Wittwer, 1956). Resonance interactions between $p_{\bar{u}}$ and $d_{\bar{u}}$ orbitals are, however, usually less extensive than $p_{\bar{u}} - p_{\bar{u}}$ interactions, such as occur, for example, with the nitro group (for which $\sigma_p^- = +1·27$ as compared with $\sigma_p = +0·778$), and therefore sulphonic acids and their anions are less resonance stabilized than nitrocompounds. Whether in a desulphonation reaction the sulphonic acid or the sulphonate anion is the reactive species will be not only a question of their relative reactivities, but also a question of their relative concentrations, which will be determined by the position of a fast pre-equilibrium.

The first mechanistically important studies of desulphonation go back to the first decade of this century (Crafts, 1907). These early rate measurements, which refer to the desulphonation of p-xylenesulphonic acid in aqueous hydrochloric acid at 140°C, have been shown more recently (Gold and Satchell, 1956) to follow the H_0 acidity function for 25°C in the range -0.4 to -2.4 with a slope of 1·05. Smaller slopes were found by Gold and Satchell (1956) for some other desulphonation reactions (Pinnow, 1915, 1917; Lantz, 1935) and were ascribed to the probable progressive protonation of the sulphonate anion. Some recent results on the desulphonation of o-, m- and p-toluene-sulphonic acids in 65–85 per cent sulphuric acid (Wanders and Cerfontain, 1967) gave the following slopes of the $\log k$ vs. $-H_0$ plots for rates at 140°C:

    o-toluenesulphonic acid   0·64

    m-toluenesulphonic acid   0·30

    p-toluenesulphonic acid   0·72 (up to $H_0 = -6.2$)

The latter plot shows a levelling off in rate at still higher acidities. The deviation of these slopes from unity cannot, however, be due to the progressive protonation of the sulphonate anion, because the sulphonic acids, with acid dissociation constants of the order of 10^{-3} (see above), exist as undissociated acids at these acidities. Indeed, the levelling off of the rate of desulphonation of p-toluenesulphonic acid at high acidities could be due to the protonation of the sulphonic acid itself. It is therefore likely that slopes of less than unity arise partly from the fact that high temperature rates (>100°C) are correlated with H_0 values at 25°C, and partly from differences in the activity coefficient behaviour of the substrate and the transition state for desulphonation, as compared with that of Hammett indicators. The linearity of the plots demonstrates that the rate of the reaction depends on the activity of the protons, as it does in other electrophilic substitutions by the hydrogen ion (Section 5.4.5), but the magnitude of the slopes is not of particular mechanistic significance.

Since sulphonic acids exist in the undissociated form in the acidity range in which desulphonation occurs, the mechanism of desulphonation should preferably be formulated as involving the sulphonic acid (as suggested by Spryskov and Ovsyanskina, 1953) and not as involving the sulphonate anion (as proposed by Baddely et al., 1944, and assumed by many others). The greater reactivity of the anion is probably outweighed by its negligibly small concentration.

The dependence of the rate of desulphonation on acidity is consistent either with a single-step electrophilic displacement of a sulphonic acid group or with a two-step mechanism involving an intermediate σ-complex. By analogy with other electrophilic substitutions by the hydrogen ion, the mechanism involving the formation of an intermediate can be formulated as follows:

$$\text{Ar–SO}_3\text{H} + \text{H}_3\text{O}^+ \xrightarrow{\text{slow}} \text{Ar}\overset{+}{\underset{\text{H}}{\diagdown}}\!\!\!\overset{\text{SO}_3\text{H}}{\diagup} \xrightarrow{\text{fast}} \text{Ar–H} + \text{H}_3\text{SO}_4^+$$

Like the acid, the intermediate σ-complex may also undergo a prototropic change:

$$\text{Ar}\overset{+}{\underset{H}{\diagdown}}{}^{SO_3H} \underset{+H^+}{\overset{-H^+}{\rightleftharpoons}} \text{Ar}\overset{+}{\underset{H}{\diagdown}}{}^{SO_3^-}$$

It has been argued that, owing to its possession of a positive charge, this intermediate should be a very strong acid (Melander, 1950), probably stronger than ordinary sulphonic acids and sulphuric acid. Kort and Cerfontain (1969a) also assume that the anion plays a part in the mechanism of sulphonation–desulphonation reactions. It may be pointed out, however, that the sulphonic acid group in the intermediate is carried on a saturated carbon atom (**5.74**). The delocalized positive charge cannot be more effective in increasing its acid strength than a localized positive pole, such as is found for example in 1-aminoethyl-sulphonic acid (**5.75**) in acid solution. The pK_1 of this cation acid has

(5.74)   CH$_3$—CH—SO$_3$H
                  |
                  NH$_3^+$
             (5.75)

been estimated as −0·33 (King, 1953), and an earlier estimate indicated a value of about 0 (Bjerrum, 1923). There is no reason why the acidity of the intermediate (**5.74**) should be much greater than this. The zwitterionic form of the intermediate is therefore probably not very important at high acidities at which desulphonation and sulphonation take place.

The rates of desulphonation increase with increasing acidity of solutions, but the rates of sulphonation, which are proportional to the activities of the sulphonating species ($H_3SO_4^+$ and $H_2S_2O_7$, see below), increase still more sharply with acid concentration, so that in concentrated sulphuric acid sulphonation is not reversible under the conditions of kinetic experiments, and it is even less so in oleum solutions. At high temperatures and at moderate sulphuric acid concentrations substantial isomerization of sulphonic acids may occur by a desulphonation–resulphonation mechanism (e.g. toluene-sulphonic acids isomerize (Wanders et al., 1967) and so do o- and m-xylene-sulphonic acids (Koeberg-Telder et al., 1969)). Complications due to desulphonation and isomerization in sulphonation studies are minimized in concentrated sulphuric acid and at lower temperatures (Cerfontain, 1968).

Earlier work on the sulphonation of benzene and alkylbenzenes (Gold and Satchell, 1956; Kilpatrick et al., 1960 and 1961) has been repeated and extended by Cerfontain and his collaborators (Kaandorp et al., 1962; Cerfontain et al., 1963; Cerfontain and Kaandorp, 1963; Kaandorp and Cerfontain, 1969;

Prinsen and Cerfontain, 1969). These reactive substrates can be sulphonated by sulphuric acid of concentration <85 per cent at 25°C. Halogeno-substituted benzenes are sulphonated at acidities >85 per cent sulphuric acid (Kort and Cerfontain, 1967, 1969a and 1969b), whereas strongly deactivated substrates (nitrobenzene, p-nitrotoluene, p-halogenotrimethylphenyl ammonium ion and benzene- and toluene-sulphonic acids) require oleum solutions as reaction media to achieve measurable rates of sulphonation (Cowdrey and Davies, 1949; Brand and Horning, 1952; Cerfontain, 1961b). Thus the general trend of substituent effects on the rates of sulphonation is the same as in other electrophilic substitutions, but an analysis in terms of a linear free energy

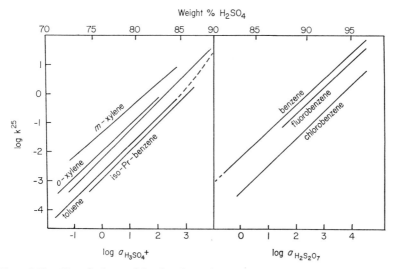

FIG. 5.10. Correlations of $\log k_{25}$ for sulphonation with the logarithms of the activities of two possible sulphonating agents (after Kort and Cerfontain, 1968).

relationship is possible only if rates of reactions involving the same electrophile are known. The identification of the nature of the electrophilic species has been a matter of some difficulty. The most reactive sulphonating electrophile is sulphur trioxide, and it has been assumed to be the active agent in the sulphonation of p-nitrotoluene in >90 per cent sulphuric acid (Cowdrey and Davies, 1949) and even in the sulphonation of benzene in <85 per cent sulphuric acid (Gold and Satchell, 1956). The activity of sulphur trioxide is, however, extremely low in sulphuric acid of concentration <85 per cent, and a more recent examination of sulphonation rates for a number of substrates over a wide acid concentration range in terms of the possibility that $H_3SO_4^+$ or $H_2S_2O_7$ may be the sulphonating agents has yielded good linear plots of near unit slope (Fig. 5.10), which show that $H_3SO_4^+$ is probably the sulphonating agent at acid concentrations <85 per cent and that $H_2S_2O_7$ is the sulphonating agent at acid concentrations >85 per cent (Kort and Cerfontain, 1968). The activity of $H_2S_2O_7$ is proportional to $a_{H_2SO_4}^2/a_{H_2O}$, both available from the work of Giauque et al. (1960)—see Chapter 1—but the activity of the $H_3SO_4^+$ ions could not be estimated without extra-thermodynamic assumptions (Kort and Cerfontain, 1968). The estimate involves also the use of the Raman data of Young et al. (1959) for the concentration of HSO_4^- ions, which are rather scanty and uncertain in the concentration region of interest (70–85 per cent sulphuric acid)—see Fig. 1.1, p. 19. Therefore the activities of $H_3SO_4^+$ ions in the sulphonation media remain uncertain and so does the conclusion that they are the sulphonating agents in <85 per cent acid. A further observation has been made that isomer ratios in the sulphonation of toluene (Cerfontain et al., 1963), naphthalene (Cerfontain and Telder, 1967b) and o- and m-xylene (Prinsen and Cerfontain, 1969) show a change between 75 per cent and 95 per cent sulphuric acid. This is particularly pronounced for the o/p substitution ratio for toluene, which increases sharply with increasing acid concentration, but a smaller increase is found also in the m/p ratio. These changes in isomer ratios are compatible with the idea of a change in the nature of the sulphonating agent, and are difficult to explain otherwise.

The two mechanisms may be formulated as follows:

$$ArH + H_2S_2O_7 \xrightarrow{slow} Ar{\overset{+}{\underset{H}{<}}}^{SO_3H} + HSO_4^- \xrightarrow{fast} ArSO_3H + H_2SO_4 \quad (I)$$

$$ArH + H_3SO_4^+ \xrightarrow{slow} Ar{\overset{+}{\underset{H}{<}}}^{SO_3H} + H_2O \xrightarrow{fast} ArSO_3H + H_3O^+ \quad (II)$$

The possibility of a prototropic change in the intermediate which, as explained above, appears remote has been ignored. Mechanism (II) is the reverse of the desulphonation mechanism formulated above, and the principle of microscopic reversibility is hence satisfied. Mechanism (I) shows the HSO_4^- ion as the active nucleophile in either the forward or the reverse reaction of the intermediate (Kort and Cerfontain, 1969a), whereas in mechanism (II) water molecules play that part. The smaller percentage of *ortho* substitution in the sulphonation of toluene in the more dilute acid is ascribed to greater steric requirements for sulphonation with $H_3SO_4^+$ ions as compared with $H_2S_2O_7$ molecules.

An analysis of the sulphonation rates in Fig. 5.10 in terms of the two sulphonation mechanisms has enabled Kort and Cerfontain (1968) to obtain the Hammett reaction constants for the two mechanisms: $\rho_I = -6.1$ and $\rho_{II} = -9.3$. The more reactive of the sulphonating agents is accordingly the less selective of the two.

Finally, the question of the relative rates of the two steps in mechanisms

(I) and (II) has been tackled by studying isotope effects on the rates of sulphonation. The earlier results of Melander (1950), which showed that benzene-t and bromobenzene-4-t are sulphonated more slowly than protium compounds, have recently been supplemented by results on deuterium isotope effect on the sulphonation of chlorobenzene in concentrated sulphuric acid (Kort and Cerfontain, 1967). It is in the concentrated sulphuric acid region (>95 per cent) that an isotope effect on the rate should be most easily detectable, because the concentrations of HSO_4^- ions decrease sharply in that region, and since this ion is involved in the removal of the proton (mechanism I), the second step of the reaction could become slow. A small deuterium isotope effect was, in fact, found ($k_H/k_D = 1\cdot 25$ in 95 per cent acid, increasing to 3·3 in 96·7 per cent acid–Kort and Cerfontain, 1967), so that it may be concluded that the second step in the sulphonation mechanism (I) becomes partly rate determining at these acid concentrations. This is not unexpected in a reaction in which the reaction product is not very stable and which can be made reversible.

Unlike the sulphonations so far discussed, the sulphonation of aniline follows an indirect route, via the formation and rearrangement of aniline-N-sulphonic acid. A comparison of the rates of sulphonation of aniline and of the rearrangement of its N-sulphonic acid has shown that the latter is much faster (Vrba and Allan, 1968a and 1968b). The products of both reactions are the same (the ratio *para*:*ortho* sulphonic acid being 85:15). This means that the formation of aniline-N-sulphonic acid is the rate determining reaction in the sulphonation of aniline. However, rapid exchange of the ^{35}S-labelled sulphonic acid group of aniline-N-sulphonic acid with the sulphuric acid medium conflicts with this conclusion, but it does show that the rearrangement is intramolecular, as postulated by Hughes and Ingold (1952). Vrba and Allan (1968a) suggest a mechanism involving a deformation of the benzene ring in the transition state.

Sulphonation in dilute oleum also leads to the formation of sulphonic acids (Brand, 1950; Brand and Horning, 1952; Brand *et al.*, 1959; Cerfontain, 1961b and 1965) and disulphuric acid is thought to be the sulphonating agent (Kort and Cerfontain, 1969b). At higher sulphur trioxide concentrations the rate is correlated with $h_0 \cdot p_{SO_3}$, which has been ascribed to HSO_3^+ as the sulphonating agent (Brand, 1950), or with $h_0 \cdot a_{H_2S_2O_7}$, suggesting that $H_3S_2O_7^+$ may be the sulphonating species (Kort and Cerfontain, 1969b). In addition to sulphonic acids, the products in this region may be sulphonic anhydrides, sulphones and sulphonated sulphones (Cerfontain, 1968).

5.4.7. MISCELLANEOUS REACTIONS INVOLVING CARBONIUM IONS

This final section will be devoted to a brief discussion of some reactions of carbonium ions, which are not easily classed under a single name or title, although they are all manifestations of the high electrophilicity of carbonium ions. This term is being used here again in its broadest sense, i.e. to embrace oxocarbonium ions and carboxonium ions as well. Some of the reactions to be discussed are not restricted to sulphuric acid media and occur in a variety of other solvents and with a variety of other catalysts, whereas for others (the Koch reaction and the Schmidt reaction) sulphuric acid is one of the most important, or the most important reaction medium and catalyst. The mechanisms of the reactions to be discussed in this section are closely related to the mechanisms of some reactions already discussed in preceding sections.

5.4.7.1. *Hydride Transfers*

Concentrated sulphuric acid is a medium in which carbonium ions can be generated from saturated hydrocarbons by hydride abstraction. This reaction may be represented by the following equation:

$$R'R''R'''CH \rightleftharpoons R'R''R'''C^+ + H^+ + 2e \qquad (5.84)$$

It can thus be seen to be an oxidation. Saturated hydrocarbons undergo this reaction only if they contain a tertiary carbon atom, i.e. if the resulting carbonium ions are stable. Aromatic systems which can yield more stable carbonium ions react more readily and at lower sulphuric acid concentrations. So, for example, while only >90 per cent sulphuric acid will convert isobutane to *t*-butyl cations, xanthene is converted to xanthyl cations at a measurable rate by 55 per cent acid and the conversion is instantaneous in 85 per cent acid (Deno *et al.*, 1962b).

A carbonium ion produced in a strongly acid medium by whatever means may act as a hydride ion acceptor with respect to other saturated hydrocarbons or aromatic systems and convert them to carbonium ions by hydride transfer. While hydride transfer reactions have been well known since the work of Bartlett *et al.* (1944), systematic studies of these reactions in sulphuric acid media are of recent date. In a hydride transfer reaction between two carbonium ion centres the rate of the reaction should be determined by the relative stabilities of the carbonium ions. A measure of these stabilities are the pK_{R^+} values (see Chapter 4) and Deno *et al.* (1962b) quote a linear correlation between $\log k$ and ΔpK_{R^+} values (due to H. Dauben and J. M. McDonough) for hydride transfers between triarylmethyl cations and several hydride donors (cycloheptatriene, trianisylmethane, triarylmethane). A similar correlation was found to be valid for xanthene as the hydride donor and four triarylmethyl cations: $\log k = 0.76(\Delta pK_{R^+}) - 2.86$ (Deno *et al.*, 1962b). The transition state for intermolecular hydride transfers ($\overset{\delta+}{C}$—H—$\overset{\delta+}{C}$) must be linear, and steric interactions between two large carbonium ions, such as triarylmethyl, may be considerable, especially in view of their non-planar, propeller-like structure.

This is illustrated by the fact that the reaction between cycloheptatriene and triarylmethyl ions is about a hundred times faster than that between two triarylmethyl systems. Also replacing an aryl group by a methyl group to give a diarylmethyl cation is estimated to increase the rate four-fold, whereas replacement by a hydrogen leads to a 3600-fold increase, after allowance is made for differences in basicity (Deno et al., 1962b). Similarly, methylcyclopentane has a much smaller activation energy for hydride abstraction than methylcyclohexane, the difference being largely due to steric effects (Kramer, 1965).

Many alcohols have been shown to be very effective hydride ion donors to acceptors such as triphenylmethyl cations (Bartlett and McCollum, 1956). The reaction can be followed in sulphuric acid/water mixtures by the disappearance of colour of the triphenylmethyl cations. Isotropic labelling has shown that it is the α-hydrogen atoms that are removed. With increasing acid concentration the rate of the reaction decreases, owing to the conversion of alcohols to oxonium ions. Ethers behave similarly. The hydride transfer reaction shows a hydrogen isotope effect, $k_H/k_D = 1\cdot 8 - 2\cdot 6$ (Bartlett and McCollum, 1956). The carbonium ions that result from these reactions of alcohols (R—$\overset{+}{\text{CH}}$—O—H) are in fact protonated aldehydes, whereas those from the reactions of ethers (R—$\overset{+}{\text{CH}}$—OR′) are the same as those that are obtained by the protonation of α,β-unsaturated ethers. They owe their considerable stability to the resonance spreading of the positive charge between the adjacent carbon and oxygen atoms.

Methanol and dioxane, however, do not react in this way. Deno et al. (1962b) showed that this was due to their protonation in sulphuric acid media. Hydride abstraction can be favoured over protonation if polyphosphoric acid is used as the reaction medium instead of sulphuric acid. Polyphosphoric acid solutions combine lower acidity with lower water activity and therefore favour carbonium ion formation according to

$$\text{ROH} + \text{H}^+ \rightleftharpoons \text{R}^+ + \text{H}_2\text{O} \qquad (5.85)$$

over protonation. With the diphenylmethyl cation as hydride acceptor, reaction rates in 72 per cent P_2O_5 are at least 10^4 times higher for a number of substrates (2-propanol, cyclopentanol, cyclohexanol, allyl alcohol, 1,2-dimethoxyethane, dioxane, ethyl ether) than in 83 per cent sulphuric acid, which has approximately the same H_R value. Even under these conditions borneol, some ethers and lactones show no detectable reaction (Deno et al., 1962b).

In superacid media (HSO_3F—SbF_5) the related reaction of methide ion abstraction has also been observed in the example of the ionization of neopentane:

$$(\text{CH}_3)_4\text{C} \rightarrow (\text{CH}_3)_3\text{C}^+ + \text{CH}_4$$

Methane was identified by mass spectroscopy and the *t*-butyl cation by n.m.r. spectra (Olah and Lukas, 1967).

5.4.7.2. Dimerization and Polymerization of Olefins

When carbonium ions are produced in a reaction medium by the protonation of olefins, there is always the possibility of the carbonium ions attacking the unchanged olefin to produce dimers, trimers and higher polymers. The general equation for the dimerization may be given as

$$-\underset{\underset{H}{|}}{\overset{\overset{H}{|}}{C}}-\overset{+}{C}{\Big\langle} + {\overset{H}{}}{\Big\rangle}C = C{\Big\langle} \longrightarrow -\underset{\underset{H}{|}}{\overset{\overset{H}{|}}{C}}-\underset{|}{\overset{\overset{H}{|}}{C}}-\underset{|}{C}-\overset{+}{C}{\Big\langle} \qquad (5.86)$$

(where alkyl and aryl groups are not shown). The dimeric carbonium ion may either lose a proton slowly by reaction with solvent nucleophiles or it may attack a further molecule of olefin leading to trimers, etc. As these are bimolecular reactions, the relative rates of the two possible paths for the dimeric carbonium ion will depend on the concentration of the olefin, the acidity of the solution and the stability of the dimer. Since the solubilities of olefins in sulphuric acid/water mixtures are low, dimerizations tend to be slow and intramolecular rearrangements of carbonium ions take place prior to this reaction, so that mixtures of products may in general be expected. It has also been shown that the higher polymeric carbonium ions may undergo scission to regenerate the original ions (or to form some other fragments), so that polymerizations are reversible (Hofmann and Schriesheim, 1962b).

Few systematic studies of dimerizations in aqueous strong acids are available. Comparisons of rates of dimerization of different substrates at a constant acidity of the medium (van der Zanden and Rix, 1956a and 1956b) are of limited value, because at low acidities it is differences in the extents of protonation of the olefins that will determine the difference in rate, rather than differences in the reactivities of the species involved. So, for example, *iso*-propenylbenzene (**5.76**) dimerizes in 43 per cent sulphuric acid much more readily than propenylbenzene (**5.77**), probably primarily owing to a higher

$$\underset{(5.76)}{C_6H_5-\underset{|}{\overset{\overset{CH_3}{|}}{C}}=CH_2} \qquad \underset{(5.77)}{C_6H_5-CH=CH-CH_3}$$

degree of its conversion into the carbonium ion. The same behaviour is found for the *p*-methoxy derivatives of (**5.76**) and (**5.77**) (van der Zanden and Rix, 1956a). Deno *et al.* (1963a) have shown that the dimerization rate of olefins goes through a maximum in the region of half-protonation of the olefins, i.e. when the product of the concentrations of the two reactants, the carbonium ion and the olefin, is a maximum. Certain allylic cations, obtained from cyclic dienes, for example, which are highly stable in concentrated sulphuric acid, disappear rapidly in more dilute acid owing to the alkylation of the unchanged olefin:

$$\text{[cyclopentenyl cation]} + \text{[cyclopentene]} \longrightarrow \text{[cyclopentyl]}-CH_2-\text{[cyclopentenyl cation]} \quad (5.87)$$

The half-life of this reaction in 35 per cent sulphuric acid, at a concentration of 0·1 M, is only 0·02 s (Deno et al., 1963a).

The dimerization of 1,1-diphenylethylene in 64–80 per cent sulphuric acid is second order in the initial concentration of the olefin. The rate decreases with increasing acid concentration, falling almost to zero in 82–83 per cent acid (Kazanskii and Entelis, 1962; Entelis and Kazanskii, 1963). This is probably due to the increasing conversion of the olefin to the carbonium ion in this acidity region. The dimerization of styrene shows a different acidity dependence from that of 1,1-diphenylethylene (Entelis et al., 1963). Entelis and Kazanskii (1963) ascribe the difference to the incursion of a mechanism involving the carbinol in the dimerization of 1,1-diphenylethylene, but this suggestion seems rather speculative.

5.4.7.3. Alkylations

Alkylations of olefins with isoparaffins in the presence of acid catalysts were first described by Ipatieff (1936). Sulphuric acid is only one of the many possible catalysts and often not the most effective one. Some information about its role in the reaction has been obtained by Hofmann and Schriesheim (1962a and 1962b), who showed the catalytic process to be very complicated. The acid does not serve merely to initiate the reaction by protonating the olefin and to provide a medium for the ionic reactions that follow. The catalytic properties of the acid change with the progress of the process. There is an induction period with fresh acid which is probably the time needed for the protonation of the olefin and for the hydride abstraction from the isoparaffin to take place in order to yield the active carbonium ions. During this initial period sulphuric acid darkens due to the formation of unsaturated hydrocarbons, most probably cyclic dienes. They are probably partly protonated and also take part in hydride transfer reactions, so that sulphuric acid improves as a catalyst with use. A spent sulphuric acid catalyst contains a sludge, whose composition is not known.

Alkylation of a mixture of butenes with isobutane leads to a high percentage of octanes in the product, in particular 2,2,4-trimethylpentane, because the protonation of both butene-1 and butene-2 yields the same carbonium ion, which rearranges to the t-butyl cation. A study of ^{14}C distribution in the reaction products has shown that the initial product results primarily from olefin polymerization reactions, while alkylation becomes predominant only under steady-state conditions (Hofmann and Schriesheim, 1962b).

Alkylation of nitriles using alcohols as alkylating agents is a well-known reaction that occurs in a homogeneous acid medium. Alkenes may also be used as alkylating agents (Ritter and Minieri, 1948). The alkylation of hydrogen cyanide with t-butyl cations yields N-t-butylformamide (Ritter and Kalish, 1948) and the analogous alkylation of acrylonitrile yields N-t-butylacrylamide. The kinetics of this latter reaction in 20–69 per cent sulphuric acid were studied by Deno et al. (1957). The rate is proportional to the concentrations of the nitrile and t-butanol, and to the acidity of the medium, h_0. Since the nitrile is not extensively protonated at these acidities, the acidity dependence of the rate reflects progressive protonation of t-butyl alcohol, which was therefore assumed to be the alkylating agent. t-Butyl cation does not seem to be free under these conditions, since it does not abstract hydride ions from xanthene (Deno et al., 1962b). It may be regarded as strongly solvated.

5.4.7.4. *Carbonylation of Carbonium Ions* (*the Koch Reaction*)

The carbonylation of aliphatic and alicyclic carbonium ions is the reverse of the decarbonylation reaction of carboxylic acids, which was discussed in Section 5.4.4. This reaction is a manifestation of the electrophilic reactivity of carbonium ions, like the other reactions discussed in the present section. The reaction is an industrially important synthetic route to branched chain aliphatic carboxylic acids. The fullest account of the basic conditions of the synthesis has been given by Koch (1955) and some recent developments have been described by Möller (1966).

The basic step of the synthesis, the carbonylation itself, is an attack of carbonium ions on carbon monoxide which leads to the formation of oxo-

$$RR'R''C^+ + CO \rightarrow RR'R''C\text{---}CO^+$$

carbonium ions. The strongly acid media required for the reaction are essential both to maintain sufficiently high concentrations of the reactant carbonium ions, which are commonly produced from olefins, and to stabilize the product oxocarbonium ions. Concentrated sulphuric acid is a particularly suitable catalyst and solvent for this reaction, but other highly acidic solvents are also used (hydrogen fluoride and phosphoric acid, all with or without addition of boron trifluoride). The concentration of sulphuric acid must be greater than 90 per cent, and 96 per cent represents an optimum concentration (Koch, 1955). Carbon monoxide is led into the reaction vessel under pressure (1–100 atm) at ordinary temperatures (0–70°C). Strong stirring is essential because the system is heterogeneous, the liquid phase itself consisting of two layers. After dilution of the reaction mixtures with water carboxylic acids are obtained, and after dilution with methanol the corresponding methyl esters. The reactions (5.88) and (5.89) are the reverse of the complex ionizations discussed

$$R_3CO^+ + H_2O \rightleftharpoons R_3COOH + H^+ \quad (5.88)$$
$$R_3CO^+ + CH_3OH \rightleftharpoons R_3COOCH_3 + H^+ \quad (5.89)$$

in Chapter 4 and are relatively simple processes.

The products of the reaction are as a rule complex mixtures of carboxylic acids owing to carbonium ion rearrangements prior to carbonylation. For example, the number of isomeric carboxylic acids from n-alkenes is $z = N - 2$, where N is the number of carbon atoms in the chain. n-Decene gives four secondary and four tertiary carboxylic acids (Möller, 1963). It is clear that carbonylation is slow compared with carbonium ion rearrangements, but whether the products are determined kinetically or thermodynamically under synthetic conditions is not known with certainty. Boron fluoride containing catalysts have a smaller isomerizing action than concentrated sulphuric acid and favour the formation of secondary acids (Möller, 1966).

A kinetic study of the isomerization of some trialkylacetic acids in concentrated sulphuric acid, which proceeds by a decarbonylation–carbonylation process, has shown that the same equilibrium mixture is obtained from different isomeric acids (Lundeen, 1960). With a half-life of only 51 min, 2,2-dimethylvaleric acid (I) and 2-methyl-2-ethylbutyric acid (II) convert into an equilibrium mixture in the ratio I/II = 2·84. The thermodynamically most stable species appears thus to predominate in isomerized mixtures (Lundeen, 1960).

The decarbonylation of formic acid is used as the source of carbon monoxide in a laboratory variant of the Koch synthesis (Koch and Haaf, 1958; Haaf, 1964). Numerous sources of carbonium ions for the reaction, apart from the olefins, have been investigated, e.g. alcohols (Eidus et al., 1962) and saturated tertiary hydrocarbons in the presence of hydride ion acceptors (Puzitskii et al., 1963).

5.4.7.5. *Reactions with Hydrazoic Acid (the Schmidt Reaction)*

The state of hydrazoic acid in sulphuric acid media was discussed in Chapter 3. Hydrazoic acid is fully protonated in concentrated sulphuric acid and the concentration of the unprotonated compound must be negligible. Its reactivity as a nucleophile is certainly much higher than that of the protonated compound. Hydrazide anion is a very strong nucleophile. Reactions of a variety of carbonium ions with hydrazoic acid in sulphuric acid solution are immediately followed by a rearrangement of the products, both steps being known as the Schmidt reaction. While azides can be prepared by other methods and their rearrangement can be studied independently (see Section 5.4.3), the formation of azides in sulphuric acid solution cannot be separated from the subsequent rearrangement. As the Schmidt reaction is normally carried out in a two-phase system, by mixing a solution of sodium azide in chloroform with a solution of the organic compounds (alcohols, olefins, aldehydes, ketones and carboxylic acids) in sulphuric acid at moderate temperatures, the first stage of the process is the transfer of the hydrazide into the sulphuric acid layer. In this way the supply of the hydrazide is slow and controlled. Otherwise the reaction might be too violent. The organic compounds are present in protonated or carbonium ion form in sulphuric acid solution. The electrophilic carbonium ion centre reacts with hydrazoic acid. It is probable that the reaction is fast, and possibly diffusion controlled. It is not known whether hydrazoic acid or protonated hydrazoic acid is the main nucleophile, but both may contribute to the overall rate. The fact that the rate of decomposition of hydrazoic acid at 40°C is reduced by the addition of m- and p-nitrobenzoic acid (Briggs and Littleton, 1943) may be taken to mean that hydrazoic acid reacts fast with these acids to form the acyl azides, which then decompose at their own rate (Smith, 1963).

The order of reactivity of ketones in the Schmidt reaction is aliphatic ketones > acetophenone > benzophenone, which is essentially the order of their basicities (Smith, 1963). Ketones react much more readily than alcohols (at 0°C, while alcohols require higher temperature), which is also a reflection of the relative ease of carbonium ion formation from the two classes of compound. Carbonium ions from olefins and alcohols may undergo extensive rearrangement prior to the reaction with hydrazoic acid, and the Schmidt reaction with such compounds yields a mixture of products.

The suggestion that in a Schmidt reaction the formation of azides is the first step is due to Olivieri-Mandalà (1925). The route from the carbonium ion to the azide appears to be a straightforward addition when olefins and alcohols are concerned:

$$RR'R''C^+ + HN_3 \rightarrow RR'R''CN_3 + H^+ \qquad (5.90)$$

The reactions of aldehydes and ketones may be more complicated. If the formation of the hypothetical iminodiazonium ions from aldehydes and ketones actually takes place, the addition of hydrazoic acid must be followed by the loss of the elements of water (Smith, 1963):

$$R_2\overset{+}{C}=OH + HN_3 \rightarrow R_2\underset{|}{\overset{OH}{C}}-N_3 + H^+ \rightarrow R_2C=\overset{+}{N}-N_2 + H_2O \qquad (5.91)$$

This step is not, however, essential for the mechanism of the subsequent rearrangement, which has also been formulated in terms of the primary adduct (Newman and Gildenhorn, 1948), although it is then not easy to account for the close similarity of the products of the Schmidt reaction with those of the Beckmann rearrangement of the corresponding oximes (see Section 5.4.3).

5. REACTION MECHANISMS IN SULPHURIC ACID SOLUTIONS

The reaction of carboxylic acids with hydrazoic acid probably takes place via the equilibrium concentrations of oxocarbonium ions:

$$R-\overset{+}{C}\equiv O + HN_3 \rightarrow R-CO-N_3 + H^+ \qquad (5.92)$$

Some support for this view comes from a recent demonstration by McNamara and Stothers (1964) that negatively substituted benzoic acids, which normally give poor yields of anilines in the usual Schmidt method using concentrated sulphuric acid, give almost quantitative yields when 100 per cent sulphuric acid is used. If the protonated acids are the reactive species, however, a mechanistic step involving the loss of the elements of water must intervene prior to the rearrangement.

Finally, it may be mentioned that certain azides which arise from very stable carbonium ions, such as triarylmethyl, dissociate reversibly in sulphuric acid (Coombs, 1958b) and undergo no rearrangement. Clearly the heterolysis of the C—N bond is easier in such compounds than the heterolysis of the N—N_2 bond, owing to the great stability of the resulting carbonium ions. This reversible dissociation results in abnormal products in the Schmidt reaction of 9-methyl-fluoren-9-ol arising through dimerization of the cation (Coombs, 1958a).

References

Abraham, R. J., Bullock, E., and Mitra, S. S. (1959). *Can. J. Chem.* **37**, 1859.
Adams, E. Q., and Rosenstein, L. (1914). *J. Amer. Chem. Soc.* **36**, 1452.
Akand, M. A., and Wyatt, P. A. H. (1967). *J. Chem. Soc. B*, 1326.
Anderson, A. G., and Harrison, W. F. (1964). *J. Amer. Chem. Soc.* **86**, 708.
Angus, W. R., and Leckie, A. H. (1935). *Proc. Roy. Soc. (London) Ser. A*, **150**, 615.
Archer, G., and Bell, R. P. (1959). *J. Chem. Soc.* 3228.
Arcus, C. L., and Evans, J. V. (1958). *J. Chem. Soc.* 789.
Arcus, C. L., and Lucken, E. A. (1955). *J. Chem. Soc.* 1634.
Armstrong, V. C., and Moodie, R. B. (1968). *J. Chem. Soc. B*, 275.
Armstrong, V. C., Farlow, D. W., and Moodie, R. B. (1968a). *J. Chem. Soc. B*, 1099.
Armstrong, V. C., Farlow, D. W., and Moodie, R. B. (1968b). *Chem. Commun.* 1362.
Arnand, R. (1967). *Bull. Soc. Chim. Fr.* 4541.
Arnesen, R. T., and Langmyhr, F. J. (1964). *Acta Chem. Scand.* **18**, 2400.
Arnett, E. M. (1963). *Progr. Phys. Org. Chem.* **1**, 223.
Arnett, E. M., and Anderson, J. N. (1963). *J. Amer. Chem. Soc.* **85**, 1542.
Arnett, E. M., and Bushick, R. D. (1962). *J. Org. Chem.* **27**, 111.
Arnett, E. M., and Bushick, R. D. (1964). *J. Amer. Chem. Soc.* **86**, 1564.
Arnett, E. M., and Mach, G. W. (1964). *J. Amer. Chem. Soc.* **86**, 2671.
Arnett, E. M., and Wu, C. Y. (1960a). *J. Amer. Chem. Soc.* **82**, 4999.
Arnett, E. M., and Wu, C. Y. (1960b). *J. Amer. Chem. Soc.* **82**, 5660.
Arnett, E. M., and Wu, C. Y. (1962). *J. Amer. Chem. Soc.* **84**, 1684.
Arnett, E. M., Wu, C. Y., Anderson, J. N., and Bushick, R. D. (1962). *J. Amer. Chem. Soc.* **84**, 1674.
Arotsky, J., Mishra, H. C., and Symons, M. C. R. (1961). *J. Chem. Soc.* 12.
Arotsky, J., Mishra, H. C., and Symons, M. C. R. (1962). *J. Chem. Soc.* 2582.
Arotsky, J., Butler, R., and Darby, A. C. (1966). *Chem. Commun.* 650.
Aschan, O. (1913). *Ber.* **46**, 2162.
Bachmann, W. E., and Barton, M. X. (1938). *J. Org. Chem.* **3**, 300.
Back, T. A., and Praestgaard, E. L. (1957). *Acta. Chem. Scand.* **11**, 901.
Baddely, G., Holt, G., and Kenner, J. (1944). *Nature (London)*, **154**, 361.
Baird, R. L., and Aboderin, A. (1963). *Tetrahedron Lett.* 235.
Baird, R. L., and Aboderin, A. (1964). *J. Amer. Chem. Soc.* **86**, 252.
Baker, J. W., Cooper, K. E., and Ingold, C. K. (1928). *J. Chem. Soc.* 426.
Bakule, R., and Long, F. A. (1963). *J. Amer. Chem. Soc.* **85**, 2309, 2313.
Bancroft, K. C. C., Bott, R. W., and Eaborn, C. (1964). *Chem. Ind. (London)*, 1951.
Banholzer, K., and Schmid, H. (1956). *Helv. Chim. Acta.* **39**, 548.
Banthorpe, D. V., and Thomas, J. A. (1965). *J. Chem. Soc.* 7149.
Banthorpe, D. V., Hughes, E. D., and Williams, D. L. H. (1964). *J. Chem. Soc.* 5349.
Banthorpe, D. V., Thomas, J. A., and Williams, D. L. H. (1965). *J. Chem. Soc.* 6135.

Barr, J., Gillespie, R. J., and Robinson, E. A. (1961). *Can. J. Chem.* **39**, 1266.
Bartlett, P. D., and McCollum, J. D. (1956). *J. Amer. Chem. Soc.* **78**, 1441.
Bartlett, P. D., Condon, F. E., and Schneider, A. (1944). *J. Amer. Chem. Soc.* **66**, 1531.
Bascombe, K. N., and Bell, R. P. (1957). *Discuss. Faraday Soc.* No. 24, 158.
Bascombe, K. N., and Bell, R. P. (1959). *J. Chem. Soc.* 1096.
Bass, S. J., and Gillespie, R. J. (1960). *J. Chem. Soc.* 814.
Bass, S. J., Flowers, R. H., Gillespie, R. J., Robinson, E. A., and Solomons, C. (1960a). *J. Chem. Soc.* 4315.
Bass, S. J., Gillespie, R. J., and Robinson, E. A. (1960b). *J. Chem. Soc.* 821.
Bates, R. (1966). *In* "The Chemistry of Non-aqueous Solvents" (Ed. J. J. Lagowski), Chapter 3. Academic Press, New York and London.
Bayliss, N. S., and Watts, D. W. (1955). *Chem. Ind. (London)*, 1353.
Bayliss, N. S., and Watts, D. W. (1956). *Aust. J. Chem.* **9**, 319.
Bean, G. P., and Katritzky, A. R. (1968). *J. Chem. Soc. B*, 864.
Bell, R. P. (1941). "Acid-Base Catalysis". Oxford University Press, London.
Bell, R. P. (1959). "The Proton in Chemistry", Methuen, London.
Bell, R. P., and Brown, A. H. (1954). *J. Chem. Soc.* 774.
Bell, R. P., and Clunie, J. C. (1952). *Proc. Roy. Soc. (London), Ser. A*, **212**, 16.
Bell, R. P., Gold, V., Hilton, J., and Rand, M. H. (1954). *Discuss. Faraday Soc.* No. 17, 151.
Bell, R. P., Dowdling, A. L., and Noble, J. A. (1955). *J. Chem. Soc.* 3106.
Bell, R. P., Bascombe, K. N., and McCoubrey, J. C. (1956). *J. Chem. Soc.* 1286.
Bell, R. P., Preston, J., and Whitney, R. B. (1962). *J. Chem. Soc.* 1166.
Bellamy, L. J. (1955). *J. Chem. Soc.* 4221.
Bellamy, L. J., and Pace, R. J. (1963). *Spectrochim. Acta.* **19**, 1831.
Bellamy, L. J., and Williams, R. L. (1957). *J. Chem. Soc.* 863.
Bender, M. L. (1951). *J. Amer. Chem. Soc.* **73**, 1626.
Bender, M. L., and Chen, M. C. (1963). *J. Amer. Chem. Soc.* **85**, 37.
Bender, M. L., and Ginger, R. D. (1955). *J. Amer. Chem. Soc.* **77**, 348.
Bender, M. L., Ginger, R. D., and Unik, J. P. (1958). *J. Amer. Chem. Soc.* **80**, 1044.
Bender, M. L., Ladenheim, H., and Chen, M. C. (1961). *J. Amer. Chem. Soc.* **83**, 123.
Bennett, G. M., Brand, J. C. D., and Williams, G. (1946). *J. Chem. Soc.* 869.
Bennett, G. M., Brand, J. C. D., James, D. M., Saunders, T. J., and Williams, G. (1947). *J. Chem. Soc.* 474.
Berger, A., Loewenstein, A., and Meiboom, S. (1959), *J. Amer. Chem. Soc.* **81**, 62.
Bergmann, M., and Radt, F. (1921). *Ber.* **54**, 1652.
Bertoli, V., and Plesh, P. H. (1966). *Chem. Commun.* 625.
Bethell, D., and Gold, V. (1967). "Carbonium Ions". Academic Press, London and New York.
Bigeleisen, J. (1949). *J. Chem. Phys.* **17**, 675.
Bigeleisen, J. (1952). *J. Phys. Chem.* **56**, 823.
Bigeleisen, J. (1964). *Pure Appl. Chem.* **8**, 217.
Bigeleisen, J., Haschemeyer, R. H., Wolfsberg, M., and Yankwich, P. E. (1962). *J. Amer. Chem. Soc.* **84**, 1813.
Biggs, A. I. (1961). *J. Chem. Soc.* 2572.
Bingham, E. C., and Stone, B. (1923). *J. Phys. Chem.* **27**, 701.
Birchall, T., and Gillespie, R. J. (1963). *Can. J. Chem.* **41**, 2642.
Birchall, T., and Gillespie, R. J. (1965). *Can. J. Chem.* **43**, 1045.
Birchall, T., Bourns, A. N., Gillespie, R. J., and Smith, P. J. (1964). *Can. J. Chem.* **42**, 1433.

Bistrzycki, A., and Fellmann, M. (1910). *Ber.* **43**, 772.
Bistrzycki, A., and Reintke, E. (1905). *Ber.* **38**, 839.
Bistrzycki, A., and Ryncki, L. (1912). *Chem. Ztg.* **36**, 403.
Bistrzycki, A., and Siemiradzki, B. (1906). *Ber.* **39**, 51.
Bjerrum, N. (1923). *Z. Phys. Chem. Abt. A*, **104**, 147.
Blangley, L. (1938). *Helv. Chim. Acta.* **21**, 1579.
Bolton, P. D. (1966). *Aust. J. Chem.* **19**, 1013.
Bonhoeffer, K. F. (1934). *Z. Elektrochem.* **40**, 469.
Bonhoeffer, K. F., and Reitz, O. (1937). *Z. Phys. Chem. Abt. A*, **179**, 135.
Bonner, T. G., and Barnard, M. (1958). *J. Chem. Soc.* 4181.
Bonner, T. G., and Brown, F. (1966). *J. Chem. Soc. B*, 658.
Bonner, T. G., and Lockhart, J. C. (1957). *J. Chem. Soc.* 364.
Bonner, T. G., and Phillips, J. (1966). *J. Chem. Soc.* 650.
Bonner, T. G., Bowyer, F., and Williams, G. (1953). *J. Chem. Soc.* 2650.
Bonner, T. G., Thorne, M. P., and Wilkins, J. M. (1955). *J. Chem. Soc.* 2351.
Borovikov, Yu. Ya., and Fialkov, Yu. Ya. (1965). *Elektrokhimiya*, **1**, 1106.
Bose, H. (1941). *Indian J. Phys.* **15**, 411.
Boyd, R. H. (1961). *J. Amer. Chem. Soc.* **83**, 4288.
Boyd, R. H. (1963a). *J. Amer. Chem. Soc.* **85**, 1555.
Boyd, R. H. (1963b). *J. Phys. Chem.* **67**, 737.
Boyd, R. H. (1969). *In* "Solute–Solvent Interactions" (Eds. J. F. Coetzee and C. D. Ritchie), pp. 98–218. Marcel Dekker, New York.
Boyd, R. H., and Wang, C.-H. (1965). *J. Amer. Chem. Soc.* **87**, 430.
Boyd, R. H., Taft, R. W., Jr., Wolfe, A. P., and Christman, D. R. (1960). *J. Amer. Chem. Soc.* **82**, 4729.
Brand, J. C. D. (1946). *J. Chem. Soc.* 585.
Brand, J. C. D. (1950). *J. Chem. Soc.* 997.
Brand, J. C. D., and Horning, W. C. (1952). *J. Chem. Soc.* 3922.
Brand, J. C. D., and Rutherford, A. (1952). *J. Chem. Soc.* 3916.
Brand, J. C. D., Horning, W. C., and Thornley, M. B. (1952). *J. Chem. Soc.* 1374.
Brand, J. C. D., James, J. C., and Rutherford, A. (1953). *J. Chem. Soc.* 2447.
Brand, J. C. D., Jarvie, A. W. P., and Horning, W. C. (1959). *J. Chem. Soc.* 3844.
Brayford, J. R., and Wyatt, P. A. H. (1955). *J. Chem. Soc.* 3438.
Brayford, J. R., and Wyatt, P. A. H. (1956). *Trans. Faraday Soc.* **52**, 642.
Bredig, G., and Lichty, D. M. (1906). *Z. Elektrochem.* **12**, 459.
Brickman, M., and Ridd, J. H. (1965). *J. Chem. Soc.* 6845.
Brickman, M., Utley, J. H. P., and Ridd, J. H. (1965). *J. Chem. Soc.* 6851.
Briggs, L. H., and Littleton, J. W. (1943). *J. Chem. Soc.* 421.
Brock Robertson, E., and Dunford, H. B. (1964). *J. Amer. Chem. Soc.* **86**, 5080.
Brodskii, A. I., and Micluchin, G. P. (1941). *Dokl. Akad. Nauk. SSSR.* **32**, 588.
Brønsted, J. N., and Wynne-Jones, W. F. K. (1929). *Trans. Faraday Soc.* **25**, 59.
Bronwer, D. M., Mackor, E. L., and Maclean, C. (1966). *Rec. Trav. Chim. Pays-Bas*, **85**, 109, 114.
Brown, H. C., McDaniel, D. H., and Häfliger, O. (1955). *In* "Determination of Organic Structures by Physical Methods" (Eds. E. A. Braude and F. C. Nachod), pp. 567–662. Academic Press, New York and London.
Brownstein, S., Bunton, C. A., and Hughes, E. D. (1958). *J. Chem. Soc.* 4354.
Bunnett, J. F. (1961). *J. Amer. Chem. Soc.* **83**, 4956, 4968, 4973, 4978.
Bunnett, J. F., and Olsen, F. P. (1966). *Can. J. Chem.* **44**, 1899, 1917.
Bunton, C. A., and Carr, M. D. (1963a). *J. Chem. Soc.* 5854.

Bunton, C. A., and Carr, M. D. (1963b). *J. Chem. Soc.* 5861.
Bunton, C. A., Lewis, T. A., and Llewellyn, D. R. (1954). *Chem. Ind. (London)*, 1154.
Bunton, C. A., Konasiewicz, A., and Llewellyn, D. R. (1955a). *J. Chem. Soc.* 604.
Bunton, C. A., Lewis, T. A., and Llewellyn, D. R. (1955b). *J. Chem. Soc.* 4419.
Bunton, C. A., Hadwick, T., Llewellyn, D. R., and Pocker, Y. (1956). *Chem. Ind. (London)*, 547.
Bunton, C. A., Ley, J. B., Rhind-Tutt, A. J., and Vernon, C. A. (1957). *J. Chem. Soc.* 2327.
Bunton, C. A., Dahn, H., and Pocker, Y. (1958a). *Chem. Ind. (London)*, 1516.
Bunton, C. A., Hadwick, T., Llewellyn, D. R., and Pocker Y. (1958b). *J. Chem. Soc.* 403.
Bunton, C. A., Llewellyn, D. R., and Wilson, I. (1958c). *J. Chem. Soc.* 4747.
Burkett, H., Schubert, W. M., Schultz, F., Murphy, R. B., and Talbot, R. (1959). *J. Amer. Chem. Soc.* **81**, 3923.
Burwell, R. L. Jr. (1954). *Chem. Rev.* **54**, 615.
Burwell, R. L. Jr., Scott, R. B., Maury, L. G., and Hussey, A. S. (1954). *J. Amer. Chem. Soc.* **76**, 5822.
Campbell, A., and Kenyon, J. (1946). *J. Chem. Soc.* 25.
Campbell, A. N., Kartzmark, E. M., Bisset, D., and Bednas, M. E. (1953). *Can. J. Chem.* **31**, 303.
Campbell, H. J., and Edward, J. T. (1960). *Can. J. Chem.* **38**, 2109.
Cannizzaro, S. (1854). *Ann. Chem.* **92**, 114.
Casadevall, A., Cauquil, G., and Corriu, R. (1964a). *Bull. Soc. Chim. Fr.* 187.
Casadevall, A., Cauquil, G., and Corriu, R. (1964b). *Bull. Soc. Chim. Fr.* 204.
Carr, M. D., and England, B. D. (1958). *Proc. Chem. Soc., London*, 350.
Cerfontain, H. (1961a). *Rec. Trav. Chim. Pays-Bas*, **80**, 257.
Cerfontain, H. (1961b). *Rec. Trav. Chim, Pays-Bas*, **80**, 296.
Cerfontain, H. (1965). *Rec. Trav. Chim, Pays-Bas*, **84**, 551.
Cerfontain, H. (1968). "Mechanistic Aspects of Aromatic Sulphonation and Desulphonation", Interscience, New York.
Cerfontain, H., and Kaandorp, A. W. (1963). *Rec. Trav. Chim. Pays-Bas*, **82**, 923.
Cerfontain, H., and Telder, A. (1967a). *Rec. Trav. Chim. Pays-Bas*, **86**, 371.
Cerfontain, H., and Telder, A. (1967b). *Rec. Trav. Chim. Pays-Bas*, **86**, 527.
Cerfontain, H., Sixma, F. L. J., and Vollbracht, L. (1963). *Rec. Trav. Chim. Pays-Bas*, **82**, 659.
Challis, B. C., and Long, F. A. (1965a). *Discuss. Faraday Soc.* No. 39, 67.
Challis, B. C., and Long, F. A. (1965b). *J. Amer. Chem. Soc.* **87**, 1196.
Challis, B. C., and Ridd, J. H. (1960). *Proc. Chem. Soc. London*, 245.
Charlon, M. (1964). *J. Org. Chem.* **29**, 1222.
Chédin, J. (1935). *Compt. Rend.* **200**, 1397.
Chédin, J. (1937). *Ann. Chim. (Paris)*, **8**, 302.
Cherbuliez, E. (1923). *Helv. Chim. Acta*, **6**, 281.
Chiang, Y., and Whipple, E. B. (1963). *J. Amer. Chem. Soc.* **85**, 2763.
Chiang, Y., Hinman, R. L., Theodoropoulos, S., and Whipple, E. B. (1967). *Tetrahedron*, **23**, 745.
Chmiel, C. T., and Long, F. A. (1956). *J. Amer. Chem. Soc.* **78**, 3326.
Chute, W. J., Dunn, G. E., McKenzie, J. C., Myers, G. S., Smart, G. N. R., Suggitt, J. W., and Wright, G. F. (1948). *Can. J. Res. Sect. B*, **26**, 114.
Ciaccio, L. L., and Marcus, R. A. (1962). *J. Amer. Chem. Soc.* **84**, 1838.
Clark, J., and Perrin, D. D. (1964). *Quart. Rev. Chem. Soc.* **18**, 295.

Clerc, J. T., Štefanac, Z., and Simon, W. (1965). *Helv. Chim. Acta*, **48**, 1566.
Cobb, A. W., and Walton, J. H. (1937). *J. Phys. Chem.* **41**, 351.
Coe, J. S., and Gold, V. (1960). *J. Chem. Soc.* 4571.
Colapietro, J., and Long, F. A. (1960). *Chem. Ind. (London)*, 1056.
Collie, J. N., and Tickle, T. (1899). *J. Chem. Soc.* **75**, 710.
Collins, C. J., Rainey, W. T., Smith, W. B., and Kaye, I. A. (1959). *J. Amer. Chem. Soc.* **81**, 460.
Conley, R. T., and Lange, R. J. (1963). *J. Org. Chem.* **28**, 210, 278.
Coombs, M. M. (1958a). *J. Chem. Soc.* 4200.
Coombs, M. M. (1958b). *J. Chem. Soc.* 3454.
Coombes, R. G., Moodie, R. B., and Schofield, K. (1969). *J. Chem. Soc.* 52.
Coryell, C. D., and Fix, R. C. (1955). *J. Inorg. Nucl. Chem.* **1**, 119.
Cowdrey, W. A., and Davies, D. S. (1949). *J. Chem. Soc.* 1871.
Crafts, J. M. (1907). *Bull. Soc. Chim. Fr.* [4], **1**, 917.
Craig, R. A., Garrett, A. B., and Newman, M. S. (1950). *J. Amer. Chem. Soc.* **72**, 163.
Crowell, T. I., and Haukins, M. G. (1969). *J. Phys. Chem.* **73**, 1380.
Cruickshank, D. W. J. (1961). *J. Chem. Soc.* 5486.
Culbertson, G., and Pettit, R. (1963). *J. Amer. Chem. Soc.* **85**, 741.
Curphey, T. J., Santer, J. O., Rosenblum, M., and Richards, J. H. (1960). *J. Amer. Chem. Soc.* **82**, 5249.
Currell, D., and Fry, A. (1956). *J. Amer. Chem. Soc.* **78**, 4377.
Curtius, T., and Darapsky, S. (1901). *J. Prakt. Chem* [2], **63**, 428.
Dacre, B., and Wyatt, P. A. H. (1960). *Proc. Chem. Soc. London*, 18.
Dacre, B., and Wyatt, P. A. H. (1961). *J. Chem. Soc.* 2962.
Darling, H. E. (1964). *J. Chem. Eng. Data*, **9**, 421.
Davies, C., and Addis, H. W. (1937). *J. Chem. Soc.* 1622.
Davis, C. T., and Geissman, T. A. (1954). *J. Amer. Chem. Soc.* **76**, 3507.
Davison, A., McFarlane, W., Pratt, L., and Wilkinson, G. (1962). *J. Chem. Soc.* 3653.
Deane, C. W. (1937). *J. Amer. Chem. Soc.* **59**, 849.
Deane, C. W., and Huffman, J. R. (1943). *Ind. Eng. Chem.* **35**, 684.
Deans, F. B., and Eaborn, C. (1959). *J. Chem. Soc.* 2299.
de Fabrizio, E. (1966). *Ric. Sci.* **36**, 1321.
de Fabrizio, E. C. R., Kalatzis, E., and Ridd, J. H. (1966). *J. Chem. Soc. B*, 533.
Degani, I., Fochi, R., and Spunta, G. (1968a). *Boll. Sci. Fac. Chim. Ind. Bologna*, **26**, 3.
Degani, I., Fochi, R., and Spunta, G. (1968b). *Boll. Sci. Fac. Chim. Ind. Bologna*, **26**, 31.
de la Mare, P. B. D., and Ridd, J. H. (1959). "Aromatic Substitution—Nitration and Halogenation". Butterworths, London.
de la Mare, P. B. D., Hughes, E. D., and Vernon, C. A. (1950). *Research (London)*, **3**, 192, 242.
de la Mare, P. B. D., Ketley, D. A., and Vernon, C. A. (1954). *J. Chem. Soc.* 1290.
de la Mare, P. B. D., Dunn, T. M., and Harvey, J. T. (1957). *J. Chem. Soc.* 923.
de la Mare, P. B. D., Kiamud Din, M., and Ridd, J. H. (1960). *J. Chem. Soc.* 561.
Denney, D. B., and Klemchuk, P. P. (1958). *J. Amer. Chem. Soc.* **80**, 3285.
Deno, N. C., and Evans, W. L. (1957). *J. Amer. Chem. Soc.* **79**, 5804.
Deno, N. C., and Perizzolo, C. (1957). *J. Amer. Chem. Soc.* **79**, 1345.
Deno, N. C., and Pittman, C. U., Jr. (1964). *J. Amer. Chem. Soc.* **86**, 1744.
Deno, N. C., and Sacher, E. (1965). *J. Amer. Chem. Soc.* **87**, 5120.
Deno, N. C., and Stein, R. (1956). *J. Amer. Chem. Soc.* **78**, 578.

Deno, N. C., and Taft, R. W., Jr. (1954). *J. Amer. Chem. Soc.* **76**, 244.
Deno, N. C., and Turner, J. O. (1966). *J. Org. Chem.* **31**, 1969.
Deno, N. C., and Wisotsky, M. J. (1963). *J. Amer. Chem. Soc.* **85**, 1735.
Deno, N. C., Jaruzelski, J. J., and Schriesheim, A. (1955). *J. Amer. Chem. Soc.* **77**, 3044.
Deno, N. C., Edwards, T., and Perizzolo, C. (1957). *J. Amer. Chem. Soc.* **79**, 2108.
Deno, N. C., Groves, P. T., and Saines, G. (1959). *J. Amer. Chem. Soc.* **81**, 5790.
Deno, N. C., Groves, P. T., Jaruzelski, J. J., and Lugash, M. (1960). *J. Amer. Chem. Soc.* **82**, 4719.
Deno, N. C., Peterson, H. J., and Sacher, E. (1961). *J. Phys. Chem.* **65**, 199.
Deno, N. C., Richley, H. C., Hodge, J. D., and Wisotsky, M. J. (1962a). *J. Amer. Chem. Soc.* **84**, 1498.
Deno, N. C., Saines, G., and Spangler, M. (1962b). *J. Amer. Chem. Soc.* **84**, 3295.
Deno, N. C., Richley, H. C., Friedman, N., Hodge, J. D., Houser, J. J., and Pittman, C. U., Jr. (1963a). *J. Amer. Chem. Soc.* **85**, 2991.
Deno, N. C., Friedman, N., Hodge, J. D., and Houser, J. J. (1963b). *J. Amer. Chem. Soc.* **85**, 2995.
Deno, N. C., Bollinger, J., Friedman, N., Hafer, K., Hodge, J. D., and Houser, J. J. (1963c). *J. Amer. Chem. Soc.* **85**, 2998.
Deno, N. C., Boyd, D. B., Hodge, J. D., Pittman, C. U., Jr., and Turner, J. O. (1964a). *J. Amer. Chem. Soc.* **86**, 1744.
Deno, N. C., Pittman, C. U., Jr., and Wisotsky, M. J. (1964b). *J. Amer. Chem. Soc.* **86**, 4370.
Deno, N. C., Kish, F. A., and Peterson, H. J. (1965). *J. Amer. Chem. Soc.* **87**, 2157.
Deno, N. C., Gaugler, R. W., and Schulze, T. (1966a). *J. Org. Chem.* **31**, 1968.
Deno, N. C., Gaugler, R. W., and Wisotsky, M. J. (1966b). *J. Org. Chem.* **31**, 1967.
Deno, N. C., La Vietes, D., Mockus, J., and Scholl, P. S. (1968). *J. Amer. Chem. Soc.* **90**, 6457.
Derbyshire, D. H., and Waters, W. A. (1950a). *J. Chem. Soc.* 564.
Derbyshire, D. H., and Waters, W. A. (1950b). *J. Chem. Soc.* 573.
Derbyshire, D. H., and Waters, W. A. (1950c). *J. Chem. Soc.* 3694.
Derbyshire, D. H., and Waters, W. A. (1951). *J. Chem. Soc.* 73.
De Right, R. E. (1933). *J. Amer. Chem. Soc.* **55**, 4761.
De Right, R. E. (1934). *J. Amer. Chem. Soc.* **56**, 618.
Deschamps, J. M. (1957). *Compt. Rend.* **245**, 1432.
Dewar, M. J. S. (1963). *In* "Molecular Rearrangements" (Ed. P. de Mayo). Vol. I. Chapter 5. John Wiley, New York.
Dewar, M. J. S., and Maitlis, P. M. (1957). *J. Chem. Soc.* 2521.
Dittmar, H. R. (1929). *J. Phys. Chem.* **33**, 533.
Dolman, D., and Stewart, R. (1967). *Can. J. Chem.* **45**, 903.
Dostrowsky, I., and Klein, F. S. (1955). *J. Chem. Soc.* 791.
Drenth, W., and Hogeveen, H. (1960). *Rec. Trav. Chim. Pays-Bas*, **79**, 1002.
Drucker, C. (1937). *Trans. Faraday Soc.* **33**, 660.
Drude, P. (1897). *Z. Phys. Chem. Abt. A*, **23**, 270.
Drude, P. (1902). *Z. Phys. Chem. Abt. A*, **40**, 635.
Duffy, J. A., and Leisten, J. A. (1960). *J. Chem. Soc.* 545, 853.
Dunstan, A. E. (1914). *Proc. Chem. Soc., London*, **30**, 104.
Dunstan, A. E., and Wilson, R. W. (1907). *J. Chem. Soc.* **91**, 83.
Dunstan, A. E., and Wilson, R. W. (1908). *J. Chem. Soc.* **93**, 2179.

Durand, J.-P., Davidson, M., Hallin, M., and Coussemant, F. (1966). *Bull. Soc. Chim. Fr.* 43, 52.
Durie, R. A., and Shannon, J. S. (1958). *Aust. J. Chem.* **11**, 168.
Eaborn, C. (1953). *J. Chem. Soc.* 3148.
Eaborn, C. (1956). *J. Chem. Soc.* 4858.
Eaborn, C., and Pande, K. C. (1960). *J. Chem. Soc.* 1566.
Eaborn, C., and Pande, K. C. (1961). *J. Chem. Soc.* 3715.
Eaborn, C., and Taylor, R. (1959). *Chem. Ind. (London)*, 949.
Eaborn, C., and Taylor, R. (1960). *J. Chem. Soc.* 3301.
Eaborn, C., and Taylor, R. (1961a). *J. Chem. Soc.* 247.
Eaborn, C., and Taylor, R. (1961b). *J. Chem. Soc.* 1012.
Eaborn, C., and Waters, J. A. (1961). *J. Chem. Soc.* 542.
Eberz, W. F., and Lucas, H. J. (1934). *J. Chem. Soc.* **56**, 1230.
Edward, J. T. (1963). *Chem. Ind. (London)*, 489.
Edward, J. T. (1964). *Trans. Roy. Soc. Can.* **2**, 313.
Edward, J. T., and Meacock, S. C. R. (1957a). *J. Chem. Soc.* 2000.
Edward, J. T., and Meacock, S. C. R. (1957b). *J. Chem. Soc.* 2007.
Edward, J. T., and Meacock, S. C. R. (1957c). *J. Chem. Soc.* 2009.
Edward, J. T., and Stollar, H. (1963). *Can. J. Chem.* **41**, 721.
Edward, J. T., and Wang, I. C. (1962). *Can. J. Chem.* **40**, 966.
Edward, J. T., Hutchinson, H. P., and Meacock, S. C. R. (1955). *J. Chem. Soc.* 2520.
Edward, J. T., Chang, H. S., Yates, K., and Stewart, R. (1960). *Can. J. Chem.* **38**, 1518.
Edward, J. T., Leane, J. B., and Wang, I. C. (1962). *Can. J. Chem.* **40**, 1521.
Eidus, Ya. T., Puzitskii, K. V., and Ryabova, K. G. (1962). *Zh. Obshch. Khim.* **32**, 3198.
Eistert, B., Merkel, E., and Reiss, W. (1954). *Chem. Ber.* **87**, 1513.
Ellefsen, P. R., and Gordon, L. (1967). *Talanta*, **14**, 409.
Elliott, W. W., and Hammick, D. Ll. (1951). *J. Chem. Soc.* 3402.
Entelis, S. G., and Kazanskii, K. S. (1963). *Kinet. i Katal.* **4**, 713.
Entelis, S. G., Kazanskii, K. S., and Kogan, G. A. (1963). *Kinet. i Katal.* **4**, 277, 589.
Exner, O. (1963). *Collect. Czech. Chem. Commun.* **28**, 935.
Exner, O. (1966). *Collect. Czech. Chem. Commun.* **31**, 65.
Eyring, H., and Cagle, F. Wm., Jr. (1952). *J. Phys. Chem.* **56**, 889.
Fabbri, G., and Roffia, S. (1959). *Proc. Int. Meet. Mol. Spectrosc.* 4th, Bologna, **3**, 963 (Published 1962).
Fajans, E., and Goodeve, C. F. (1936). *Trans. Faraday Soc.* **32**, 511.
Fee, J. A., and Fife, T. H. (1966). *J. Org. Chem.* **31**, 2343.
Filler, R., Wang, C.-S., McKinney, M. A., and Miller, F. N. (1967). *J. Amer. Chem. Soc.* **89**, 1026.
Fischer, A., Grigor, B. A., Packer, J., and Vaughan J. (1961). *J. Amer. Chem. Soc.* **83**, 4208.
Flexser, L. A., Hammett, L. P., and Dingwall, A. (1935). *J. Amer. Chem. Soc.* **57**, 2103.
Flowers, R. H., Gillespie, R. J., and Wasif, S. (1956). *J. Chem. Soc.* 607.
Flowers, R. H., Gillespie, R. J., Oubridge, J. V., and Solomons, C. (1958). *J. Chem. Soc.* 667.
Flowers, R. H., Gillespie, R. J., Robinson, E. A., and Solomons, C. (1960a). *J. Chem. Soc.* 4327.
Flowers, R. H., Gillespie, R. J., and Robinson, E. A. (1960b). *Can. J. Chem.* **38**, 1363.
Flowers, R. H., Gillespie, R. J., and Robinson, E. A. (1963). *Can. J. Chem.* **41**, 2464.

Fraenkel, G., and Niemann, C. (1958). *Proc. Nat. Acad. Sci. U.S.A.* **44**, 688.
Fraenkel, G., and Franconi, C. (1960). *J. Amer. Chem. Soc.* **82**, 4478.
Friedman, H. B., and Elmore, G. V. (1941). *J. Amer. Chem. Soc.* **63**, 864.
Fry, A., and Calvin, M. (1952). *J. Phys. Chem.* **56**, 897.
Gardner, J. N., and Katritzky, A. R. (1957). *J. Chem. Soc.* 4375.
Gastaminza, A., Modro, T. A., Ridd, J. H., and Utley, J. H. P. (1968). *J. Chem. Soc. B*, 534.
Gel'bshtein, A. I., Shcheglova, G. G., and Temkin, M. I. (1956). *Zh. Neorg. Khim.* **1**, 506.
Gerrard, W., and Macklen, E. D. (1959). *Chem. Rev.* **59**, 1105.
Giauque, W. F., Kunzler, J. E., and Hornung, E. W. (1956). *J. Amer. Chem. Soc.* **78**, 5482.
Giauque, W. F., Hornung, E. W., Kunzler, J. E., and Rubin, T. R. (1960). *J. Amer. Chem. Soc.* **82**, 62.
Giguère, P. A., and Savoie, R. (1960). *Can. J. Chem.* **38**, 2467.
Gilbert, E. E. (1965). "Sulphonation and Related Reactions." Interscience, New York.
Gillespie, R. J. (1950a). *J. Chem. Soc.* 2493.
Gillespie, R. J. (1950b). *J. Chem. Soc.* 2516.
Gillespie, R. J. (1950c). *J. Chem. Soc.* 2542.
Gillespie, R. J. (1950d). *J. Chem. Soc.* 2997.
Gillespie, R. J. (1954). *J. Chem. Soc.* 1851.
Gillespie, R. J. (1963). *In* "Friedel-Crafts and Related Reactions" (Ed. G. A. Olah), Vol. I, p. 169. Interscience, New York.
Gillespie, R. J., and Birchall, T. (1963). *Can. J. Chem.* **41**, 148.
Gillespie, R. J., and Cole, R. H. (1956). *Trans. Faraday Soc.* **52**, 1325.
Gillespie, R. J., and Leisten, J. A. (1954a). *Quart. Rev. Chem. Soc.* **8**, 40.
Gillespie, R. J., and Leisten, J. A. (1954b). *J. Chem. Soc.* **1**, 7.
Gillespie, R. J., and Millen, D. J. (1948). *Quart. Rev. Chem. Soc.* **2**, 277.
Gillespie, R. J., and Norton, D. G. (1953). *J. Chem. Soc.* 971.
Gillespie, R. J., and Oubridge, J. V. (1956). *J. Chem. Soc.* 80.
Gillespie, R. J., and Passerini, R. C. (1956). *J. Chem. Soc.* 3850.
Gillespie, R. J., and Robinson, E. A. (1957a). *Proc. Chem. Soc.* (*London*), 145.
Gillespie, R. J., and Robinson, E. A. (1957b). *J. Chem. Soc.*, 4233.
Gillespie, R. J., and Robinson, E. A. (1959). *Advan. Inorg. Chem. Radiochem.* **1**, 386.
Gillespie, R. J., and Robinson, E. A. (1962a). *Can. J. Chem.* **40**, 644.
Gillespie, R. J., and Robinson, E. A. (1962b). *Can. J. Chem.* **40**, 658.
Gillespie, R. J., and Robinson, E. A. (1962c). *Can. J. Chem.* **40**, 784.
Gillespie, R. J., and Robinson, E. A. (1964). *J. Amer. Chem. Soc.* **86**, 5676.
Gillespie, R. J., and Robinson, E. A. (1965). *In* "Non-aqueous Solvent Systems" (Ed. T. C. Waddington), Chapter IV, p. 117. Academic Press, London and New York.
Gillespie, R. J., and Robinson, E. A. (1968). *In* "Carbonium Ions" (Eds. G. A. Olah and P. von R. Schleyer), Vol. I. p. 111. Interscience, New York.
Gillespie, R. J., and Senior, J. B. (1964a). *Inorg. Chem.* **3**, 440.
Gillespie, R. J., and Senior, J. B. (1964b). *Inorg. Chem.* **3**, 972.
Gillespie, R. J., and Solomons, C. (1957). *J. Chem. Soc.* 1796.
Gillespie, R. J., and Solomons, C. (1960). *J. Chem. Educ.* **37**, 202.
Gillespie, R. J., and Wasif, S. (1953a). *J. Chem. Soc.* 209.
Gillespie, R. J., and Wasif, S. (1953b). *J. Chem. Soc.* 215.
Gillespie, R. J., and Wasif, S. (1953c). *J. Chem. Soc.* 221.

Gillespie, R. J., and Wasif, S. (1953d). *J. Chem. Soc.* 964.
Gillespie, R. J., and White, R. F. M. (1958). *Trans Faraday Soc.* **54**, 1846.
Gillespie, R. J., and White, R. F. M. (1960). *Can. J. Chem.* **38**, 1371.
Gillespie, R. J., Hughes, E. D., and Ingold, C. K. (1950a). *J. Chem. Soc.* 2473.
Gillespie, R. J., Hughes, E. D., Ingold, C. K., and Peeling, E. R. A. (1950b). *J. Chem. Soc.* 2504.
Gillespie, R. J., Oubridge, J. V., and Solomons, C. (1957). *J. Chem. Soc.* 1804.
Gillespie, R. J., Robinson, E. A., and Solomons, C. (1960). *J. Chem. Soc.* 4320.
Gillespie, R. J., Kapoor, R., and Robinson, E. A. (1966). *Can. J. Chem.* **44**, 1197.
Glasstone, S., Laidler, K. J., and Eyring, H. (1941). "The Theory of Rate Processes". McGraw-Hill Book Co., New York.
Gleason, A. H., and Dougherty, G. (1929). *J. Amer. Chem. Soc.* **51**, 310.
Gleghorn, J. T., Moodie, R. B., Qureshi, E. A., and Schofield, K. (1968). *J. Chem. Soc. B*, 312.
Gmitro, J. J., and Vermeulen, Th. (1964). *Amer. Inst. Chem. Eng. J.* **10**, 740.
Goering, H. L., and Dilgren, R. E. (1960). *J. Amer. Chem. Soc.* **82**, 5744.
Gold, V. (1955). *J. Chem. Soc.* 1263.
Gold, V. (1964). *In* "Friedel-Crafts and Related Reactions" (Ed. G. A. Olah), Vol. II, p. 1252. Interscience, New York.
Gold, V., and Gruen, L. C. (1966). *J. Chem. Soc. B*, 600.
Gold, V., and Hawes, B. W. V. (1951). *J. Chem. Soc.* 2102.
Gold, V., and Hilton, J. (1955). *J. Chem. Soc.* 838, 343.
Gold, V., and Kessick, M. A. (1965). *Discuss. Faraday Soc.* No. 39, 84.
Gold, V., and Satchell, D. P. N. (1955a). *J. Chem. Soc.* 3609.
Gold, V., and Satchell, D. P. N. (1955b). *J. Chem. Soc.* 3619.
Gold, V., and Satchell, D. P. N. (1956). *J. Chem. Soc.* 1635, 2743.
Gold, V., and Satchell, R. S. (1963). *J. Chem. Soc.* 1930.
Gold, V., and Tye, F. L. (1950). *J. Chem. Soc.* 2932.
Gold, V., and Tye, F. L. (1952a). *J. Chem. Soc.* 2172.
Gold, V., and Tye, F. L. (1952b). *J. Chem. Soc.* 2181.
Gold, V., and Tye, F. L. (1952c). *J. Chem. Soc.* 2184.
Gold, V., Lambert, R. W., and Satchell, D. P. N. (1959). *Chem. Ind. (London)*, 1312.
Gold, V., Lambert, R. W., and Satchell, D. P. N. (1960). *J. Chem. Soc.* 2461.
Goldfarb, A. R., Mele, A., and Gutstein, N. (1955). *J. Amer. Chem. Soc.* **77**, 6194.
Gore, P. H. (1957). *J. Chem. Soc.* 1437.
Grabovskaya, Zh. E., and Vinnik, M. I. (1966). *Zh. Fiz. Khim.* **40**, 2272.
Grabovskaya, Zh. E., Andreeva, L. R., and Vinnik, M. I. (1967). *Zh. Fiz. Khim.* **41**, 918.
Grace, J. A., and Symons, M. C. R. (1959). *J. Chem. Soc.* 961.
Graebe, C., and Aubin, C. (1888). *Ann. Chem.* **247**, 261.
Greenewalt, C. H. (1925). *Ind. Eng. Chem.* **17**, 522.
Greenwood, N. N., and Thompson, A. (1959). *J. Chem. Soc.* 3474.
Greenwood, N. N., and Thompson A. (1960). *Inorg. Syn.* **6**, 121.
Gregory, B. J., Moodie, R. B., and Schofield, K. (1968). *Chem. Commun.* 1380.
Gruen, L. C., and Long, F. A. (1967). *J. Amer. Chem. Soc.* **89**, 1287.
Grunwald, E., Loewenstein, A., and Meiboom, S. (1957a). *J. Chem. Phys.* **27**, 630.
Grunwald, E., Loewenstein, A., and Meiboom, S. (1957b). *J. Chem. Phys.* **27**, 641.
Grunwald, E., Heller, A., and Klein, F. S. (1957c). *J. Chem. Soc.* 2604.
Gudmundsen, C. H., and McEwen, W. E. (1957). *J. Amer. Chem. Soc.* **79**, 329.
Guggenheim, E. A. (1926). *Phil. Mag.*, 7th Ser. **2**, 538.

Gutowsky, H. S., and Saika, A. (1953). *J. Chem. Phys.* **21**, 1688.
Haaf, W. (1964). *Brennst.-Chem.* **45**, 209.
Haake, P., and Cook, R. D. (1968). Tetrahedron Lett. 427.
Haake, P., and Hurst, G. H. (1966). *J. Amer. Chem. Soc.* **88**, 2544.
Haake, P., Cook, R. D., and Hurst, G. H. (1967). *J. Amer. Chem. Soc.* **89**, 2650.
Haase, R., Sauermann, P. F., and Dücker, K.-H. (1966). *Z. Phys. Chem. (Frankfurt)*, **48**, 206.
Hahn, C.-S., and Jaffé, H. H. (1962). *J. Amer. Chem. Soc.* **84**, 949.
Halevi, E. A., Ron, A., and Seiser, S. (1965). *J. Chem. Soc.* 2560.
Hall, N. F., and Spengeman, W. F. (1940). *J. Amer. Chem. Soc.* **62**, 2489.
Hall, S. K., and Robinson, E. A. (1964). *Can. J. Chem.* **42**, 1113.
Hammett, L. P. (1935). *Chem. Rev.* **16**, 67.
Hammett, L. P. (1937). *J. Amer. Chem. Soc.* **59**, 96.
Hammett, L. P. (1940). "Physical Organic Chemistry". McGraw-Hill, New York and London.
Hammett, L. P., and Chapman, R. P. (1934). *J. Amer. Chem. Soc.* **56**, 1282.
Hammett, L. P., and Deyrup, A. J. (1932). *J. Amer. Chem. Soc.* **54**, 2721.
Hammett, L. P., and Deyrup, A. J. (1933). *J. Amer. Chem. Soc.* **55**, 1900.
Hammett, L. P., and Lowenheim, F. A. (1934). *J. Amer. Chem. Soc.* **56**, 2620.
Hammett, L. P., and Paul, M. A. (1934). *J. Amer. Chem. Soc.* **56**, 830.
Handa, T. (1955). *Bull. Chem. Soc. Jap.* **28**, 483.
Hantzsch, A. (1908). *Z. Phys. Chem.* **61**, 257.
Hantzsch, A. (1909). *Z. Physik. Chem.* **65**, 41.
Hantzsch, A. (1921). *Ber.* **54 B**, 2573.
Hantzsch, A. (1922). *Ber.* **55 B**, 953.
Hantzsch, A. (1930). *Ber.* **63 B**, 1782.
Hart, H., and Fish, R. W. (1958). *J. Amer. Chem. Soc.* **80**, 5894.
Hart, H., and Fish, R. W. (1960). *J. Amer. Chem. Soc.* **82**, 5419.
Hart, H., and Fish, R. W. (1961). *J. Amer. Chem. Soc.* **83**, 4460.
Hart, H., and Sedor, E. A. (1967). *J. Amer. Chem. Soc.* **89**, 2342.
Hartshorn, S. R., and Ridd, J. H. (1968). *J. Chem. Soc. B*, 1063, 1068.
Hartwell, E. J., Richards, R. E., and Thompson, H. W. (1948). *J. Chem. Soc.* 1436.
Hass, H. B., and Riley, E. F. (1943). *Chem. Rev.* **32**, 373.
Hassel, O., and Roemming, Chr. (1960). *Acta Chem. Scand.* **14**, 398.
Hearne, G., Tamele, M., and Converse, W. (1941). *Ind. Eng. Chem.* **33**, 805.
Heck, R., and Winstein, S. (1957). *J. Amer. Chem. Soc.* **79**, 3432.
Hekkert, G. L., and Drenth, W. (1961). *Rec. Trav. Chim. Pays-Bas.* **80**, 1285.
Herbert, R. A., Goren, M. B., and Vernon, A. A. (1952). *J. Amer. Chem. Soc.* **74**, 5779.
Hetherington, G., Nichols, M. J., and Robinson, P. L. (1955a). *J. Chem. Soc.* 3141.
Hetherington, G., Hub, D. R., Nichols, M. J., and Robinson, P. L. (1955b). *J. Chem. Soc.* 3300.
Hetherington, G., Hub, D. R., and Robinson, P. L. (1955c). *J. Chem. Soc.* 4041.
Hill, R. K., Conley, R. T., and Chortyk, O. T. (1965). *J. Amer. Chem. Soc.* **87**, 5646.
Hinman, R. L., and Lang, J. (1964). *J. Amer. Chem. Soc.* **86**, 3796.
Hinman, R. L., and Whipple, E. B. (1962). *J. Amer. Chem. Soc.* **84**, 2534.
Hofmann, J. E., and Schriesheim, A. (1962a). *J. Amer. Chem. Soc.* **84**, 953.
Hofmann, J. E., and Schriesheim, A. (1962b). *J. Amer. Chem. Soc.* **84**, 957.
Hogeveen, H. (1966). *Rec. Trav. Chim. Pays-Bas*, **85**, 1072.
Hogeveen, H. (1967). *Rec. Trav. Chim. Pays-Bas*, **86**, 1288.

Hogeveen, H., and Bickel, A. F. (1967). *Rec. Trav. Chim. Pays-Bas*, **86**, 1313.
Hogeveen, H., and Drenth, W. (1963a). *Rec. Trav. Chim. Pays-Bas*, **82**, 375.
Hogeveen, H., and Drenth, W. (1963b). *Rec. Trav. Chim. Pays-Bas*, **82**, 410.
Hogeveen, H., Bickel, A. F., Hilbers, C. W., Mackor, E. L., and MacLean, C. (1966). *Chem. Commun.* 898.
Högfeldt, E. (1963). *Acta Chem. Scand.* **17**, 785.
Högfeldt, E., and Bigeleisen, J. (1960). *J. Amer. Chem. Soc.* **82**, 15.
Holstead, C., Lamberton, A. H., and Wyatt, P. A. H. (1953). *J. Chem. Soc.* 3341.
Hood, G. C., and Reilly, C. A. (1957). *J. Chem. Phys.* **27**, 1126.
Hoop, G. M., and Tedder, J. M. (1961). *J. Chem. Soc.* 4685.
Hopkinson, A. C. (1969). *J. Chem. Soc. B*, 203.
Hornung, E. W., and Giauque, W. F. (1955). *J. Amer. Chem. Soc.* **77**, 2744.
Hoshino, S., Hosoya, H., and Nagakura, S. (1966). *Can. J. Chem.* **44**, 1961.
Hosoya, H., and Nagakura, S. (1961). *Spectrochim. Acta*, **17 A**, 324.
Hughes, E. D., and Ingold, C. K. (1952). *Quart. Rev. Chem. Soc.* **6**, 34.
Hughes, E. D., and Jones, G. T. (1950). *J. Chem. Soc.* 2678.
Hughes, E. D., Ingold, C. K., and Pearson, R. B. (1958a). *J. Chem. Soc.* 4357.
Hughes, E. D., Ingold, C. K., and Ridd, J. H. (1958b). *J. Chem. Soc.* 58 and accompanying papers.
Huisgen, R., Brade, H., Walz, H., and Glogger, J. (1957). *Chem. Ber.* **90**, 1437.
Ingold, C. K. (1953). "Structure and Mechanism in Organic Chemistry", G. Bell and Sons. London.
Ingold, C. K., and Shaw, F. R. (1927). *J. Chem. Soc.* 2918.
Ingold, C. K., and Wilson, C. L. (1934.) *J. Chem. Soc.* 773.
Ingold, C. K., Raisin, C. G., and Wilson, C. L. (1934), *Nature (London)*, **134**, 734.
Ingold, C. K., Raisin, C. G., and Wilson, C. L. (1936a). *J. Chem. Soc.* 915.
Ingold, C. K., Raisin, C. G., and Wilson, C. L. (1936b). *J. Chem. Soc.* 1637.
Ingold, C. K., Millen, D. J., and Poole, H. G. (1950). *J. Chem. Soc.* 2576.
Ipatieff, V. N. (1936). "Catalytic Reactions at High Pressures and Temperatures", p. 673–701, The Macmillan Co., New York.
Jaffé, H. H. (1953). *Chem. Rev.* **53**, 191.
Jaffé, H. H., and Doak, G. O. (1955). *J. Amer. Chem. Soc.* **77**, 4441.
Jaffé, H. H., and Gardner, R. W. (1958). *J. Amer. Chem. Soc.* **80**, 320.
Jaques, D. (1965). *J. Chem. Soc.* 3874.
Jaques, D., and Leisten, J. A. (1961). *J. Chem. Soc.* 4963.
Jaques, D., and Leisten, J. A. (1964). *J. Chem. Soc.* 2683.
Jellinek, H. H. G., and Gordon, A. (1949). *J. Phys. Colloid. Chem.* **53**, 996.
Jellinek, H. H. G., and Urwin, J. R., (1953). *J. Phys. Chem.* **57**, 900.
Johnson, C. D., Katritzky, A. R., Ridgewell, B. J., and Shakir, N. (1965). *Tetrahedron*, **21**, 1055.
Johnson, C. D., Katritzky, A. R., and Shapiro, S. A. (1969). *J. Amer. Chem. Soc.* **91**, 6654.
Johnson, E. I., and Partington, J. R. (1931). *J. Chem. Soc.* 86.
Jörgenson, M. J., and Hartter, D. R. (1963). *J. Amer. Chem. Soc.* **85**, 878.
Kaandorp, A. W., and Cerfontain, H. (1969). *Rec. Trav. Chim. Pays-Bas*, **88**, 725.
Kaandorp, A. W., Cerfontain, H., and Sixma, F. L. J. (1962). *Rec. Trav. Chim. Pays-Bas*, **81**, 969.
Kabachnik, M. I. (1956). *Dokl. Akad. Nauk. SSSR.* **110**, 393.
Kalatzis, E., and Ridd, J. H. (1966). *J. Chem. Soc. B*, 529.
Karasch, N., Buess, C. M., and King, W. (1953). *J. Amer. Chem. Soc.* **75**, 6035.

Katritzky, A. R., and Jones, R. A. Y. (1961). *Chem. Ind. (London)*, 722.
Katritzky, A. R., and Kingsland, M. (1968). *J. Chem. Soc. B*, 862.
Katritzky, A. R., Waring, A. J., and Yates, K. (1963). *Tetrahedron*, **19**, 465.
Kazanskii, K. S., and Entelis, S. G. (1962). *Kinet. i Katal.* **3**, 36.
Kendall, J., and Carpenter, C. D. (1914). *J. Amer. Chem. Soc.* **36**, 2498.
Kershaw, D. N., and Leisten, J. A. (1960). *Proc. Chem. Soc. (London)*, 84.
Kiamud Din, M., and Choudhury, A. K. (1963). *Chem. Ind. (London)*, 1840.
Kilpatrick, M. (1963). *J. Amer. Chem. Soc.* **85**, 1036.
Kilpatrick, M., and Hyman, H. H. (1958). *J. Amer. Chem. Soc.* **80**, 87.
Kilpatrick, M., Meyer, M. W., and Kilpatrick, M. L. (1960). *J. Phys. Chem.* **64**, 1433.
Kilpatrick, M., Meyer, M. W., and Kilpatrick, M. L. (1961). *J. Phys. Chem.* **65**, 530, 1189, 1312.
King, E. J. (1953). *J. Amer. Chem. Soc.* **75**, 2204.
Kipping, F. S. (1907). *J. Chem. Soc.* 209.
Kirkbride, B. J., and Wyatt, P. A. H. (1958). *Trans. Faraday Soc.* **54**, 483.
Klofutar, C., Krašovec, F., and Kušar, M. (1968). *Croat. Chim. Acta*, **40**, 23.
Knietsch, R. (1901). *Ber*, **34**, 4069.
Knowles, J. R., and Norman, R. O. C. (1961). *J. Chem. Soc.* 3885.
Knowles, J. R., Norman, R. O. C., and Rada, G. K. (1960). *J. Chem. Soc.* 4885.
Koch, H. (1955). *Brennst.-Chem.* **36**, 321.
Koch, H., and Haaf, W. (1958). *Ann. Chem.* **618**, 251.
Koeberg-Telder, A., Prinsen, A. J., and Cerfontain, H. (1969). *J. Chem. Soc. B*, 1004.
Koepp, H.-M., Wendt, H., and Strehlow, H., (1960). *Z. Elektrochem.* **64**, 483.
Kohlrausch, F. (1876). *Pogg. Ann.* **159**, 233, 240.
Kohlrausch, W. (1882). *Ann. Phys.* **17**, 69.
Kohnstam, G. (1967). *Adv. Phys. Org. Chem.* **5**, 121.
Kort, C. W. F., and Cerfontain, H. (1967). *Rec. Trav. Chim. Pays-Bas*, **86**, 865.
Kort, C. W. F., and Cerfontain, H. (1968). *Rec. Trav. Chim. Pays-Bas*, **87**, 24.
Kort, C. W. F., and Cerfontain, H. (1969a). *Rec. Trav. Chim. Pays-Bas*, **88**, 860.
Kort, C. W. F., and Cerfontain, H. (1969b). *Rec. Trav. Chim. Pays-Bas*, **88**, 1298.
Kortüm, G., Vogel, W., and Andrussow, K. (1961). "Dissociation Constants of Organic Acids in Aqueous Solution". Butterworths, London.
Koski, W. S. (1949). *J. Amer. Chem. Soc.* **71**, 4042.
Koskikallio, J., and Whalley, E. (1959a). *Can. J. Chem.* **37**, 788.
Koskikallio, J., and Whalley, E. (1959b). *Trans. Faraday Soc.* **55**, 809.
Koskikallio, J., and Whalley, E. (1959c). *Trans. Faraday Soc.* **55**, 815.
Kovacic, P., and Bennett, R. P. (1961). *J. Amer. Chem. Soc.* **83**, 221, 743.
Kovacic, P., Russel, R. L., and Bennett, R. P. (1964). *J. Amer. Chem. Soc.* **86**, 1588.
Krafft, F. (1875). *Ber.* **8**, 1044.
Kramer, G. M. (1965). *J. Org. Chem.* **30**, 2671.
Kreevoy, M. M., and Taft, R. W., Jr. (1955). *J. Amer. Chem. Soc.* **77**, 3146, 5590.
Kresge, A. J., and Chiang, Y. (1959). *J. Amer. Chem. Soc.* **81**, 5509.
Kresge, A. J., and Chiang, Y. (1961a). *Proc. Chem. Soc. London*, 81.
Kresge, A. J., and Chiang, Y. (1961b). *J. Amer. Chem. Soc.* **83**, 2877.
Kresge, A. J., and Chiang, Y. (1962). *J. Amer. Chem. Soc.* **84**, 3976.
Kresge, A. J., and Chiang, Y. (1967). *J. Amer. Chem. Soc.* **89**, 4411.
Kresge, A. J., and Hakka, L. E. (1966). *J. Amer. Chem. Soc.* **88**, 3868.
Kresge, A. J., Barry, G. W., Charles, K. R., and Chiang, Y. (1962). *J. Amer. Chem. Soc.* **84**, 4343.

Kresge, A. J., Hakka, L. E., Mylonakis, S., and Sato, Y. (1965). *Discuss. Faraday Soc.* No. 39, 75.
Kresge, A. J., Chiang, Y., and Sato, Y. (1967). *J. Amer. Chem. Soc.* **89**, 4418.
Krieble, V. K., and Holst, K. A. (1938). *J. Amer. Chem. Soc.* **60**, 2976.
Krieble, V. K., and Noll, C. I. (1939). *J. Amer. Chem. Soc.* **61**, 560.
Kuhn, L. P. (1949). *J. Amer. Chem. Soc.* **71**, 1575.
Kuhn, L. P., and Corwin, A. H. (1948). *J. Amer. Chem. Soc.* **70**, 3370.
Kuhn, S. J., and Olah, G. A. (1961). *J. Amer. Chem. Soc.* **83**, 4564.
Kuivila, H. G., and Nahabedian, K. V. (1961a). *J. Amer. Chem. Soc.* **83**, 2159.
Kuivila, H. G., and Nahabedian, K. V. (1961b). *J. Amer. Chem. Soc.* **83**, 2164.
Kunzler, J. E. (1953). *Anal. Chem.* **26**, 93.
Kunzler, J. E., and Giauque, W. F. (1952a). *J. Amer. Chem. Soc.* **74**, 804.
Kunzler, J. E., and Giauque, W. F. (1952b). *J. Amer. Chem. Soc.* **74**, 3472.
Kunzler, J. E., and Giauque, W. F. (1952c). *J. Amer. Chem. Soc.* **74**, 5271.
Kwart, H., and Weisfeld, L. B. (1958). *J. Amer. Chem. Soc.* **80**, 4670.
Ladenheim, H., and Bender, M. L. (1960). *J. Amer. Chem. Soc.* **82**, 1895.
Laidler, K. J., and Landskroener, P. A. (1956). *Trans. Faraday Soc.* **52**, 200.
Landini, D., Modena, G., Scorrano, G., and Taddei, F. (1969). *J. Amer. Chem. Soc.* **91**, 6703.
Lane, C. A. (1964). *J. Amer. Chem. Soc.* **86**, 2521.
Lange, W. (1933). *Z. Anorg. Allg. Chem.* **215**, 321.
Lantz, R. (1935). *Bull. Soc. Chim. Fr.* **2**, 2092.
La Planche, L. A., and Rogers, M. T. (1964). *J. Amer. Chem. Soc.* **86**, 337.
Laughlin, R. G. (1967). *J. Amer. Chem. Soc.* **89**, 4268.
Lavrushin, V. F., Verkhovod, N. N., and Movchan, P. K. (1955). *Dokl. Akad. Nauk. SSSR*, **105**, 723.
Layne, W. S., Jaffé, H. H., and Zimmer, H. (1963). *J. Amer. Chem. Soc.* **85**, 1816.
Le Fave, G. M. (1949). *J. Amer. Chem. Soc.* **71**, 4148.
Leisten, J. A. (1955). *J. Chem. Soc.* 298.
Leisten, J. A. (1956). *J. Chem. Soc.* 1572.
Leisten, J. A. (1959). *J. Chem. Soc.* 765.
Leisten, J. A. (1961). *J. Chem. Soc.* 2191.
Leisten, J. A., and Walton, P. R. (1964). *J. Chem. Soc.* 3180.
Leisten, J. A., and Wright, K. L. (1964). *J. Chem. Soc.* 3173.
Levy, J. B., Taft, R. W., Jr., Aaron, D., and Hammett, L. P. (1951). *J. Amer. Chem. Soc.* **73**, 3792.
Levy, J. B., Taft, R. W., Jr., and Hammett, L. P. (1953). *J. Amer. Chem. Soc.* **75**, 1253.
Lewis, G. N., and Bigeleisen, J. (1943). *J. Amer. Chem. Soc.* **65**, 1144.
Librovich, N. B., and Vinnik, M. I. (1966). *Dokl. Akad. Nauk. SSSR*, **166**, 647.
Lichty, D. M. (1907). *J. Phys. Chem.* **11**, 225.
Lichty, D. M. (1908). *J. Amer. Chem. Soc.* **30**, 1834.
Liler, M. (1959). *In* "Hydrogen Bonding", Proceedings of a Symposium (Ljubljana) (Ed. D. Hadži), p. 519, Pergamon Press, London.
Liler, M. (1962). *J. Chem. Soc.* 4272.
Liler, M. (1963). *J. Chem. Soc.* 3106.
Liler, M. (1965a). *Chem. Commun.* 244.
Liler, M. (1965b). *J. Chem. Soc.* 4300.
Liler, M. (1966a). *Nature. (London).* **211**, 523.
Liler, M. (1966b). *J. Chem. Soc.* 205.

Liler, M. (1967). *Spectrochim. Acta*, **23A**, 139.
Liler, M. (1969). *J. Chem. Soc. B*, 385.
Liler, M., and Kosanović, Dj. (1958). *J. Chem. Soc.* 1084.
Ling, A. C., and Kendall, F. H. (1967). *J. Chem. Soc. B*, 440.
Livingston, R. L., and Rao, C. N. R. (1960). *J. Phys. Chem.* **64**, 756.
Lobry de Bruyn, C. A., and Sluiter, C. H. (1904). *Proc. Kon. Ned. Akad. Wetensch.* **6**, 773.
Lockhart, J. C. (1966). *J. Chem. Soc. B*, 1174.
Loewenstein, A., and Connor, T. M. (1963). *Ber. Bunsenges. Phys. Chem.* **67**, 280.
Long, F. A. (1968). *In* "Hydrogen-bonded Solvent Systems" (Eds. A. K. Covington and P. Jones), p. 285. Taylor and Francis, London.
Long, F. A., and Bigeleisen, J. (1959). *Trans. Faraday Soc.* **55**, 2077.
Long, F. A., and Paul, M. A. (1957). *Chem. Rev.* **57**, 935.
Long, F. A., and Pritchard, J. G. (1956a). *J. Amer. Chem. Soc.* **78**, 2663.
Long, F. A., and Pritchard, J. G. (1956b). *J. Amer. Chem. Soc.* **78**, 6008.
Long, F. A., and Purchase, M. (1950). *J. Amer. Chem. Soc.* **72**, 3267.
Long, F. A., and Schulze, J. (1964). *J. Amer. Chem. Soc.* **86**, 327.
Long, F. A., and Varker, A. (1957). *Chem. Rev.* **57**, 990.
Long, F. A., Dunkle, F. B., and McDevit, W. F. (1951). *J. Phys. Colloid Chem.* **55**, 829, 813.
Long, F. A., Pritchard, J. G., and Stafford, F. E. (1957). *J. Amer. Chem. Soc.* **79**, 2362.
Longridge, J. L., and Long, F. A. (1967). *J. Amer. Chem. Soc.* **89**, 1292.
Longridge, J. L., and Long, F. A. (1968a). *J. Amer. Chem. Soc.* **90**, 3088.
Longridge, J. L., and Long, F. A. (1968b). *J. Amer. Chem. Soc.* **90**, 3092.
Lowen, A. M., Murray, M. A., and Williams, G. (1950). *J. Chem. Soc.* 3321.
Lucas, H. J., and Eberz, W. F. (1934). *J. Amer. Chem. Soc.* **56**, 460.
Lucas, H. J., and Liu, Y. P. (1934). *J. Amer. Chem. Soc.* **56**, 2138.
Lucas, H. J., Stewart, W. T., and Pressman, D. (1944). *J. Amer. Chem. Soc.* **66**, 1818.
Luder, W. F., and Zuffanti, S. (1944). *J. Amer. Chem. Soc.* **66**, 524.
Lundeen, A. (1960). *J. Amer. Chem. Soc.* **82**, 3228.
Lutskii, A. E., and Dorofeev, V. V. (1963). *Zh. Obshch. Khim.* **33**, 2331.
Mackor, E. L., Hofstra, A., and Van der Waals, J. H. (1958). *Trans. Faraday Soc.* **54**, 66, 186.
Manassen, J., and Klein, F. S. (1960). *J. Chem. Soc.* 4203.
Marburg, S., and Jencks, W. P. (1962). *J. Amer. Chem. Soc.* **84**, 232.
March, D. M., and Henshall, T. (1962). *J. Phys. Chem.* **66**, 840.
Markham, A. E., and Kobe, K. A. (1941). *J. Amer. Chem. Soc.* **63**, 1165.
Martinsen, H. (1905). *Z. Phys. Chem.* **50**, 385.
Martinsen, H. (1907). *Z. Phys. Chem.* **59**, 605.
Masson, I. (1931). *J. Chem. Soc.* 3200.
Masson, I., and Argument, C. (1938). *J. Chem. Soc.* 1702.
Masters, B. J., and Norris, T. H. (1952). *J. Amer. Chem. Soc.* **74**, 2395.
Maury, L. G., Burwell, R. L., Jr., and Tuxworth, R. H. (1954). *J. Amer. Chem. Soc.* **76**, 5831.
Mavel, M. G. (1961). *J. Chim. Phys.* **58**, 545.
McCulloch, L. (1946). *J. Amer. Chem. Soc.* **68**, 2735.
McIntyre, D., and Long, F. A. (1954). *J. Amer. Chem. Soc.* **76**, 3240.
McLaren, D. A., and Schachat, R. E. (1949). *J. Org. Chem.* **14**, 254.
McLean, J. D., Rabinovich, B. S., and Winkler, C. A. (1942). *Can. J. Res.* **20B**, 168.
McNamara, A. J., and Stothers, J. B. (1964). *Can. J. Chem.* **42**, 2354.

McNulty, P. J., and Pearson, D. E. (1959). *J. Amer. Chem. Soc.* **81**, 612.
McTigue, P. T., and Gruen, L. C. (1963). *Aust. J. Chem.* **16**, 177.
Melander, L. (1950). *Ark. Kemi*, **2**, 211.
Melander, L. (1960). "Isotope Effects on Reaction Rates". Ronald Press, New York.
Miles, F. D., Niblock, H., and Wilson, G. L. (1940). *Trans. Faraday Soc.* **36**, 345.
Millen, D. J. (1950). *J. Chem. Soc.* 2589.
Mocek, M. M., and Stewart, R. (1963). *Can. J. Chem.* **41**, 1641.
Möller, K. E. (1963). *Angew. Chem.* **75**, 1098. (International edition **2**, 719).
Möller, K. E. (1966). *Brennst.-Chem.* **47**, 10.
Moodie, R. B., Wale, P. D., and White, T. J. (1963a). *J. Chem. Soc.* 4237.
Moodie, R. B., Schofield, K., and Williamson, M. J. (1963b). *Chem. Ind. (London)*, 1283.
Moodie, R. B., Schofield, K., and Williamson, M. J. (1964). "Nitrocompounds" (Proceedings of the International Symposium, Warsaw 1963). Pergamon Press, London.
Moore, W. (1932). *J. Econ. Entomol.* **25**, 729.
Morgan, J. L. R., and Davis, C. E. (1916). *J. Amer. Chem. Soc.* **38**, 555.
Mountford, G. A., and Wyatt, P. A. H. (1966). *Trans. Faraday Soc.* **62**, 3201.
Murray, M. A., and Williams, G. (1950). *J. Chem. Soc.* 3322.
Nagakura, S., Minegishi, A., and Stanfield, K. (1957). *J. Amer. Chem. Soc.* **79**, 1033.
Nahabedian, K. V., and Kuivila, H. G. (1961). *J. Amer. Chem. Soc.* **83**, 2167.
Nelson, J. H., Garvey, R. G., and Ragsdale, R. O. (1967). *J. Heterocycl. Chem.* **4**, 591.
Nenitzescu, C. D. (1968). *In* "Carbonium Ions" (Eds. G. A. Olah and P. von R. Scheyer), Vol. I, p. 23. Interscience, New York.
Newman, M. S. (1941). *J. Amer. Chem. Soc.* **63**, 2431.
Newman, M. S. (1942). *J. Amer. Chem. Soc.* **64**, 2324.
Newman, M. S., and Deno, N. C. (1951a). *J. Amer. Chem. Soc.* **73**, 3644.
Newman, M. S., and Deno, N. C. (1951b). *J. Amer. Chem. Soc.* **73**, 3651.
Newman, M. S., and Gildenhorn, H. L. (1948). *J. Amer. Chem. Soc.* **70**, 317.
Newman, M. S., Kuivila, H. G., and Garrett, A. B. (1945). *J. Amer. Chem. Soc.* **67**, 704.
Newman, M. S., Craig, R. A., and Garrett, A. B. (1949). *J. Amer. Chem. Soc.* **71**, 869.
Nikolov, K., and Mikhailov, M. (1965). *Izv. Fiz. Inst. ANEB, Bulg. Akad. Nauk*, **13**, 25; *C.A.* **64**, 1558c.
Norris, T. H. (1950). *J. Amer. Chem. Soc.* **72**, 1220.
Noyce, D. S., and Avarbock, H. S. (1962). *J. Amer. Chem. Soc.* **84**, 1644.
Noyce, D. S., and Heller, R. A. (1965). *J. Amer. Chem. Soc.* **87**, 4325.
Noyce, D. S., and Jörgenson, M. J. (1961). *J. Amer. Chem. Soc.* **83**, 2525.
Noyce, D. S., and Jörgenson, M. J. (1962). *J. Amer. Chem. Soc.* **84**, 4312.
Noyce, D. S., and Jörgenson, M. J. (1963). *J. Amer. Chem. Soc.* **85**, 2420, 2427.
Noyce, D. S., and Kittle, P. A. (1965a). *J. Org. Chem.* **30**, 1896.
Noyce, D. S., and Kittle, P. A. (1965b). *J. Org. Chem.* **30**, 1899.
Noyce, D. S., and Lane, C. A. (1962). *J. Amer. Chem. Soc.* **84**, 1635.
Noyce, D. S., and Pryor, W. A. (1955). *J. Amer. Chem. Soc.* **77**, 1397.
Noyce, D. S., and Reed, W. L. (1958). *J. Amer. Chem. Soc.* **80**, 5539.
Noyce, D. S., Pryor, W. A., and King, P. A. (1959). *J. Amer. Chem. Soc.* **81**, 5423.
Noyce, D. S., Woo, G. L., and Jörgenson, M. J. (1961). *J. Amer. Chem. Soc.* **83**, 1160.

Noyce, D. S., King, P. A., Kirby, F. B., and Reed, W. L. (1962a). *J. Amer. Chem. Soc.* **84**, 1632.
Noyce, D. S., Avarbock, H. S., and Reed, W. L. (1962b). *J. Amer. Chem. Soc.* **84**, 1647.
Noyce, D. S., Matesich, M. A., Schiavelli, O. P. M. D., and Peterson, P. E. (1965). *J. Amer. Chem. Soc.* **87**, 2295.
Nylén, P. (1941). *Z. Anorg. Allg. Chem.* **246**, 227.
Oae, S., Kitao, T., and Kitaoka, Y. (1965a). *Bull. Chem. Soc. Jap.* **38**, 543.
Oae, S., Kitao, T., Kitaoka, Y., and Kawamura, S. (1965b). *Bull. Chem. Soc. Jap.* **38**, 546.
Oddo, G., and Casalino, A. (1917a). *Gazz. Chim. Ital.* **47**, II, 200.
Oddo, G., and Casalino, A. (1917b). *Gazz. Chim. Ital.* **47**, II, 232.
Oddo, G., and Scandola, E. (1908). *Gazz. Chim. Ital.* **38**, I, 608.
Oddo, G., and Scandola, E. (1909a). *Gazz. Chim. Ital.* **39**, I, 569.
Oddo, G., and Scandola, E. (1909b). *Gazz. Chim. Ital.* **39**, II, 1.
Oddo, G., and Scandola, E. (1909c). *Z. Phys. Chem.* **66**, 138.
Oddo, G., and Scandola, E. (1910). *Gazz. Chim. Ital.* **40**, II, 163.
O'Brien, J. L., and Niemann, C. (1951). *J. Amer. Chem. Soc.* **73**, 4264.
O'Brien, J. L., and Niemann, C. (1957). *J. Amer. Chem. Soc.* **79**, 1386.
O'Ferrall, R. A. M., and Kouba, J. (1967). *J. Chem. Soc. B*, 985.
O'Gorman, J. M., and Lucas, H. J. (1950). *J. Amer. Chem. Soc.* **72**, 5489.
Olivieri-Mandalà, E. (1925). *Gazz. Chim. Ital.* **55**, I, 271.
Olsson, S., and Russell, M. (1969). *Ark. Kemi*, **31**, 439.
Olah, G. A., and Bollinger, J. M. (1967). *J. Amer. Chem. Soc.* **89**, 2993.
Olah, G. A., and Calin, M. (1968). *J. Amer. Chem. Soc.* **90**, 401.
Olah, G. A., and Kiovsky, T. E. (1968a). *J. Amer. Chem. Soc.* **90**, 4666.
Olah, G. A., and Kiovsky, T. E. (1968b). *J. Amer. Chem. Soc.* **90**, 6461.
Olah, G. A., and Kreienbühl, P. (1967). *J. Amer. Chem. Soc.* **89**, 4756.
Olah, G. A., and Lukas, J. (1967). *J. Amer. Chem. Soc.* **89**, 2227.
Olah, G. A., and Namanworth, E. (1966). *J. Amer. Chem. Soc.* **88**, 5327.
Olah, G. A., and O'Brien, D. H. (1967). *J. Amer. Chem. Soc.* **89**, 1725.
Olah, G. A., and von R. Schleyer, P., eds. (1968). "Carbonium Ions", Vol. I. Interscience, New York.
Olah, G. A., and White, A. M. (1967a). *J. Amer. Chem. Soc.* **89**, 3591.
Olah, G. A., and White, A. M. (1967b). *J. Amer. Chem. Soc.* **89**, 4752.
Olah, G. A., and White, A. M. (1967c). *J. Amer. Chem. Soc.* **89**, 7072.
Olah, G. A., and White, A. M. (1968). *J. Amer. Chem. Soc.* **90**, 6087.
Olah, G. A., Kuhn, S. J., and Mlinko, A. (1956). *J. Chem. Soc.* 4256.
Olah, G. A., O'Brien, D. H. and Pittman, C. U., Jr. (1967a). *J. Amer. Chem. Soc.* **89**, 2996.
Olah, G. A., Sommer, J., and Namanworth, E. (1967b). *J. Amer. Chem. Soc.* **89**, 3576.
Olah, G. A., O'Brien, D. H., and Calin, M. (1967c). *J. Amer. Chem. Soc.* **89**, 3582.
Olah, G. A., Calin, M., and O'Brien, D. H. (1967d). *J. Amer. Chem. Soc.* **89**, 3586.
Olah, G. A., O'Brien, D. H., and White, A. M. (1967e). *J. Amer. Chem. Soc.* **89**, 5694.
Orékhov, A., and Tiffeneau, M. (1926). *Compt. Rend.* **182**, 67.
Osborn, A. R., Mak, T. C.-W., and Whalley, E. (1961). *Can. J. Chem.* **39**, 1101.
Oulevey, G., and Susz, B. P. (1965). *Helv. Chim. Acta*, **48**, 630.
Paddock, N. L. (1964). *Quart. Rev. Chem. Soc.* **18**, 168.
Pal'm, V. A. (1956). *Dokl. Akad. Nauk. SSSR*, **108**, 270.

Pal'm, V. A. (1961). *Russ. Chem. Rev.* **30**, No. 9, 471.
Pascard, R. (1955). *Compt. Rend.* **240**, 2162.
Pascard-Billy, C. (1965). *Acta Crystallogr.* **18**, 827.
Paternò, E., and Spallino, R. (1907). *Gazz. Chim. Ital.* **37**, I, 111.
Paul, M. A. (1952). *J. Amer. Chem. Soc.* **74**, 141.
Paul, M. A. (1954). *J. Amer. Chem. Soc.* **76**, 3236.
Paul, M. A., and Long, F. A. (1957). *Chem. Rev.* **57**, 1.
Pearson, D. E., and Ball, F. (1949). *J. Org. Chem.* **14**, 118.
Pearson, D. E., and Cole, W. E. (1955). *J. Org. Chem.* **20**, 488.
Pedersen, K. J. (1934). *J. Phys. Chem.* **38**, 581.
Perrin, C. (1964). *J. Amer. Chem. Soc.* **86**, 256.
Perrin, D. D. (1965). "Dissociation Constants of Organic Bases in Aqueous Solution". Butterworths. London.
Pinnow, J. (1915). *Z. Elektrochem.* **21**, 380.
Pinnow, J. (1917). *Z. Elektrochem.* **23**, 243.
Plattner, P. A., Heilbronner, E., and Weber, S. (1949). *Helv. Chim. Acta*, **32**, 574.
Plattner, P. A., Heilbronner, E., and Weber, S. (1950). *Helv. Chim. Acta*, **33**, 1663.
Pocker, Y. (1959). *Chem. Ind.* (*London*), 332.
Polanyi, M., and Szabo, A. (1934). *Trans. Faraday Soc.* **30**, 508.
Popiel, R. (1964). *J. Chem. Eng. Data*, **9**, 269.
Pople, J. A., Schneider, W. G., and Bernstein, H. J. (1959). "High-Resolution Nuclear Magnetic Resonance". McGraw-Hill Book Co., New York.
Pospišil, L., Tomanová, J., and Kůta, J. (1968). *Collect. Czech. Chem. Commun.* **33**, 594.
Potts, H. A., and Smith, G. F. (1957). *J. Chem. Soc.* 4018.
Pracejus, H. (1959). *Chem. Ber.* **92**, 988.
Pracejus, H., Kehlen, M., Kehlen, H., and Matschiner, H. (1965). *Tetrahedron*, **21**, 2257.
Pressman, D., and Lucas, H. J. (1939). *J. Amer. Chem. Soc.* **61**, 2271.
Pressman, D., Brewer, L., and Lucas, H. J. (1942). *J. Amer. Chem. Soc.* **64**, 1122.
Price, F. P. (1948). *J. Amer. Chem. Soc.* **70**, 871.
Prinsen, A. J., and Cerfontain, H. (1969). *Rev. Trav. Chim. Pay-Bas.* **88**, 833.
Pritchard, J. G., and Long, F. A. (1956). *J. Amer. Chem. Soc.* **78**, 2667.
Pritchard, J. G., and Long, F. A. (1958). *J. Amer. Chem. Soc.* **80**, 4162.
Purlee, E. L., and Taft, R. W., Jr. (1956). *J. Amer. Chem. Soc.* **78**, 5811.
Puzitskii, K. V., Eidus, Ya. T., and Ryabova, K. G. (1963). *Zh. Obshch. Khim.* **33**, 3278.
Raaen, V. F., and Collins, C. J. (1958). *J. Amer. Chem. Soc.* **80**, 1409.
Rabinovich, B. S., and Winkler, C. A. (1942). *Can. J. Res.* **20B**, 221.
Rabinovich, B. S., and Winkler, C. A. (1952). *Can. J. Res.* **20B**, 73.
Rabinovich, B. S., Winkler, C. A., and Stewart, A. R. P. (1942). *Can. J. Res.* **20B**, 121.
Ramsey, B. G. (1966). *J. Amer. Chem. Soc.* **88**, 5358.
Rao, N. R., and Ramanaiah, K. V. (1966). *Indian J. Pure Appl. Phys.* **4**, 206.
Reagan, M. T. (1969). *J. Amer. Chem. Soc.* **91**, 5506.
Reeves, R. L. (1966). *J. Amer. Chem. Soc.* **88**, 2240.
Reid, E. E. (1899). *Amer. Chem. J.* **21**, 284.
Reid, E. E. (1900). *Amer. Chem. J.* **24**, 397.
Reitz, O. (1936). *Naturwissenschaften*, **24**, 814.
Reitz, O. (1938). *Z. Elektrochem.* **44**, 693.
Riesz, P., Taft, R. W., Jr., and Boyd, R. H. (1957). *J. Amer. Chem. Soc.* **79**, 3724.

Ritter, J. J., and Kalish, J. F. (1948). *J. Amer. Chem. Soc.* **70**, 4048.
Ritter, J. J., and Minieri, P. P. (1948). *J. Amer. Chem. Soc.* **70**, 4045.
Roberts, J. D., and Moreland, W. T. (1953). *J. Amer. Chem. Soc.* **75**, 2167.
Robinson, E. A., and Ciruna, J. A. (1964). *J. Amer. Chem. Soc.* **86**, 5677.
Robinson, E. A., and Quadri, S. A. A. (1967a). *Can. J. Chem.* **45**, 2385.
Robinson, E. A., and Quadri, S. A. A. (1967b). *Can. J. Chem.* **45**, 2391.
Robinson, E. A., and Zaidi, S. A. A. (1968). *Can. J. Chem.* **46**, 3927.
Robinson, R. A., and Stokes, R. H. (1955). "Electrolyte Solutions". Butterworths, London.
Roebuck, A. K., and Evering, B. L. (1953). *J. Amer. Chem. Soc.* **75**, 1631.
Ropp, G. A. (1958). *J. Amer. Chem. Soc.* **80**, 6691.
Ropp, G. A. (1960). *J. Amer. Chem. Soc.* **82**, 842.
Ropp, G. A., Weinberg, A. J., and Neville, O. K. (1951). *J. Amer. Chem. Soc.* **73**, 5573.
Rördam, H. N. K. (1915). *J. Amer. Chem. Soc.* **37**, 557.
Rosenthal, D., and Taylor, T. I. (1957). *J. Amer. Chem. Soc.* **79**, 2684.
Rothrock, T. S., and Fry, A. (1958). *J. Amer. Chem. Soc.* **80**, 4349.
Roughton, J. E. (1951). *J. Appl. Chem. (London)*, Suppl. Issue, No. 2, S 141.
Rubin, T. R., and Giauque, W. F. (1952). *J. Amer. Chem. Soc.* **74**, 800.
Russell, M. (1969). *Ark. Kemi*, **31**, 455.
Rutherford, K. G., and Newman, M. S. (1957). *J. Amer. Chem. Soc.* **79**, 213.
Ryabova, R. S., and Vinnik, M. I. (1963). *Zh. Fiz. Khim.* **37**, 2529.
Ryabova, R. S., Medvetskaya, I. M., and Vinnik, M. I. (1966). *Zh. Fiz. Khim.* **40**, 339.
Sackur, O. (1902). *Z. Elektrochem.* **8**, 77; Zentr. 1902. I. 554.
Saika, A. (1960). *J. Amer. Chem. Soc.* **82**, 3540.
Salomaa, P., and Kankaanpera, A. (1961). *Acta Chem. Scand.* **15**, 871.
Salomaa, P., and Keisala, H. (1966). *Acta Chem. Scand.* **20**, 902.
Salomaa, P., Schaleger, L. L., and Long, F. A. (1964). *J. Amer. Chem. Soc.* **86**, 1.
Samuel, D., and Silver, B. L. (1965). *Advan. Phys. Org. Chem.* **3**, 123.
Satchell, D. P. N. (1956). *J. Chem. Soc.* 3911.
Satchell, D. P. N. (1959). *J. Chem. Soc.* 463.
Schaleger, L. L., and Long, F. A. (1963). *Advan. Phys. Org. Chem.* **1**, 1.
Schierz, E. R. (1923). *J. Amer. Chem. Soc.* **45**, 447.
Schierz, E. R., and Ward, H. T. (1928). *J. Amer. Chem. Soc.* **50**, 3240.
Schmid, H. (1946). *Helv. Chim. Acta*, **29**, 1144.
Schmidt, K. F. (1924). *Ber.* **57**, 704.
Schonberg, A., and Azzam, R. C. (1958). *J. Org. Chem.* **23**, 286.
Schubert, W. M. (1949). *J. Amer. Chem. Soc.* **71**, 2639.
Schubert, W. M., and Burkett, H. (1956). *J. Amer. Chem. Soc.* **78**, 64.
Schubert, W. M., and Latourette, H. K. (1952). *J. Amer. Chem. Soc.* **74**, 1829.
Schubert, W. M., and Myhre, P. C. (1958). *J. Amer. Chem. Soc.* **80**, 1755.
Schubert, W. M., and Quacchia, R. H. (1962). *J. Amer. Chem. Soc.* **84**, 3778.
Schubert, W. M., and Zahler, R. E. (1954). *J. Amer. Chem. Soc.* **76**, 1.
Schubert, W. M., Donohue, J., and Gardner, J. D. (1954). *J. Amer. Chem. Soc.* **76**, 9.
Schubert, W. M., Zahler, R. E., and Robbins, J. (1955). *J. Amer. Chem. Soc.* **77**, 2293.
Schubert, W. M., Lamm, B., and Keefe, J. R. (1964). *J. Amer. Chem. Soc.* **86**, 4727.
Schulze, J., and Long, F. A. (1964a). *J. Amer. Chem. Soc.* **86**, 322.
Schulze, J., and Long, F. A. (1964b). *J. Amer. Chem. Soc.* **86**, 331.

Schwarzenbach, G., and Wittwer, C. (1947). *Helv. Chim. Acta*, **30**, 659.
Seel, F., and Sauer, H. (1957). *Z. Anorg. Allg. Chem.* **292**, 1.
Sekuur, Th. J., and Kranenburg, P. (1966). *Tetrahedron Lett.* **39**, 4793.
Senderens, J. B. (1927). *Compt. Rend.* **184**, 856.
Setkina, V. N., and Kursanov, D. N. (1958). *Dokl. Akad. Nauk SSSR*, **120**, 801.
Setlik, B. (1889). *Chem. Ztg.* **13**, 1670.
Shankman, S., and Gordon, A. R. (1939). *J. Amer. Chem. Soc.* **61**, 2370.
Shaw, W. H. R., and Walker, D. G. (1958). *J. Amer. Chem. Soc.* **80**, 5337.
Shilov, E. A., and Kaniaev, N. (1939). *Dokl. Akad. Nauk SSSR*, **24**, 890.
Shorter, J. (1969). *Chem. Brit.* **5**, 269.
Sidhu, G. S., Thyagarajan, G., and Bhalerao, U. T. (1966). *Chem. Ind. (London)*, 1301.
Silbermann, W. E., and Henshall, T. (1957). *J. Amer. Chem. Soc.* **79**, 4107.
Simamura, O., and Mizuno, Y. (1957). *Bull. Chem. Soc. Jap.* **30**, 196.
Singer, K., and Vamplew, P. A. (1956). *J. Chem. Soc.* 3971.
Skrabal, A., and Zahorka, A. (1933). *Monatsh. Chem.* **63**, 1.
Smith, P. A. S. (1963). In "Molecular Rearrangements" (Ed. P. de Mayo), Vol. 1, Chapter 8. John Wiley, New York.
Smith, P. A. S., and Antoniades, E. P. (1960). *Tetrahedron*, **9**, 210.
Smith, P. A. S., and Brown, B. B. (1951). *J. Amer. Chem. Soc.* **73**, 2438.
Smith, P. A. S., and Horwitz, J. P. (1950). *J. Amer. Chem. Soc.* **72**, 3718.
Somiya, T. (1927). *Proc. Imp. Acad. Tokyo*, **3**, 76.
Sommer, L. H., Pietrusza, E. W., Kerr, G. T., and Whitmore, F. C. (1946). *J. Amer. Chem. Soc.* **68**, 156.
Spryskov, A. A., and Ovsyanskina, N. A. (1953). *Sb. Statei Obshch. Khim.* **2**, 882; C. A. **49**, 6894 (1955).
Stamhuis, E. J., and Drenth, W. (1963). *Rec. Trav. Chim. Pays-Bas.* **82**, 385.
Steigman, J., and Shane, N. (1965). *J. Phys. Chem.* **69**, 968.
Stewart, R., and Granger, M. R. (1961). *Can. J. Chem.* **39**, 2508.
Stewart, R., and Yates, K. (1958). *J. Amer. Chem. Soc.* **80**, 6355.
Stewart, R., and Yates, K. (1959). *Can. J. Chem.* **37**, 664.
Stewart, R., and Yates, K. (1960). *J. Amer. Chem. Soc.* **82**, 4059.
Stewart, R., Granger, M. R., Moodie, R. B., and Muenster, L. J. (1963). *Can. J. Chem.* **41**, 1065.
Stieglitz, J., and Leech, P. N. (1914). *J. Amer. Chem. Soc.* **36**, 272.
Stiles, M., and Mayer, R. P. (1959). *J. Amer. Chem. Soc.* **81**, 1497.
Stock, L. M., and Brown, H. C. (1963). *Advan. Phys. Org. Chem.* **1**, 35.
Stopperka, K. (1966a). *Z. Anorg. Allg. Chem.* **344**, 263.
Stopperka, K. (1966b). *Z. Anorg. Allg. Chem.* **345**, 264.
Strehlow, R., and Wendt, H. (1961). *Z. Physik. Chem. (Frankfurt)*, **30**, 141.
Suarez, C. (1966). *Int. J. Appl. Radiat. Isotop.* **17**, 491.
Surfleet, B., and Wyatt, P. A. H. (1965). *J. Chem. Soc.* 6524.
Swain, C. G., and Rosenberg, A. S. (1961). *J. Amer. Chem. Soc.* **83**, 2154.
Swain, C. G., Stivers, E. C., Reuwer, J. F., Jr., and Shaad, L. J. (1958). *J. Amer. Chem. Soc.* **80**, 5888.
Sweeting, L. M., and Yates, K. (1966). *Can. J. Chem.* **44**, 2395.
Szmant, H. H., and Brost, G. A. (1951). *J. Amer. Chem. Soc.* **73**, 4175.
Szmant, H. H., Devlin, O. M., and Brost, G. A. (1951). *J. Amer. Chem. Soc.* **73**, 3059.
Taft, R. W., Jr. (1952a). *J. Amer. Chem. Soc.* **74**, 3120.
Taft, R. W., Jr. (1952b). *J. Amer. Chem. Soc.* **74**, 5372.

Taft, R. W., Jr. (1956). *In* "Steric Effects in Organic Chemistry" (Ed. M. S. Newman). John Wiley, New York and London.
Taft, R. W., Jr. (1960a). *J. Phys. Chem.* **64**, 1805.
Taft, R. W., Jr. (1960b). *J. Amer. Chem. Soc.* **82**, 2965.
Taft, R. W., Jr., and Levins, P. L. (1962). *Anal. Chem.* **34**, 436.
Taft, R. W., Jr., and Riesz, P. (1955). *J. Amer. Chem. Soc.* **77**, 902.
Taft, R. W., Jr., Levy, J. B., Aaron, D., and Hammett, L. P. (1952). *J. Amer. Chem. Soc.* **74**, 4735.
Taft, R. W., Jr., Purlee, E. L., Riesz, P., and De Fazio, C. A. (1955). *J. Amer. Chem. Soc.* **77**, 1584.
Takeda, M., and Stejskal, E. O. (1960). *J. Amer. Chem. Soc.* **82**, 25.
Tchelintzev, B., and Kozlov, N. (1914). *Zh. Russ. Fiz. Khim. Obscchest.* **46**, 708.
Tedder, J. M. (1957). *J. Chem. Soc.* 4003.
Terada, H. (1959). *Nippon Kagaku Zasshi*, **80**, 1053.
Tietz, R. F., and McEwen, W. E. (1955). *J. Amer. Chem. Soc.* **77**, 4007.
Tiffeneau, M. (1907). *Ann. Chim. Phys.* 8th Ser. **10**, 328.
Tiffeneau, M., and Dorlencourt, (1909). *Ann. Chim. Phys.* 8th Ser. **16**, 237.
Tillett, J. G. (1962). *J. Chem. Soc.* 5142.
Timm, E. W., and Hinshelwood, C. N. (1938). *J. Chem. Soc.* 862.
Traficante, D. D., and Maciel, G. E. (1966). *J. Phys. Chem.* **70**, 1314.
Traube, W., and Reubke, E. (1921). *Ber.* **54B**, 1618.
Treffers, H. P., and Hammett, L. P. (1937). *J. Amer. Chem. Soc.* **59**, 1708.
Tsukerman, S. V., Kutulya, L. A., Surov, Yu. N., Lavrushin, V. F., and Yurev, Yu. K. (1965). *Dokl. Akad. Nauk SSSR*, **164**, 354.
Tutundžić, P. S., and Liler, M. (1953). *Bull. Soc. Chim. Belgrade*, **18**, 521.
Tutundžić, P. S., Liler, M., and Kosanović, Dj. (1954a). *Bull. Soc. Chim. Belgrade*, **19**, 225.
Tutundžić, P. S., Liler, M., and Kosanović, Dj. (1954b). *Bull. Soc. Chim. Belgrade*, **19**, 549.
Tutundžić, P. S., Liler, M., and Kosanović, Dj. (1955). *Bull. Soc. Chim. Belgrade*, **20**, 497.
Utley, J. H. P., and Vaughan, T. A. (1968). *J. Chem. Soc.* 196.
van der Heijde, H. B. (1955). *Chem. Weekbl.* **51**, 823.
van der Zanden, J. M., and Rix, Th. R. (1956a). *Rec. Trav. Chim. Pays-Bas*, **75**, 1166.
van der Zanden, J. M., and Rix, Th. R. (1956b). *Rec. Trav. Chim. Pays-Bas*, **75**, 1343.
Vedel, J. (1967). *Ann. Chim. (Paris)*, **2**, 336.
Veibel, S. (1930). *Ber.* **63**, 1577.
Verhulst, J. (1931). *Bull. Soc. Chim. Belg.* **40**, 475.
Vetešnik, P., Bielavský, J., and Večera, M. (1968). *Collect. Czech. Chem. Commun.* **33**, 1687.
Vinnik, M. I. (1966). *Russ. Chem. Rev.* 802.
Vinnik, M. I., and Ryabova, R. S. (1962). *Zh. Fiz. Khim.* **36**, 2601.
Vinnik, M. I., and Ryabova, R. S. (1964). *Zh. Fiz. Khim.* **38**, 606.
Vinnik, M. I., and Zarakhani, N. G. (1960). *Zh. Fiz. Khim.* **34**, 2671.
Vinnik, M. I., and Zarakhani, N. G. (1963). *Dokl. Akad. Nauk SSSR*, **152**, 1147, and subsequent papers.
Vinnik, M. I., Ryabova, R. S., and Chirkov, N. M. (1959a). *Zh. Fiz. Khim.* **33**, 1992.
Vinnik, M. I., Ryabova, R. S., and Chirkov, N. M. (1959b). *Zh. Fiz. Khim.* **33**, 2677.
Vinnik, M. I., Ryabova, R. S., and Belova, G. V. (1962). *Zh. Fiz. Khim.* **36**, 942.

Vinnik, M. I., Ryabova, R. S., Grabovskaya, Zh. E., Koslov, K., and Kubar, I. (1963). *Zh. Fiz. Khim.* **37**, 94.
Virtanen, P. O. I., and Maikkula, M. (1968). *Tetrahedron Lett.* 4855.
Virtanen, P. O. I., and Södervall, T. (1967). *Suom. Kemistilehti B*, **40**, 337.
von Dobeneck, H., and Kiefer, R. (1965). *Ann. Chem.* **684**, 115.
von Kothner, P. (1901). *Ann. Chem.* **319**, 1.
Vrba, Z., and Allan, Z. J. (1968a). *Tetrahedron Lett.* **43**, 4507.
Vrba, Z., and Allan, Z. J. (1968b). *Collect. Czech. Chem. Commun.* **33**, 2502.
Walden, P. (1901). *Ber.* **34**, 4185.
Walden, P. (1902). *Z. Anorg. Allg. Chem.* **29**, 371.
Walden, P. (1903). *Z. Phys. Chem.* **46**, 182.
Walrafen, G. E. (1964). *J. Chem. Phys.* **40**, 2326.
Walrafen, G. E., and Dodd, D. M. (1961). *Trans. Faraday Soc.* **57**, 1286.
Walrafen, G. E., and Young, T. F. (1960). *Trans. Faraday Soc.* **56**, 1419.
Walsh, A. D. (1946). *Trans. Faraday Soc.* **42**, 56.
Wanders, A. C. M., and Cerfontain, H. (1967). *Rec. Trav. Chim. Pays-Bas*, **86**, 1199.
Wanders, A. C. M., Cerfontain, H., and Kort, C. W. F. (1967). *Rec. Trav. Chim. Pays-Bas*, **86**, 301.
Wasif, S. (1955). *J. Chem. Soc.* 372.
Webster, B. M. (1952). *Rec. Trav. Chim. Pays-Bas*, **71**, 1159, 1171.
Welch, C. M., and Smith, H. A. (1950). *J. Amer. Chem. Soc.* **72**, 4748.
Wells, P. R. (1963). *Chem. Rev.* **63**, 171.
Wells, P. R. (1968). "Linear Free Energy Relationships". Academic Press, London and New York.
Westheimer, F. H. (1961). *Chem. Rev.* **61**, 265.
Westheimer, F. H., and Kharasch, M. S. (1946). *J. Amer. Chem. Soc.* **68**, 1871.
Whalley, E. (1964). *Advan. Phys. Org. Chem.* **2**, 93.
Whipple, E. B., Chiang, Y., and Hinman, R. L. (1963). *J. Amer. Chem. Soc.* **85**, 26.
White, W. N., Klink, J. R., Lazdins, D., Hathaway, C., Golden, J. T., and White, H. S. (1961). *J. Amer. Chem. Soc.* **83**, 2024.
Whitford, E. L. (1925). *J. Amer. Chem. Soc.* **47**, 953.
Wiberg, K. (1955). *Chem. Rev.* **55**, 713.
Wiig, E. O. (1930a). *J. Amer. Chem. Soc.* **52**, 4729.
Wiig, E. O. (1930b). *J. Amer. Chem. Soc.* **52**, 4737.
Wiles, L. A. (1953). *J. Chem. Soc.* 996.
Wiles, L. A. (1956). *Chem. Rev.* **56**, 329.
Wiles, L. A., and Baughan, E. C. (1953). *J. Chem. Soc.* 933.
Wiley, R. H., and Moyer, A. N. (1954). *J. Amer. Chem. Soc.* **76**, 5706.
Willi, A. V. (1965). "Säurekatalytische Reaktionen der Organischen Chemie", Friedrich Vieweg & Sohn, Braunschweig.
Williams, G., and Hardy, M. L. (1953). *J. Chem. Soc.* 2560.
Williams, G., and Simkins, R. J. (1952). *J. Chem. Soc.* 3086.
Williams, G., and Simkins, R. J. (1953). *J. Chem. Soc.* 1386.
Williams, J. F. A. (1962). *Tetrahedron*, **18**, 1487.
Winstein, S., and Henderson, R. B. (1950). *In* "Heterocyclic Compounds" (Ed. R. C. Elderfield), Vol. 1, Chapter 1. John Wiley, New York.
Winstein, S., and Lucas, H. J. (1937). *J. Amer. Chem. Soc.* **59**, 1461.
Wolfenden, R., and Jencks, W. P. (1961). *J. Amer. Chem. Soc.* **83**, 2763.
Woodward, L. A., and Horner, R. G. (1934). *Proc. Roy. Soc. London, A*, **144**, 129.
Wurtz, C.-A. (1884). *Bull. Soc. Chim. Fr.* [2], **42**, 286.

Wyatt, P. A. H. (1957). *Discuss. Faraday Soc.* No. 24, 163.
Wyatt, P. A. H. (1960). *Trans. Faraday Soc.* **56**, 490.
Wyatt, P. A. H. (1961). *Trans. Faraday Soc.* **57**, 773.
Wyatt, P. A. H. (1969). *Trans. Faraday Soc.* **65**, 585.
Wynne-Jones, W. F. K. (1936). *Trans. Faraday Soc.* **32**, 1397.
Wynne-Jones, W. F. K., and Eyring, H. (1935). *J. Chem. Phys.* **3**, 492.
Yates, K. (1964). *Can. J. Chem.* **42**, 1239.
Yates, K., and McClelland, R. A. (1967). *J. Amer. Chem. Soc.* **89**, 2686.
Yates, K., and Riordan, J. C. (1965). *Can. J. Chem.* **43**, 2328.
Yates, K., and Stevens, J. B. (1965). *Can. J. Chem.* **43**, 529.
Yates, K., and Thompson, A. A. (1967). *Can. J. Chem.* **45**, 2997.
Yates, K., and Wai, H. (1964). *J. Amer. Chem. Soc.* **86**, 5408.
Yates, K., and Wai, H. (1965). *Can. J. Chem.* **43**, 2131.
Yates, K., Stevens, J. B., and Katritzky, A. R. (1964). *Can. J. Chem.* **42**, 1957.
Yeh, S.-J., and Jaffé, H. H. (1959a). *J. Amer. Chem. Soc.* **81**, 3274.
Yeh, S.-J., and Jaffé, H. H. (1959b). *J. Amer. Chem. Soc.* **81**, 3279.
Yeh, S.-J., and Jaffé, H. H. (1959c). *J. Amer. Chem. Soc.* **81**, 3283.
Young, T. F., and Walrafen, G. E. (1961). *Trans. Faraday Soc.* **57**, 34.
Young, T. F., Maranville, L. F., and Smith, H. M. (1959). "The Structure of Electrolyte Solutions" (Ed. W. J. Hamer). John Wiley, New York.
Yukawa, Y., and Tsuno, Y. (1959). *Bull. Chem. Soc. Jap.* **32**, 971.
Zarakhani, N. G., and Vinnik, M. I. (1962). *Zh. Fiz. Khim.* **36**, 916.
Zarakhani, N. G., and Vinnik, M. I. (1963). *Zh. Fiz. Khim.* **37**, 503.
Zarakhani, N. G., Budylina, V. V. and Vinnik, M. I., (1965). *Zh. Fiz. Khim.* **39**, 1863.
Zollinger, H., and Wittwer, C. (1956). *Helv. Chim. Acta*, **39**, 347.
Zucker, L., and Hammett, L. P. (1939). *J. Amer. Chem. Soc.* **61**, 2779, 2785, 2791.

Author Index

Numbers in italics indicate the pages on which the references are listed

A

Aaron, D., 210, *319*, *326*
Aboderin, A., 245, *307*
Abraham, R. J., 98, 99, *307*
Adams, E. Q., 96, *307*
Addis, H. W., 112, *311*
Akand, M. A., 285, *307*
Allan, Z. J., 297, *327*
Anderson, A. G., 142, *307*
Anderson, J. N., 49, 66, 119, 120, 121, *307*
Andreeva, L. R., 48, *315*
Andrussow, K., 256, 292, *318*
Angus, W. R., 162, *307*
Antoniades, E. P., 252, *325*
Archer, G., 234, *307*
Arcus, C. L., 249, 252, *307*
Argument, C., 163, *320*
Armstrong, V. C., 64, 65, 105, 108, 193, *307*
Arnand, R., 124, *307*
Arnesen, R. T., 127, *307*
Arnett, E. M., 26, 38, 41, 45, 49, 60, 62, 64, 66, 86, 96, 97, 105, 112, 118, 119, 120, 121, 122, 127, 152, 153, *307*
Arotsky, J., 163, 164, 289, *307*
Aschan, O., 133, *307*
Aubin, C., 227, *315*
Avarbock, H. S., 221, 237, *321*, *322*
Azzam, R. C., 253, *324*

B

Bachmann, W. E., 248, *307*
Back, T. A., 144, *307*
Baddely, G., 292, 293, *307*
Baird, R. L., 245, *307*
Baker, J. W., 281, *307*
Bakule, R., 233, 234, 235, *307*
Ball, F., 247, *323*

Bancroft, K. C. C., 265, *307*
Banholzer, K., 254, *307*
Banthorpe, D. V., 286, *307*
Barnard, M., 229, 230, *309*
Barr, J., 84, 138, 164, *308*
Barry, G. W., 123, *318*
Bartlett, P. D., 298, 299, *308*
Barton, M. X., 248, *307*
Bascombe, K. N., 33, 34, 36, 37, 52, 53, 102, 188, *308*
Bass, S. J., 10, 11, 12, 13, 68, 69, 71, 73, 78, *308*
Bates, R., 35, *308*
Baughan, E. C., 126, 127, *327*
Bayliss, N. S., 162, *308*
Bean, G. P., 271, *308*
Bednas, M. E., 10, *310*
Bell, R. P., 33, 34, 36, 37, 52, 53, 102, 177, 181, 222, 233, 234, 270, *307*, *308*
Bellamy, L. J., 91, 92, 134, *308*
Belova, G. V., 226, *327*
Bender, M. L., 194, 197, 198, 199, 204, 288, *308*, *319*
Bennett, G. M., 161, 277, 284, *308*
Bennett, R. P., 291, *318*
Berger, A., 108, *308*
Bergmann, M., 133, *308*
Bernstein, H. J., 24, *323*
Bertoli, V., 141, *308*
Bethell, D., 86, 139, 140, 143, 153, 237, *308*
Bhalerao, U. T., 291, *325*
Bickel, A. F., 131, 140, *317*
Bielavský, J., 43, 44, *326*
Bigeleisen, J., 38, 39, 178, 179, 180, 258, 271, *308*, *317*, *319*, *320*
Biggs, A. I., 37, *308*
Bingham, E. C., 7, *308*
Birchall, T., 109, 110, 122, 128, 131, 132, *308*, *314*

Bisset, D., 10, *310*
Bistrzycki, A., 253, *309*
Bjerrum, N., 294, *309*
Blangley, L., 288, *309*
Bollinger, J., 142, *312*
Bollinger, J. M., 189, *322*
Bolton, P. D., 192, 193, *309*
Bonhoeffer, K. F., 178, *309*
Bonner, T. G., 29, 38, 43, 125, 229, 230, 284, 285, *309*
Borovikov, Yu. Ya., 91, *309*
Bose, H., 22, *309*
Bott, R. W., 265, *307*
Bourns, A. N., 122, *308*
Bowyer, F., 284, *309*
Boyd, D. B., 244, *312*
Boyd, R. H., 26, 29, 37, 44, 47, 48, 49, 211, 212, 217, 218, 220, *309*, *323*
Brade, H., 109, 110, *317*
Brand, J. C. D., 3, 9, 22, 29, 38, 43, 49, 57, 70, 78, 115, 161, 277, 284, 295, 297, *308*, *309*
Brayford, J. R., 22, 69, 70, *309*
Bredig, G., 254, *309*
Brewer, L., 223, *323*
Brickman, M., 281, 283, *309*
Briggs, L. H., 144, 253, 304, *309*
Brock Robertson, E., 55, *309*
Brodskii, A. I., 246, *309*
Brønsted, J. N., 187, 189, *309*
Bronwer, D. M., 122, *309*
Brost, G. A., 123, 165, *325*
Brown, A. H., 188, *308*
Brown, B. B., 251, *325*
Brown, F., 285, *309*
Brown, H. C., 87, 93, 271, 276, 283, 290, *309*, *325*
Brownstein, S., 286, *309*
Budylina, V. V., 209, *328*
Buess, C. M., 124, *317*
Bullock, E., 98, 99, *307*
Bunnett, J. F., 175, *309*
Bunton, C. A., 40, 188, 197, 198, 200, 208, 217, 238, 239, 240, 243, 286, *309*, *310*
Burkett, H., 261, *310*, *324*
Burwell, R. L., Jr., 183, 244, 245, *310*, *320*
Bushick, R. D., 45, 49, 66, 121, 127, 152, 153, *307*
Butler, R., 289, *307*

C

Cagle, F. Wm., Jr., 258, *313*
Calin, M., 108, 127, 128, 183, *322*
Calvin, M., 258, *314*
Campbell, A., 251, *310*
Campbell, A. N., 10, *310*
Campbell, H. J., 124, *310*
Cannizzaro, S., 155, *310*
Carpenter, C. D., 118, 132, *318*
Carr, M. D., 243, 290, *309*, *310*
Casadevall, A., 131, 133, 159, *310*
Casalino, A., 128, 132, 158, *322*
Cauquil, G., 131, 133, 159, *310*
Cerfontain, H., 16, 22, 280, 292, 293, 294, 295, 296, 297, *310*, *317*, *318*, *323*, *327*
Challis, B. C., 270, 272, 288, *310*
Chang, H. S., 104, *313*
Chapman, R. P., 47, 48, 49, 62, 63, *316*
Charles, K. R., 123, *318*
Charlton, M., 94, 95, *310*
Chédin, J., 161, *310*
Chen, M. C., 204, *308*
Cherbuliez, E., 115, *310*
Chiang, Y., 31, 98, 99, 123, 265, 266, 269, 270, *310*, *318*, *319*, *327*
Chirkov, N. M., 226, 255, 256, *326*, *327*
Chmiel, C. T., 204, *310*
Chortyk, O. T., 249, *316*
Choudhury, A. K., 290, *318*
Christman, D. R., 212, 217, 218, 220, *309*
Chute, W. J., 285, *310*
Ciaccio, L. L., 277, *310*
Ciruna, J. A., 160, *324*
Clark, J., 87, 91, 108, 118, *310*
Clerc, J. T., 35, *311*
Clunie, J. C., 181, *308*
Cobb, A. W., 207, 208, *311*
Coe, J. S., 212, *311*
Colapietro, J., 265, *311*
Cole, R. H., 9, *314*
Cole, W. E., 246, 248, *323*
Collie, J. N., 118, 136, *311*
Collins, C. J., 242, *311*, *323*
Condon, F. E., 298, *308*
Conley, R. T., 249, *311*, *316*
Connor, T. M., 169, *320*
Converse, W., 238, 239, *316*
Cook, R. D., 105, 114, 117, 137, *316*
Coombs, M. M., 249, 305, *311*

Coombes, R. G., 209, 210, *311*
Cooper, K. E., 281, *307*
Corpin, A. H., 132, *319*
Corriu, R., 131, 133, 159, *310*
Coryell, C. D., 38, *311*
Coussemant, F., 213, *313*
Cowdrey, W. A., 295, *311*
Crafts, J. M., 292, *311*
Craig, R. A., 70, 151, 157, 165, *311*, *321*
Crowell, T. I., 208, *311*
Cruickshank, D. W. J., 22, *311*
Culbertson, G., 124, *311*
Curphey, T. J., 35, *311*
Currell, D., 238, *311*
Curtius, T., 249, 250, *311*

D

Dacre, B., 17, 22, *311*
Dahn, H., 238, *310*
Darapsky, S., 249, 250, *311*
Darby, A. C., 289, *307*
Darling, H. E., 10, *311*
Davidson, M., 213, *313*
Davies, C., 112, *311*
Davies, D. S., 295, *311*
Davis, C. E., 5, 6, 8, *321*
Davis, C. T., 32, 33, 34, 136, *311*
Davison, A., 144, *311*
Deane, C. W., 226, 227, *311*
Deans, F. B., 275, *311*
de Fabrizio, E., 45, 288, *311*
De Fazio, C. A., 211, 218, *326*
Degani, I., 136, 137, *311*
de la Mare, P. B. D., 277, 281, 284, 287, 288, 289, 290, *311*
Denney, D. B., 228, 229, *311*
Deno, N. C., 20, 29, 30, 31, 45, 46, 49, 64, 65, 67, 96, 104, 117, 119, 120, 121, 124, 128, 140, 141, 142, 145, 152, 153, 154, 155, 157, 161, 162, 207, 213, 215, 221, 243, 244, 245, 257, 258, 277, 278, 281, 298, 299, 300, 301, 302, *311*, *312*, *321*
Derbyshire, D. H., 289, 290, *312*
De Right, R. E., 255, *312*
Deschamps, J. M., 162, *312*
Devlin, O. M., 165, *325*
Dewar, M. J. S., 286, 291, *312*
Deyrup, A. J., 26, 27, 28, 32, 36, 41, 57, 58, 68, 70, 78, 125, 127, 171, 226, 246, 247, *316*
Dilgren, R. E., 238, *315*
Dingwall, A., 33, 118, 124, 128, 129, *313*
Dittmar, H. R., 255, *312*
Doak, G. O., 112, 113, *317*
Dodd, D. M., 18, 19, 20, 21, *327*
Dolman, D., 40, 98, *312*
Donohue, J., 158, 262, *324*
Dorlencourt, 240, *326*
Dorofeev, V. V., 124, *320*
Dostrowsky, I., 217, *312*
Dougherty, G., 226, *315*
Dowdling, A. L., 181, *308*
Drenth, W., 224, *312*, *316*, *317*, *325*
Drucker, C., 39, *312*
Drude, P., 9, *312*
Dücker, K.-H., 5, 10, *316*
Duffy, J. A., 194, 195, 196, *312*
Dunford, H. B., 55, *309*
Dunkle, F. B., 205, *320*
Dunn, G. E., 285, *310*
Dunn, T. M., 290, *311*
Dunstan, A. E., 7, 8, *312*
Durand, J.-P., 213, *313*
Durie, R. A., 127, *313*

E

Eaborn, C., 265, 264, 266, 274, 275, 276, *307*, *311*, *313*
Eberz, W. F., 211, 213, *313*, *320*
Edward, J. T., 26, 34, 42, 51, 52, 54, 56, 64, 104, 105, 109, 110, 119, 121, 124, 128, 192, 193, 209, *310*, *313*
Edwards, T., 207, 302, *312*
Eidus, Ya. T., 303, *313*, *323*
Eistert, B., 126, *313*
Ellefsen, P. R., 112, *313*
Elliott, W. W., 254, 255, 256, *313*
Elmore, G. V., 201, *314*
England, B. D., 290, *310*
Entelis, S. G., 301, *313*, *318*
Evans, J. V., 252, *307*
Evans, W. L., 153, 154, *311*
Evering, B. L., 244, *324*
Exner, O., 92, 93, *313*
Eyring, H., 171, 258, *313*, *315*, *328*

F

Fabbri, G., 18, *313*
Fajans, E., 22, *313*
Farlow, D. W., 108, 193, *307*
Fee, J. A., 193, *313*
Fellmann, M., 254, *309*
Fialkov, Yu. Ya., 9, *309*
Fife, T. H., 193, *313*
Filler, R., 153, *313*
Fischer, A., 125, *313*
Fish, R. W., 160, *316*
Fix, R. C., 38, *311*
Flexser, L. A., 33, 118, 124, 128, 129, *313*
Flowers, R. H., 12, 24, 25, 69, 70, 73, 74, 75, 77, 84, 85, 159, 165, *308*, *313*
Fochi, R., 136, 137, *311*
Fraenkel, G., 109, *314*
Franconi, C., 109, *314*
Friedman, H. B., 201, *314*
Friedman, N., 142, 243, 300, 301, *312*
Fry, A., 238, 240, 258, *311*, *314*, *324*

G

Gardner, J. D., 158, 262, *324*
Gardner, J. N., 114, *314*
Gardner, R. W., 101, *317*
Garrett, A. B., 70, 151, 157, 158, 165, 226, *311*, *321*
Garvey, R. G., 114, *321*
Gastaminza, A., 283, *314*
Gaugler, R. W., 117, *312*
Geissman, T. A., 32, 33, 34, 136, *311*
Gel'bshtein, A. I., 37, *314*
Gerrard, W., 118, *314*
Giauque, W. F., 2, 3, 11, 12, 13, 14, 15, 16, 17, 296, *314*, *317*, *319*, *324*
Giguère, P. A., 18, *314*
Gilbert, E. E., 292, *314*
Gildenhorn, H. L., 251, 304, *321*
Gillespie, R. J., 2, 7, 8, 9, 10, 11, 12, 13, 18, 20, 21, 23, 24, 25, 38, 39, 57, 71, 67, 68, 69, 70, 71, 72, 73, 74, 75, 76, 77, 78, 79, 80, 81, 83, 84, 85, 98, 108, 109, 110, 112, 115, 116, 117, 122, 123, 126, 128, 131, 132, 137, 138, 145, 150, 151, 152, 158, 159, 160, 161, 163, 164, 165, 166, 254, 277, 278, 284, *308*, *313*, *314*, *315*
Ginger, R. D., 194, 197, 198, 199, *308*
Glasstone, S., 171, *315*
Gleason, A. H., 226, *315*
Gleghorn, J. T., 284, *315*
Glogger, J., 109, 110, *317*
Gmitro, J. J., 14, *315*
Goering, H. L., 238, *315*
Gold, V., 29, 30, 86, 139, 140, 143, 153, 154, 178, 179, 181, 208, 211, 212, 218, 220, 237, 263, 264, 265, 266, 267, 268, 270, 271, 293, 294, 295, *308*, *311*, *316*
Golden, J. T., 286, *327*
Goldfarb, A. R., 104, 128, *315*
Goodeve, C. F., 22, *313*
Gordon, A., 192, *317*
Gordon, A. R., 14, *325*
Gordon, L., 112, *313*
Gore, P. H., 280, *315*
Goren, M. B., 126, *316*
Grabovskaya, Zh. E., 48, 49, 226, *315*, *327*
Grace, J. A., 155, 156, *315*
Graebe, C., 227, *315*
Granger, M. R., 124, 129, 130, *325*
Greenewalt, C. H., 14, *315*
Greenwood, N. N., 6, 7, 8, 24, 25, *315*
Gregory, B. J., 247, 248, *315*
Grigor, B. A., 125, *313*
Groves, P. T., 30, 31, 46, 104, 140, 141, 155, *312*
Gruen, L. C., 218, 234, 265, *315*, *321*
Grunwald, E., 63, 64, 217, *315*
Gudmundsen, C. H., 251, *315*
Guggenheim, E. A., 181, *315*
Gutowsky, H. S., 23, 24, *316*
Gutstein, N., 104, 128, *315*

H

Haaf, W., 303, *316*, *318*
Haake, P., 105, 114, 117, 137, *316*
Haase, R., 5, 10, *316*
Hadwick, T., 239, 240, *310*
Hafer, K., 142, *312*
Häfliger, O., 87, 93, *309*
Hahn, C.-S., 115, *316*
Hakka, L. E., 123, 265, *318*, *319*
Halevi, E. A., 286, *316*
Hall, N. F., 40, *316*
Hall, S. K., 138, 139, 143, *316*
Hallin, M., 213, *313*

Hammett, L. P., 11, 26, 27, 28, 32, 33, 36, 41, 47, 48, 49, 57, 58, 62, 63, 68, 70, 71, 68, 88, 118, 124, 125, 127, 128, 129, 131, 135, 145, 146, 157, 158, 168, 171, 175, 205, 210, 213, 226, 233, 234, 246, 247, 254, 256, *313*, *316*, *319*, *326*, *328*
Hammick, D. W., 254, 255, 256, *313*
Handa, T., 126, 142, *316*
Hantzsch, A., 103, 115, 118, 121, 125, 128, 144, 145, 146, 150, 151, 152, 161, *316*
Hardy, M. L., 95, 128, *327*
Harrison, W. F., 142, *307*
Hart, H., 160, 231, *316*
Hartshorn, S. R., 283, *316*
Hartter, D. R., 36, 38, *317*
Hartwell, E. J., 134, *316*
Harvey, J. T., 290, *311*
Haschemeyer, R. H., 258, *308*
Hass, H. B., 117, *316*
Hassel, O., 118, *316*
Hathaway, C., 286, *327*
Haukins, M. G., 208, *311*
Hawes, B. W. V., 29, 30, *315*
Hearne, G., 238, 239, *316*
Heck, R., 241, *316*
Heilbronner, E., 142, *323*
Hekkert, G. L., 224, *316*
Heller, A., 217, *315*
Heller, R. A., 215, *321*
Henderson, R. B., 187, *327*
Henshall, T., 227, 230, *325*
Herbert, R. A., 127, *316*
Hetherington, G., 10, 161, 162, *316*
Hilbers, C. W., 131, *316*
Hill, R. K., 249, *316*
Hilton, J., 181, 208, *308*, *315*
Hinman, R. L., 42, 43, 98, 99, 100, *310*, *316*, *327*
Hinshelwood, C. N., 197, *326*
Hodge, J. D., 142, 243, 300, 301, *312*
Hofmann, J. E., 300, 301, 302, *316*
Hofstra, A., 140, 143, *320*
Hogeveen, H., 104, 122, 131, 140, 224, *312*, *316*, *317*
Högfeldt, E., 39, 55, 178, *316*
Holst, K. A., 189, *319*
Holstead, C., 108, *317*
Holt, G., 292, 293, *307*
Hood, G. C., 23, *317*
Hoop, G. M., 144, 291, *317*

Hopkinson, A. C., 204, 205, *317*
Horner, R. G., 19, *327*
Horning, W. C., 29, 38, 43, 49, 58, 70, 78, 115, 295, 297, *309*
Hornung, E. W., 12, 13, 14, 15, 16, 17, 296, *314*, *317*
Horwitz, J. P., 252, *325*
Hoshino, S., 131, *317*
Hosoya, H., 131, *317*
Houser, J. J., 142, 243, 300, 301, *312*
Hub, D. R., 10, 162, *316*
Huffman, J. R., 227, *311*
Hughes, E. D., 13, 15, 68, 69, 70, 145, 161, 285, 286, 287, 288, 289, 297, *307*, *309*, *311*, *315*, *317*
Huisgen, R., 109, 110, *317*
Hurst, G. H., 105, 114, 117, *316*
Hussey, A. S., 244, 245, *310*
Hutchinson, H. P., 192, 193, *313*
Hyman, H. H., 160, *318*

I

Ingold, C. K., 13, 15, 68, 69, 70, 145, 161, 187, 189, 196, 203, 204, 234, 263, 278, 281, 285, 287, 288, 291, 297, *307*, *315*, *317*
Ipatieff, V. N. 301, *317*

J

Jaffé, H. H., 40, 87, 101, 102, 111, 112, 113, 115, 191, *316*, *317*, *319*, *328*
James, D. M., 277, 284, *308*
James, J. C., 9, *309*
Jaques, D., 67, 121, 183, 184, 185, 186, 202, *317*
Jaruzelski, J. J., 29, 31, 45, 46, 104, 152, 153, 154, 155, *312*
Jarvie, A. W. P., 49, 58, 70, 78, 297, *309*
Jellinek, H. H. G., 192, *317*
Jencks, W. P., 103, 193, *320*, *327*
Johnson, C. D., 36, 37, 112, 113, 114, *317*
Johnson, E. I., 136, *317*
Jones, G. T., 286, 287, *317*
Jones, R. A. Y., 105, *318*
Jörgenson, M. J., 36, 38, 40, 125, 236, *317*, *321*

K

Kaandorp, A. W., 294, *310*, *317*
Kabachnik, M. I., 88, *317*
Kalatzis, E., 287, *311*, 217
Kalish, J. F., 302, *324*
Kaniaev, N., 289, *325*
Kankaanpera, A., 188, *324*
Kapoor, R., 165, *315*
Karasch, N., 124, *317*
Kartzmark, E. M., 10, *310*
Katrizky, A. R., 33, 34, 36, 37, 41, 42, 105, 112, 113, 114, 271, 284, *308*, *314*, *317*, *318*, *328*
Kawamura, S., 138, *322*
Kaye, I. A., 242, *311*
Kazanskii, K. S., 301, *313*, *318*
Keefe, J. R., 211, 212, 213, 215, *324*
Kehlen, H., 107, *323*
Kehlen, M., 107, *323*
Keisala, H., 124, *324*
Kendall, F. H., 271, *320*
Kendall, J., 118, 132, *318*
Kenner, J., 292, 293, *307*
Kenyon, J., 251, *310*
Kerr, G. T., 165, *325*
Kershaw, D. N., 204, *318*
Kessick, M. A., 178, 179, 211, 212, *315*
Ketley, D. A., 289, *311*
Kharasch, M. S., 161, 277, *327*
Kiamud Din, M., 290, *311*, *318*
Kiefer, R., 259, *326*
Kilpatrick, M., 160, 188, 294, *318*
Kilpatrick, M. L., 294, *318*
King, E. J., 294, *318*
King, P. A., 36, 215, 224, 236, 237, *321*, *322*
King, W., 124, *317*
Kingsland, M., 284, *318*
Kiovsky, T. E., 104, 112, 117, *322*
Kipping, F. S., 274, *318*
Kirby, F. B., 215, 237, *322*
Kirkbride, B. J., 17, 78, *318*
Kish, F. A., 213, 215, 221, *312*
Kitao, T., 137, 138, *322*
Kitaoka, Y., 137, 138, *322*
Kittle, P. A., 131, 158, *321*
Klein, F. S., 217, 221, *312*, *315*, *320*
Klemchuk, P. P., 228, 229, *311*
Klink, J. R., 286, *327*
Klofutar, C., 112, 114, *318*

Knietsch, R., 7, 11, 12, 15, *318*
Knowles, J. R., 282, 283, *318*
Kobe, K. A., 143, *320*
Koch, H., 302, 303, *318*
Koeberg-Telder, A., 294, *318*
Koepp, H.-M., 34, *318*
Kogan, G. A., 301, *313*
Kohlrausch, F., 9, *318*
Kohlrausch, W., 12, *318*
Kohnstam, G., 178, *318*
Konasiewicz, A., 217, *310*
Kort, C. W. F., 16, 294, 295, 296, 297, *318*, *337*
Kortüm, G., 256, 292, *318*
Kosanović, Dj., 103, 104, 118, 133, 149, 182, 207, *320*, *326*
Koski, W. S., 25, *318*
Koskikallio, J., 183, 187, 189, *318*
Koslov, K., 226, *327*
Kouba, J., 271, *322*
Kovacic, P., 291, *318*
Kozlov, N., 118, *326*
Krafft, F., 290, *318*
Kramer, G. M., 299, *318*
Kranenburg, P., 128, *325*
Krášovec, F., 122, 114, *318*
Kreevoy, M. M., 188, *318*
Kreienbühl, P., 103, *322*
Kresge, A. J., 31, 123, 265, 266, 269, 270, *318*, *319*
Krieble, V. K., 189, 206, *319*
Kubar, I., 226, *327*
Kuhn, L. P., 132, 203, *319*
Kuhn, S. J., 277, *319*, *322*
Kuivila, H. G., 158, 226, 273, 274, *319*, *321*
Kunzler, J. E., 2, 3, 4, 11, 13, 14, 15, 16, 17, 296, *314*, *319*
Kursanov, D. N., 218, *325*
Kušar, M., 112, 114, *318*
Kůta, J., *323*
Kutulya, L. A., 126, *326*
Kwart, H., 216, *319*

L

Ladenheim, H., 204, 288, *308*, *319*
Laidler, K. J., 171, 202, *315*, *319*
Lambert, R. W., 270, *315*
Lamberton, A. H., 108, *317*

Lamm, B., 211, 212, 213, 215, *324*
Landini, D., 137, *319*
Landskroener, P. A., 202, *319*
Lane, C. A., 201, 202, 215, 220, 221, *319, 321*
Lang, J., 42, 43, 99, 100, *316*
Lange, R. J., 249, *311*
Lange, W., 164, *319*
Langmyhr, F. J., 127, *307*
Lantz, R., 293, *319*
La Planche, L. A., 109, *319*
Latourette, H. K., 173, 261, *324*
Laughlin, R. G., 64, 110, 111, *319*
La Vietes, D., 245, *312*
Lavrushin, V. F., 126, 152, *319, 326*
Layne, W. S., 111, *319*
Lazdins, D., 286, *327*
Leane, J. B., 64, 105, 119, 121, *313*
Leckie, A. H., 162, *307*
Leech, P. N., 246, *325*
Le Fave, G. M., 209, *319*
Leisten, J. A., 61, 67, 98, 121, 126, 145, 146, 148, 149, 150, 151, 152, 159, 160, 181, 183, 184, 185, 186, 189, 190, 191, 194, 195, 196, 202, 203, 204, 254, *312, 314, 317, 318, 319*
Levins, P. L., 64, *326*
Levy, J. B., 210, 213, *319, 326*
Lewis, G. N., 38, *319*
Lewis, T. A., 188, 197, 198, 200, 208, 217, *310*
Ley, J. B., 40, *310*
Librovich, N. B., 48, *319*
Lichty, D. M., 10, 254, 255, *309, 319*
Liler, M., 7, 49, 64, 80, 81, 82, 83, 86, 90, 91, 103, 104, 105, 106, 108, 109, 115, 116, 118, 125, 128, 129, 130, 133, 134, 135, 136, 149, 182, 200, 207, 255, 256, *319, 320, 326*
Ling, A. C., 271, *320*
Littleton, J. W., 144, 253, 304, *309*
Liu, Y. P., 213, *320*
Livingston, R. L., 250, *320*
Llewellyn, D. R., 188, 197, 198, 200, 208, 217, 239, 240, *310*
Lobry de Bruyn, C. A., 245, *320*
Lockhart, J. C., 29, 43, 285, *309, 320*
Loewenstein, A., 63, 64, 108, 169, 217, *308, 315, 320*
Long, F. A., 26, 142, 168, 174, 177, 178, 179, 183, 186, 187, 188, 196, 201, 202, 204, 205, 208, 226, 233, 234, 235, 262, 265, 269, 270, 271, 272, *307, 310, 311, 315, 320, 323, 324*
Longridge, J. L., 262, 269, 270, *320*
Lowen, A. M., 29, *320*
Lowenheim, F. A., 11, 71, *316*
Lucas, H. J., 187, 211, 213, 222, 223, *313, 320, 322, 323, 327*
Lucken, E. A., 249, 252, *307*
Luder, W. F., 227, *320*
Lugash, M., 31, 46, 104, 155, *312*
Lukas, J., 300, *322*
Lundeen, A., 303, *320*
Lutskii, A. E., 124, *320*

M

McClelland, R. A., 202, 203, 205, *328*
McCollum, J. D., 299, *308*
McCoubrey, J. C., 188, *308*
McCulloch, L., 131, *320*
McDaniel, D. H., 87, 93, *309*
McDevit, W. F., 205, *320*
McEwen, W. E., 251, 252, *315, 326*
McFarlane, W., 144, *311*
Mach, G. W., 38, 41, 96, 97, *307*
Maciel, G. E., 158, *326*
McIntyre, D., 188, *320*
McKenzie, J. C., 285, *310*
McKinney, M. A., 153, *313*
Macklen, E. D., 118, *314*
Mackor, E. L., 122, 131, 140, 143, *309, 317, 320*
McLaren, D. A., 246, *320*
Maclean, C., 122, 131, *309, 317*
McLean, J. D., 207, *320*
McNamara, A. J., 305, *320*
McNulty, P. J., 246, 247, 248, *321*
McTigue, P. T., 234, *321*
Maikkula, M., 111, *327*
Maitlis, P. M., 291, *312*
Mak, T. C.-W., 194, *322*
Manassen, J., 221, *320*
Maranville, L. F., 19, 20, 296, *328*
Marburg, S., 193, *320*
March, D. M., 227, *320*
Marcus, R. A., 277, *310*
Markham, A. E., 143, *320*
Martinsen, H., 277, 278, *320*

Masson, I., 49, 115, 163, *320*
Masters, B. J., 25, *320*
Matesich, M. A., 225, *322*
Matschiner, H., 107, *323*
Maury, L. G., 244, 245, *310, 320*
Mavel, M. G., 118, *320*
Mayer, R. P., 243, *325*
Meacock, S. C. R., 192, 193, 209, *313*
Medvetskaya, I. M., 33, 36, 37, *324*
Meiboom, S., 63, 64, 108, 217, *308, 315*
Melander, L., 271, 278, 294, 297, *321*
Mele, A., 104, 128, *315*
Merkel, E., 126, *313*
Meyer, M. W., 294, *318*
Micluchin, G. P., 246, *309*
Mikhailov, M., 25, *321*
Miles, F. D., 15, 22, *321*
Millen, D. J., 161, 277, 278, *314, 317, 321*
Miller, F. N., 153, *313*
Minegishi, A., 231, *321*
Minieri, P. P., 302, *324*
Mishra, H. C., 163, 164, *307*
Mitra, S. S., 98, 99, *307*
Mizuno, Y., 281, *325*
Mlinko, A., 277, *322*
Mocek, M. M., 154, *321*
Mockus, J., 245, *312*
Modena, G., 137, *319*
Modro, T. A., 283, *314*
Möller, K. E., 302, 303, *321*
Moodie, R. B., 64, 65, 105, 108, 125, 191, 193, 209, 210, 247, 248, 277, 283, 284, *307, 311, 315, 321, 325*
Moreland, W. T., 94, *324*
Moore, W., 207, *321*
Morgan, J. L. R., 5, 6, 8, *321*
Mountford, G. A., 25, *321*
Movchan, P. K., 152, *319*
Moyer, A. N., 118, 129, 159, *327*
Muenster, L. J., 125, *325*
Murphy, R. B., 261, *310*
Murray, M. A., 29, 152, 153, *320, 321*
Myers, G. S., 285, *310*
Myhre, P. C., 261, *324*
Mylonakis, S., 265, *319*

N

Nagakura, S., 131, 231, *317, 321*
Nahabedian, K. V., 273, 274, *319, 321*

Namanworth, E., 120, 152, 183, *322*
Nelson, J. H., 114, *321*
Nenitzescu, C. D., 118, *321*
Neville, O. K., 257, *324*
Newman, M. S., 70, 96, 128, 145, 146, 151, 152, 155, 157, 158, 165, 226, 251, 252, *304, 311, 321, 324*
Niblock, H., 15, 22, *321*
Nichols, M. J., 10, 161, *316*
Niemann, C., 104, 109, 128, 129, *314, 322*
Nikolov, K., 25, *321*
Noble, J. A., 181, *308*
Noll, C. I., 206, *319*
Norman, R. O. C., 282, 283, *318*
Norris, T. H., 25, *320, 321*
Norton, D. G., 284, *314*
Noyce, D. S., 36, 40, 125, 131, 158, 215, 220, 221, 222, 224, 225, 231, 232, 236 237, *321, 322*
Nylén, P., 112, 114, *322*

O

Oae, S., 137, 138, *322*
O'Brien, D. H., 120, 121, 122, 127, 128, 132, 133, 183, 203, 204, 205, *322*
O'Brien, J. L., 104, 128, 129, *322*
Oddo, G., 104, 118, 121, 128, 132, 151, 158, *322*
O'Ferrall, R. A. M., 271, *322*
O'Gorman, J. M., 187, *322*
Olah, G. A., 86, 103, 104, 108, 109, 112, 117, 120, 121, 122, 127, 128, 132, 133, 143, 152, 158, 183, 189, 203, 204, 205, 209, 277, 300, *319, 322*
Olivieri-Mandalà, E., 304, *322*
Olsen, F. P., 174, *309*
Olsson, S., 266, 267, *322*
Orékhov, A., 241, *322*
Osborn, A. R., 194, *322*
Oubridge, J. V., 11, 12, 24, 25, 70, 71, 78, 83, *313, 314, 315*
Oulevey, G., 157, *322*
Ovsyanskina, N. A., 293, *325*

P

Pace, R. J., 134, *308*
Packer, J., 125, *313*
Paddock, N. L., 117, *322*

Pal'm, V. A., 39, 87, 89, *322, 323*
Pande, K. C., 275, 276 *313*
Partington, J. R., 136, *317*
Pascard, R., 22, *323*
Pascard-Billy, C., 22, *323*
Passerini, R. C., 123, 137, 138, *314*
Paternò, E., 118, *323*
Paul, M. A., 26, 36, 168, 174, 183, 188, 196, 201, 205, 208, 226, 234, 265, *316, 320 323*
Pearson, D. E., 246, 247, 248, *321, 323*
Pearson, R. B., 285, 287, 288, *317*
Pedersen, K. J., 233, *323*
Peeling, E. R. A., 145, 161, *315*
Perizzolo, C., 31, 49, 207, 302, *311, 312*
Perrin, C., 53, *323*
Perrin, D. D., 87, 91, 98, 101, 108, 112, 118, *310, 323*
Peterson, H. J., 161, 162, 213, 215, 221, 278, *312*
Peterson, P. E., 225, *322*
Pettit, R., 124, *311*
Phillips, J., 38, 125, *309*
Pietrusza, E. W., 165, *325*
Pinnow, J., 293, *323*
Pittman, C. U., Jr., 120, 122, 142, 157, 183, 244, 300, 301, *311, 312, 322*
Plattner, P. A., 142, *323*
Plesh, P. H., 141, *308*
Pocker, Y., 187, 238, 239, 240, *310, 323*
Polanyi, M., 196, *323*
Poole, H. G., 161, *317*
Popiel, R., 7, 12, *323*
Pople, J. A., 24, *323*
Pospišil, L., 129, *323*
Potts, H. A., 98, *323*
Pracejus, H., 107, *323*
Praestgaard, E. L., 144, *307*
Pratt, L., 144, *311*
Pressman, D., 222, 223, *320, 323*
Preston, J., 222, *308*
Price, F. P., 165, *323*
Prinsen, A. J., 294, 295, 296, *318, 323*
Pritchard, J. G., 179, 186, 187, *320, 323*
Pryor, W. A., 36, 224, 231, 232, 236, *321*
Purchase, M., 205, *320*
Purlee, E. L., 211, 213, 218, *323, 326*
Puzitskii, K. V., 303, *313, 323*

Q

Quacchia, R. H., 123, *324*
Quadri, S. A. A., 129, 146, 149, 151, 159, *324*
Qureshi, E. A., 284, *315*

R

Raaen, V. F., 242, *323*
Rabinovich, B. S., 192, 207, *320, 323*
Rada, G. K., 283, *318*
Radt, F., 133, *308*
Ragsdale, R. O., 114, *321*
Rainey, W. T., 242, *311*
Raisin, C. G., 263, 291, *317*
Ramanaiah, K. V., 18, *323*
Ramsey, B. G., 122, 123, *323*
Rand, M. H., 181, *308*
Rao, C. N. R., 250, *320*
Rao, N. R., 18, *323*
Reagan, M. T., 46, 49, 142, *323*
Reed, W. L., 215, 221, 222, 224, 237, *321, 322*
Reeves, R. L., 100, 101, 103, *323*
Reid, E, E., 189, *323*
Reilly, C. A., 23, *317*
Reintke, E., 253, *309*
Reiss, W., 126, *313*
Reitz, O., 178, 193, 206, 234, *309, 323*
Reubke, E., 164, *326*
Reuwer, J. F. Jr., 268, *325*
Rhind-Tutt, A. J., 40, *310*
Richards, J. H., 35, *311*
Richards, R. E., 134, *316*
Richley, H. C., 142, 300, 301, *312*
Ridd, J. H., 277, 281, 283, 284, 285, 287, 288, 289, 290, *309, 310, 311, 314, 316, 317*
Ridgewell, B. J., 112, 113, 114, *317*
Riesz, P., 211, 212, 218, *323, 326*
Riley, E. F., 117, *316*
Riordan, J. C., 42, 192, 194, *328*
Ritter, J. J., 302, *324*
Rix, Th. R., 300, *326*
Robbins, J., 158, 262, *324*
Roberts, J. D., 94, *324*
Robinson, E. A., 10, 11, 12, 18, 21, 61, 67, 69, 70, 71, 72, 73, 74, 75, 76, 77, 78, 79, 84, 85, 108, 112, 115, 117, 124, 129, 138, 139, 143, 146, 149, 151, 160, 164, 165, 166, *308, 313, 314, 315, 316, 324*

Robinson, P. L., 10, 161, 162, *316*
Robinson, R. A., 73, 75, *324*
Roebuck, A. K., 244, *324*
Roemming, Chr., 118, *316*
Roffia, S., 18, *313*
Rogers, M. T., 109, *319*
Ron, A., 286, *316*
Ropp, G. A., 257, 258, *324*
Rördam, H. N. K., 136, *324*
Rosenberg, A. S., 234, *325*
Rosenblum, M., 35, *311*
Rosenstein, L., 96, *307*
Rosenthal, D., 192, 194, *324*
Rothrock, T. S., 240, *324*
Roughton, J. E., 10, *324*
Rubin, T. R., 13, 14, 16, 16, 17, 296, *314*, *324*
Russel, R. L., 291, *318*
Russell, M., 266, 267, 269, 270, *322*, *324*
Rutherford, A., 9, 22, *309*
Rutherford, K. G., 252, *324*
Ryabova, K. G., 303, *313*, *323*
Ryabova, R. S., 33, 36, 37, 38, 58, 78, 85, 226, 255, 256, 259, *324*, *326*, *327*
Ryncki, L., 254, *309*

S

Sacher, E., 161, 162, 221, 278, *311*, *312*
Saika, A., 23, 24, 193, *316*, *324*
Saines, G., 30, 31, 46, 140, 141, 298, 299, 302, *312*
Sackur, O., 15, *324*
Salomaa, P., 124, 188, 201, *324*
Samuel, D., 180, 194, 197, 216, *324*
Santer, J. O., 35, *311*
Satchell, D. P. N., **264**, 265, 266, 267, 270, 271, 293, 294, 295, *315*, *324*
Satchell, R. S., 211, 218, 220, *315*
Sato, Y., 265, 266, *319*
Sauer, H., 162, *325*
Sauermann, P. F., 5, 10, *316*
Saunders, T. J., 277, 284, *308*
Savoie, R., 18, *314*
Scandola, E., 104, 118, 121, 132, 151, *322*
Schachat, R. E., 246, *320*
Schaleger, L. L., 177, 201, 202, *324*
Schiavelli, O. P. M. D., 225, *322*
Schierz, E. R., 255, *324*
Schmid, H., 254, 290, *307*, *324*

Schmidt, K. F., 249, 291, *324*
Schneider, A., 298, *308*
Schneider, W. G., 24, *323*
Schofield, K., 191, 209, 210, 247, 248, 277, 283, 284, *311* *315*, *321*
Scholl, P. S., 245, *312*
Schonberg, A., 253, *324*
Schriesheim, A., 29, 45, 152, 153, 154, 300, 301, 302, *312*, *316*
Schubert, W. M., 123, 158, 173, 211, 212, 213, 215, 260, 261, 262 *310*, *324*
Schultz, F., 261, *210*
Schulze, J., 142, 271, *320*, *324*
Schulze, T., 117, *312*
Schwarzenbach, G., 233, *325*
Scorrano, G., 137, *319*
Scott, R. B., 244, 245, *310*
Sedor, E. A., 231, *316*
Seel, F., 162, *325*
Seiser, S., 286, *316*
Sekuur, Th. J., 128, *325*
Senderens, J. B., 254, *325*
Senior, J. B., 163, 164, *314*
Setkina, V. N., 218, *325*
Setlik, B., 3, *325*
Shaad, L. J., 268, *325*
Shakir, N., 112, 113, 114, *317*
Shane, N., 131, *325*
Shankman, S., 14, *325*
Shannon, J. S., 127, *313*
Shapiro, S. A., 36, 37, *317*
Shaw, F. R., 278, *317*
Shaw, W. H. R., 193, *325*
Shcheglova, G. G., 37, *314*
Shilov, E. A., 289, *325*
Shorter, J., 87, 89, 91, 94, *325*
Sidhu, G. S., 291, *325*
Siemiradzki, B., 253, *309*
Silbermann, W. E., 230, *325*
Silver, B. L., 180, 194, 197, 216, *324*
Simamura, O., 281, *325*
Simkins, R. J., 285, *327*
Simon, W., 35, *311*
Singer, K., 162, *325*
Sixma, F. L. J., 294, 296, *310*, *317*
Skrabal, A., 183, *325*
Sluiter, C. H., 245, *320*
Smart, G. N. R., 285, *310*
Smith, G. F., 98, *323*
Smith, H. A., 153, *327*
Smith, H. M., 19, 20, 296, *328*

AUTHOR INDEX

Smith, P. A. S., 245, 248, 249, 251, 252, 304, *325*
Smith, P. J., 122, *308*
Smith, W. B., 242, *311*
Södervall, T., 109, 110, *327*
Solomons, C., 2, 11, 12, 24, 25, 69, 70, 72, 73, 74, 75, 76, 77, 78, 79, 80, 81, 83, 85, 115, 116, *308*, *313*, *314*, *315*
Somiya, T., 4, *325*
Sommer, J., 152, 183, *322*
Sommer, L. H., 165, *325*
Spallino, R., 118, *323*
Spangler, M., 298, 299, 302, *312*
Spengeman, W. F., 40, *316*
Spryskov, A. A., 293, *325*
Spunta, G., 136, 137, *311*
Stafford, F. E., 186, 187, *320*
Stamhuis, E. J., 224, *325*
Stanfield, K., 231, *321*
Stefanac, Z., 35, *311*
Steigman, J., 131, *325*
Stein, R., 277, 281, *311*
Stejskal, E. O., 108, *326*
Stevens, J. B., 33, 41, 42, 104, 106, 192, *328*
Stewart, A. R. P., 207, *323*
Stewart, R., 40, 98, 104, 124, 125, 129, 130, 131, 154, 191, *312*, *313*, *321*, *325*
Stewart, W. T., 222, 223, *320*
Stieglitz, J., 246, *325*
Stiles, M., 243, *325*
Stivers, E. C., 268, *325*
Stock, L. M., 271, 276, 283, 290, *325*
Stokes, R. H., 73, 75, *324*
Stollar, H., 109, 110, *313*
Stone, B., 7, *308*
Stopperka, K., 18, 30, 21, 23, 24, *325*
Stothers, J. B., 305, *320*
Strehlow, H., 34, *318*
Strehlow, R., 34, *325*
Suarez, C., 25, *325*
Suggitt, J. W., 285, *310*
Surfleet, B., 285, *325*
Surov, Yu. N., 126, *326*
Susz, B. P., 157, *322*
Swain, C. G., 234, 268, *325*
Sweeting, L. M., 49, 50, *325*
Symons, M. C. R., 155, 156, 163, 164, *307*, *315*
Szabo, A., 196, *323*
Szmant, H. H., 123, 165, *325*

T

Taddei, F., 137, *319*
Taft, R. W. Jr., 20, 50, 56, 64, 88, 92, 94, 188, 193, 198, 210, 211, 212, 213, 217, 218, 220, 257, 258, *309*, *312*, *318*, *319*, *323*, *325*, *326*
Takeda, M., 108, *326*
Talbot, R., 261, *310*
Tamele, M., 238, 239, *316*
Taylor, R., 264, 265, 266, *313*
Taylor, T. I., 192, 194, *324*
Tchelintzev, B., 118, *326*
Tedder, J. M., 144, 288, 291, *317*, *326*
Telder, A., 280, 296, *310*
Temkin, M. I., 37, *314*
Terada, H., 208, *326*
Theodoropoulos, S., 99, *310*
Thomas, J. A., 286, *307*
Thompson, A., 6, 7, 8, 24, 25, *315*
Thompson, A. A., 116, *328*
Thompson, H. W., 134, *316*
Thorne, M. P., 229, *309*
Thornley, M. B., 29, 38, 43, 115, *309*
Thyagarajan, G., 291, *325*
Tickle, T., 118, 136, *311*
Tietz, R. F., 251, 252, *326*
Tiffeneau, M., 240, 241, *322*, *326*
Tillett, J. G., 277, 281, *326*
Timm, E. W., 196, *326*
Tomanová, J., *323*
Traficante, D. D., 158, *326*
Traube, W., 164, *326*
Treffers, H. P., 131, 145, 146, 157, 158, *326*
Tsukerman, S. V., 126, *326*
Tsuno, Y., 275, 276, *328*
Turner, J. O., 119, 120, 121, 244, *312*
Tutundžić, P. S., 7, 104, 118, 133, *326*
Tuxworth, R. H., 245, *320*
Tye, F. L., 140, 143, 154, *315*

U

Unik, J. P., 198, 199, *308*
Urwin, J. R., 192, *317*
Utley, J. H. P., 281, 283, *309*, *314*, *326*

V

Vamplew, P. A., 162, *325*
van der Heijde, H. B., 4, *326*

Van der Waals, J. H., 140, 143, *320*
van der Zanden, J. M., 300, *326*
Varker, A., 262, *320*
Vaughan, J., 125, *313*
Vaughan, T. A., 283, *326*
Večera, M., 43, 44, *326*
Vedel, J., 35, *326*
Veibel, S., 288, *326*
Verhulst, J., 208, *326*
Verkhovod, N. N., 152, *319*
Vermeulen, Th., 14, *315*
Vernon, A. A., 127, *316*
Vernon, C. A., 40, 289, *310*, *311*
Vetešnik, P., 43, 44, *326*
Vinnik, M. I., 19, 20, 26, 33, 34, 36, 37, 38, 39, 48, 49, 58, 78, 85, 96, 97, 209, 226, 247, 255, 256, 259, *315*, *319*, *324*, *326*, *327*, *328*
Virtanen, P. O. I., 109, 110, 111, *327*
Vogel, W., 256, 292, *318*
Vollbracht, L., 294, 296, *310*
von Dobeneck, H., 259, *326*
von Kothner, P., 132, *326*
von R. Schleyer, P., 86, 143, *322*
Vrba, Z., 297, *327*

W

Wai, H., 40, 62, 120, 122, 129, *328*
Walden, P., 8, 9, 136, *327*
Wale, P. D., 191, 283, *321*
Walker, D. G., 193, *325*
Walrafen, G. E., 18, 19, 20, 21, *327*, *328*
Walsh, A. D., 91, *327*
Walton, J. H., 207, 208, *311*
Walton, P. R., 148, 149, 151, 160, *319*
Walz, H., 109, 110, *317*
Wanders, A. C. M., 293, 294, *327*
Wang, C.-H., 37, 153, *309*
Wang, C.-S., 153, *313*
Wang, I. C., 34, 64, 105, 119, 121, 128, *313*
Ward, H. T., 255, *324*
Waring, A. J., 34, *318*
Wasif, S., 7, 8, 11, 12, 72, 76, 78, 98, 159, 161, *313*, *314*, *315*, *327*
Waters, J. A., 275, *313*
Waters, W. A., 289, 290, *312*
Watts, D. W., 162, *308*

Weber, S., 142, *323*
Webster, B. M., 93, *327*
Weinberg, A. J., 257, *324*
Weisfeld, L. B., 216, *319*
Welch, C. M., 153, *327*
Wells, P. R., 87, 88, *327*
Wendt, H., 34, *318*, *325*
Westheimer, F. H., 161, 271, 277, *327*
Whalley, E., 178, 183, 187, 188, 189, 194, 202, *318*, *322*, *327*
Whipple, E. B., 98, 99, *310*, *316*, *327*
White, A. M., 109, 132, 133, 158, 183, 203, 204, 205, 209, *322*
White, H. S., 286, *327*
White, R. F. M., 9, 23, 24, 38, *315*
White, T. J., 191, 283, *321*
White, W. N., 286, *327*
Whitford, E. L., 255, *327*
Whitmore, F. C., 165, *325*
Whitney, R. B., 222, *308*
Wiberg, K., 178, 188, 193, 201, *327*
Wiig, E. O., 255, 259, *327*
Wiles, L. A., 126, 127, 129, *327*
Wiley, R. H., 118, 129, 159, *327*
Wilkins, J. M., 229, *309*
Wilkinson, G., 144, *311*
Willi, A. V., 168, 196, 205, 210, 263, *327*
Williams, D. L. H., 286, *307*
Williams, G., 29, 95, 128, 152, 153, 161, 277, 284, 285, *308*, *309*, *320*, *321*, *327*
Williams, J. F. A., 155, 156, 157, *327*
Williams, R. L., 91, 92, *308*
Williamson, M. J., 191, 277, 283, *321*
Wilson, C. L., 234, 263, 291, *317*
Wilson, G. L., 15, 22, *321*
Wilson, I., 217, *310*
Wilson, R. W., 7, 8, *312*
Winkler, C. A., 192, 207, *320*, *323*
Winstein, S., 187, 222, 223, 241, *316*, *327*
Wisotsky, M. J., 64, 65, 67, 104, 119, 120, 121, 124, 142, 157, *312*
Wittwer, C., 233, 292, *325*, *328*
Wolfe, A. P., 212, 217, 218, 220, *309*
Wolfenden, R., 103, *327*
Wolfsberg, M., 258, *308*
Woo, G. L., 236, *321*
Woodward, L. A., 19, *327*
Wright, G. F., 285, *310*
Wright, K. L., 146, 151, 159, *319*
Wu, C. Y., 49, 66, 120, 121, 122, *307*
Wurtz, C.-A., 221, *327*

Wyatt, P. A. H., 12, 17, 20, 22, 25, 52, 58, 69, 70, 73, 74, 75, 76, 78, 108, 285, *307, 309, 311, 317, 318, 321, 325, 328*
Wynne-Jones, W. F. K., 171, 178, 187, 189, *309, 328*

Y

Yankwich, P. E., 258, *308*
Yates, K., 33, 34, 40, 41, 42, 49, 50, 51, 53, 62, 104, 106, 116, 120, 122, 124, 129, 131, 191, 192, 194, 202, 203, 205, *313, 318, 325, 328*
Yeh, S.-J. 40, 101, 102, *328*
Young, T. F., 19, 20, 21, 296, *327, 328*

Yukawa, Y., 275, 276, *328*
Yurev, Yu. K., 126, *326*

Z

Zahler, R. E., 158, 260, 261, 262, *324*
Zahorka, A., 183, *325*
Zaidi, S. A. A., 124, *324*
Zarakhani, N. G., 19, 20, 38, 209, 247, *326, 328*
Zimmer, H., 111, *319*
Zollinger, H., 292, *328*
Zucker, L., 175, 233, 234, *328*
Zuffanti, S., 227, *320*

Subject Index

A

Absorption spectra of indicators, 32
Acetaldehyde, 234
Acetamide, 105, 193–194
Acetamides, substituted, 105
Acetic acid, 72, 157
Acetic acid/sulphuric acid, H_0, 40
Acetic anhydride, 145, 150, 158
Acetone, 124, 146, 231, 234
Acetophenone, 124, 231
Acetophenone oxime, 248
Acetylacetone, 233
Acetyl chloride, 133
Acetylferrocene, 127
Acetylglycine, 193
Acid azides, 252, 305
Acid chlorides, basicities, 133
Acidity functions,
  correlations with, 171–176, 272–273
  definitions, 27
  H_A function, 41, 42
  H_C function, 46
  H_I function, 43, 43
  H_0 or H_0' function, 27, 35, 37, 56
  H_0''' function, 40, 41, 55
  H_R function, 29, 45
  H_R' function, 30, 45
  H_+ function, 43, 44
  H_- function, 44
  J_0 function, 30
  $R_0(H)$ function, 34, 35
  physical significance, 52
Acraldehyde, β,β-dimethyl, 223
Activation parameters, 176
Activity coefficients of indicators, 47
Activity coefficient postulate, 27, 28, 47
N-Acylimidazolium ions, 193
Alcohols,
  addition on to double bonds, 221
  basicities, 119–120
  dehydration, 210, 216–218
  ionization in 100 per cent H_2SO_4, 150

Alcohols—*continued*
  hydride abstraction by, 299
  oxygen exchange, 216
Aldehydes,
  basicities, 124
  cation structure, 127
  hydration, 221–224
Aldehyde cyanohydrins, 208
Aldimines, 103
Alicyclic ketones, 124
Aliphatic ketones, 124
Alkoxycarbonium ions, 189, 299
Alkylations, 301
Alkylbenzenes, 294–295
Alkyl hydrogen sulphates, 151
Alkylidene azide, 250
1-Alkynyl ethers, 224
1-Alkynyl thioethers, 224
Allyl alcohol, 238
Allylic rearrangements, 237
Amides,
  acidity function, 41, 42, 49, 55
  basicities, 104–107
  cation structure, 108
  hydrolysis, 189–196, 197, 201
Amination, 291
Amines,
  as indicators, 36, 97
  basicity, 93, 95, 97
Amine oxides, 112
1-Amino-ethyl-sulphonic acid, 294
Ammonium sulphate, 147
t-Amyl alcohol, 218
Aniline-N-sulphonic acid, 297
Anilines, substituted,
  as indicators, 36, 97
  basicities, 97
  diazotization, 288
  sulphonation, 297
Anilinium ions,
  hydration, 50
  relative activity coefficients, 50
Anils, cyclodehydrations, 229

Anthracene, 265, 279, 280
Anthraldehydes, 124
Anthraquinone, 36, 127
Aromatic aldehydes,
  basicity, 124
  decarbonylation, 259–261
Aromatic sulphides, 123
Arsines, 112
Arsine oxides, 114
Arylmethanols,
  acidity function, 45
Aryl nitramines, 286
"Asymmetric dissociation" of 100 per cent H_2SO_4, 73
Autoprotolysis constant, 57, 76
Autoprotolysis ions, 74
Azides, 249–253, 303–305
9-Azidofluorene, 252
Azobenzenes, 100–103
Azoxybenzenes, 114
Azulene-1-d, 269
Azulene-1-carboxylic acid, 262
Azulenes, protonation, 46, 141–142, 270–272

B

n-Bases, 86–7
$\bar{u}$-Bases, 86
Basicity, factors determining base strength, 87
Beckmann fission, 249
Beckmann rearrangement, 245–249
Benzalacetophenones,
  basicity, 125
  isomerization, 232, 236
Benzaldehyde,
  basicity, 124
  condensation, 231
  semicarbazone, 103
Benzaldehydes, substituted, 124, 259–261
Benzamide, 49, 104, 134–135
  diazotization, 288
  hydrolysis, 190–191, 194
Benzamides, substituted, 104, 190–191
Benzene, 267, 279, 294–295
Benzeneboronic acids, 273–274
Benzenes, substituted, 264, 266, 269, 271, 281–282
Benzenesulphonic acid, 292

Benzhydrols, 251
Benzhydryl azides, 251
Benzil, 126, 253
Benzilic acid, 259
Benzoic acid, 62, 72, 129, 131, 134, 158
Benzoic acids, substituted, 129, 253
  decarboxylation, 261–262
  di-o-substituted, 157
Benzoic anhydride, 145, 158
Benzoin, 253
Benzophenones, 125
Benzophenone oximes, 248
Benzopyrones, 136
Benzoquinuclidine, 107
Benzoylacetic acid, 225
o-Benzoylbenzoic acids,
  basicity, 131
  complex ionization, 158
  cyclodehydration, 226
Benzoyl chlorides, basicity, 133
p-Benzoyldiphenyl, 125
Benzoylformic acid, 255–256
β-Benzoylnaphthalene, 125
Benzyl phenyl ketones, 125
$\Delta^{10,10'}$-Bianthrone, 127
Boric acid, 147, 148
Bromination, 289–290
Bromine, 283, 289
N-Bromo-succinimide, 290
Brønsted coefficient, 179, 211
t-Butanol, 151, 217, 244
Butenes, hydration, 211, 222
9-$tert$-Butylfluorenol, 249

C

Carbon bases, 139
  acidity function, 46
Carbon dioxide, 143, 253
Carbonium ions,
  formation, 86, 87, 139, 151–152, 239
  reactions, 302–304
  rearrangements, 237–245
  stability, 139
Carbon monoxide, 253, 302
Carbonylations, see Koch reaction
Carbonyl bases, 134–135
Carbonyl compounds,
  basicities, 134
  hydration, 221–224

SUBJECT INDEX

Carbonyl frequency, 90, 134
Carboxylic acids,
　basicities, 129–130
　cation structure, 131–132
　decarbonylation, 254–259
　decarboxylation, 261–262
　formation, 302–303
Chalcone, see Benzalacetophenone
Chlorination, 289–291
Chlorosulphuric acid, 39, 84
Chromone, 136
Cinnamic acid, 236
Citric acid, 255–256, 259
Complex ionizations, 145
Concentration units,
　in sulphuric acid, 2
　in oleum solutions, 2
Condensations, 231
Conductometric measurements,
　in complex ionization studies, 149
　in protonation studies, 71
　in studies of reaction rate, 182
　technique, 83
Conductometric titration,
　in complex ionization studies, 149
　in protonation studies, 83
Coumarin, 132
p-Cresol-o-t, 270
Croton-aldehyde, 222, 223
Crotonic acid, 222, 223
Crotyl alcohol, 222
Cryoscopic mixtures, 146–148
Cryoscopy,
　in complex ionization studies, 146
　in protonation studies, 67
　in studies of reaction rate, 181
Cyanocarbon acids, 44
Cyclic allylic cations, 243, 301
Cyclic ethers, 121
Cyclodehydrations, 225–231
Cyclohexa-1,2-dione, 233, 234–235
Cyclohexanes, 244
Cyclohexene, 211
Cyclopropane, 245

D

Deboronation, see Protodeboronation
Decarbonylations, 253
　of acids, 254,
　of aromatic aldehydes, 253–254, 259

Decarboxylations, 253, 261–262
Dedeuteration, see Hydrogen isotope exchange
Degermylation, 275–276
Dehydration of alcohols, 216–221
Densities of sulphuric acid solutions, 5–7
Deplumbylation, 275–276
Desilylation, 274–276
Destannylation, 275–276
Desulphonation, 291–294
Detritiation, see Hydrogen isotope exchange
2-Deuterio-2'-carboxy-biphenyl, 229
Deuterium oxide, 178
1,1'-Dianthrimide, 127
9,9-Diarylfluorenes, 231
Diarylmethanols, 153–154
1,1-Diaryl olefins, 30, 46, 140–141, 301
Diazotization, 287, 288
Dibenzalacetone, 125
Dibenzoylmethane, 126
Dideuterosulphuric acid,
　acidity functions, 39
　as reaction medium, 178
　infrared spectra, 21
　preparation, 24
　properties, 24, 25
Diethylsulphide, 122
Diferrocenyl ketone, 127
Dilatometry, 181
Dimerization, 300, 305
N,N-Dimethylaniline, nitrosation, 287
1,2-Dimethylcyclohexane-1,2-diol, 243
1,2-Dimethylcyclopentane-1,2-diol, 243
Dimethyldithiophosphinic acid, 117
3,3-Dimethyl-5-ketohexanoic acid, 230
Dimethylphosphinic acid, 117
2,6-Dimethyl-4-pyrone, 118, 136
4,6-Dimethyl-2-pyrone, 118
2,2-Dimethyl-quinuclidone-6, 107
2:4-Dinitrobenzene-sulphenyl chloride, 124
1,4-Dioxane, 40, 121
2,2'-Diphenic acid, 227
Diphenylacetic acid, 256
Diphenylamines as indicators, 97, 98
1,1-Diphenylethylene, 252, 301
Diphenylmethyl cation, 154, 299
Diphenyltriketone, 253

Distribution measurements,
  in indicator studies, 48
  in protonation studies, 66
Disulphuric acid, 11, 22, 149, 294–296

E

Electrophilic aromatic substitutions,
  by the hydrogen ion, 260, 262–276, 293
  See also Nitration, etc.
Enolization, see Keto-enol isomerization
Enthalpies of activation, 176
Entropies of activation, 177, 202
Esters,
  basicity, 202
  cation structure, 132
  hydrolysis, 196–205
  protonation, 132
Ethers,
  basicity, 120–121
  cation structure, 121
  hydride abstraction, 299
  hydrolysis, 183–186
Extraction, see Distribution

F

Ferrocene-ferricinium electrode, 34
Flavones, 136
Fluorosulphuric acid, 61, 84, 109, 117, 120, 131–132, 139, 203
Formamide, 106
  t-butyl-, 302
Formic acid,
  basicity, 256
  cation structure, 131
  decarbonylation, 254–259

G

General acid catalysis, 179
Glass electrode, 35
Glutaric anhydride, 159
Glycols, 240–243
Guaiazulene-3-d, 269
Guaiazulene-2-sulphonate-3-d, 269
Guanidine, 95
Guggenheim method, 181

H

Hammett acidity function,
  see Acidity functions

Hammett indicators, 36
Hammett substituent constants, 88, 89
Heat capacities of activation, 178
Helianthrone, 127
Hexamethylbenzene, 160
Hexamethylene-tetramine, 98
Hydration,
  of olefins, 210–216
  of triple bonds, 224–225
  of α,β-unsaturated carbonyl compounds, 221
Hydrazoic acid,
  basicity, 144
  reactions, 303–305
Hydride shift, 238
Hydride transfers, 244, 298–300
Hydrocarbons,
  as bases, 139
  polycyclic, 142, 265
  rearrangements, 232
Hydrogen-bonding spectral shifts, 60
Hydrogen chloride, 164
Hydrogen cyanide, 207, 302
Hydrogen-deuterium exchange, 243
Hydrogen disulphate ion, solvolysis, 182
Hydrogen fluoride, 131, 139, 164
Hydrogen ion, see Hydronium ion
Hydrogen isotope exchange, 263–273
Hydrolysis,
  acetals, 187–188
  acid anhydrides, 208
  amides, 189–196, 197
  benzylidine trifluoride, 209
  cyanides, 205
  cyclic ethers, 186
  epoxides, 183, 186
  esters, 196–205
  ethers, 183–186
  formals, 188
  glucosides, 187, 188
  lactones, 205
  nitriles, 205–208
  oximes, 209
  trimethylene oxide, 186
  urea, 193
Hydronium ion,
  as electrophile, 262–276
  hydrogen bonding to, 62
  infrared spectrum, 20
  structure, 6

Hypobromous acid, 289
Hypochlorous acid, 289

I

Indicator measurements, 31
   accuracy, 33, 34
   medium effects, 33
Indoles,
   acidity function, 42
   basicities, 98, 100
Interconversion of alcohols and olefins, 210, 218
Iodination, 289–290
Iodine, 163
Iodosyl cation, 163
Iodyl cation, 164
Isobutene, 211, 217, 244
Isobutyraldehyde, 238
Isocyanates, 252
Isomerizations, 232
   cis-trans, 235
   keto-enol, 233
Isotopic exchange, 180, 216–217, 243, 263–273
Isotopic labelling, 179

K

Ketimines, 103
Keto-enol isomerizations, 233–235
Ketone cyanohydrins, 208
Ketones, 124–127, 221–224, 235, 304
Ketonization, see Keto-enol isomerization
Ketoximes, 245
Koch reaction, 302–303

L

Lactams, 109–110, 246
Lactones, 205, 299
Lead tetraacetate, 166

M

Maleic anhydride, 159
Malic acid, 255–256
Mechanism of conduction, see Proton-jump conduction
Mechanisms, see Reaction mechanisms
Mechanistic criteria, 170–181

Mercaptans, 122
Mesitaldehyde, 260–261
Mesitoic acid, 157, 262
Mesityl oxide, 223, 231
Metal sulphates, 71
2-Methylallyl alcohol, 238
Methyl azide, 250
Methyl benzimidate, 193, 209
Methylenecyclobutane, 211
β-Methylglutaconic anhydride, 159
Methyl mesitoate, 204
Migratory aptitudes of groups, 240
Molon unit, 2
Monoaryl alcohols, 154–157
Monoaryl carbonium ions, 156

N

Naphthaldehydes, 124
Naphthalene, 280, 296
Naphthalene-1,8-dicarboxylic anhydride, 159
meso-Naphthobianthrone, 127
N-Nitramines, 285–286
Nitration, 277–286
N-Nitration, 285
Nitric acid, 145, 161, 223, 277–278
Nitriles,
   basicities, 103
   cation structure, 104
   formation, 250
   hydrolysis, 205–208
Nitroanilines as indicators, 36, 40, 48, 97
9-Nitroanthracene, 280
Nitrocompounds,
   as indicators, 38, 48
   basicities, 115, 116
   cation structure, 117
   resonance in, 279
Nitrogen oxides, 161–162
Nitro group, 91
Nitroso group, 91
Nitromethane, 116
Nitronium ion, 161, 277–286
Nitronium tetrafluoroborate, 277
N-Nitrosamines, 111
Nitrosation, 287, 288
Nitrosyl cation, 162, 288
Nitrous acid, 161–162
Nitryl fluoride, 162
Non-electrolytes in sulphuric acid, 79

O

Nuclear magnetic resonance, see Proton magnetic resonance

Olefins, hydration, 210–216
   polymerization, 300–301
Oleum solutions,
   acidity function, 38, 39, 56
   concentration units, 2
   conductance, 11, 12
   densities, 7
   p.m.r. spectra, 24
   Raman spectra, 21
   ultraviolet spectra, 21
   viscosity, 7, 8
Oxalic acid, 182, 254–258
Oximes,
   basicities, 111–112
   cation structure, 112
   hydrolysis, 209
   rearrangement, 245–249
Oxocarbonium ions, 157–158, 254, 305
Oxygen isotope exchange, 180, 197–200, 216–217, 247

P

Pentacyanopropenide anion, 49
Pentamethylbenzanilide, 157
Perchloric acid, aqueous,
   acidity function, 40
   as protonating medium, 62, 63
   as reaction medium, 183, 190, 214, 223, 264, 266, 270, 273
9-Phenanthraldehyde, 124
Phenanthrene, 265
4-Phenanthrene-carboxylic acid, 252
Phenols, 120, 122
Phenyl acetylenes, 225
o-Phenylenediamine, 98
Phenylenediamines, acidity function, 43
Phenylpropiolic acids, 225
2-Phenyltriarylcarbinols, 231
Phloroglucinol, 123
Phosphines, 112
Phosphine oxides, 114
Phosphonitrilic chlorides, 117
Phosphoric acid,
   protonation, 117
   as reaction medium, 249, 273

Phthalic acids, 129
Phthalic anhydride, 159
Pinacolic rearrangement, 239
Polycyclic ketones, 126
Polyisobutylene, 151
Polymerization, 300
Polyphosphoric acid, 249
Potassium pyrophosphate, 117
Potentiometric titration of oleum, 4
Primary kinetic isotope effect, 179, 257–258, 268, 271
1-Propanol, 245
Propenium ion, 139
Propenylbenzenes, 300
Propionamide, 105
Propiophenone oxime, 248
Protodeboronation, 273–274
Protodesilylation, 274–276
Proton-jump conduction, 10, 71, 75
Proton magnetic resonance,
   in protonation studies, 63
   in studies of reaction rates, 183
Pyridine oxides, 112–114
Pyrones, 136
Pyrroles, 98, 99

Q

Quinones, 126
Quinuclidine, 107

R

Racemization, 180, 216–217, 234
Raman spectra,
   in protonation studies, 65
   of sulphuric acid, 18, 19
Reaction mechanisms, main types, 168–170
Rearrangements, 232
   aldehydes to ketones, 241
   allylic, 237
   Beckmann, 245–249
   carbonium ion, 237–245, 303
   diphenyltriketone-benzoin, 253
   nitramine, 286
   pinacol-pinacolone, 239
   Schmidt, 249–253
Relative activity coefficients, 49, 51

S

Scatole, 160
Schmidt reaction, 249–253, 303–305
Secondary isotope effects, 270
Selenochromone, 136
Selenoxanthone, 136
Silicon compounds, 165, 274–276
Solvent isotope effects, 178, 268
Spectrophotometric measurements, 31, 61, 150, 181
Steric effects, 180
Stibine oxides, 114
Strong bases in sulphuric acid, 75
Styrenes, hydration, 213, 214–215, 225
Substituent constants, 88, 89, 92, 95
Substituent effects, 88–95, 180
Succinic anhydride, 159
Sulphonamides,
  basicities, 110, 111
  cation structure, 110
Sulphonation, 183, 291–297
Sulphones, 137–139
Sulphonic acids, 293
Sulphonyl group, 91
Sulphoxides, 137, 139
Sulphur dioxide, 143, 159
Sulphuric acid/acetic acid, H_0, 40
Sulphuric acid, aqueous,
  acidity functions, 35, 37, 40
  activity of water, 15
  boiling points, 14
  conductance, 5, 9, 11
  densities, 4, 5
  excess volume of mixing, 5
  infrared spectra, 18
  p.m.r. spectra, 23
  Raman spectra, 18, 19
  Rayleigh scattering, 18
  relative partial molal enthalpies, 16
  surface tension, 5, 8
  vapour pressures, 14
  viscosities, 5, 7
  X-ray diffraction, 22
Sulphuric acid, 100 per cent,
  "asymmetric dissociation", 12, 73–75
  autoprotolysis, 11, 76–79
  conductance, 10
  cryoscopic constant, 13
  density, 6
  dielectric constant, 8, 9
Sulphuric acid—*continued*
  freezing point, 2, 13
  heat capacity, 16
  heat of autoprotolysis, 17
  infrared spectrum, 18
  ionic self-dehydration, 15, 17
  latent heat of fusion, 17
  p.m.r. spectrum, 23
  self-ionization, 11, 15, 76–79
  specific heat, 17
  surface tension, 6
  viscosity, 6
Sulphuric acid monohydrate,
  freezing point, 13
  latent heat of fusion, 17
Sulphuric acid, partially aqueous, H_0, 40
Sulphuric acid, solid,
  X-ray diffraction, 22
Sulphuric acid-^{35}S, 25, 297
Sulphuric acidium ion, 21, 294–296
Sulphuric acid/sulphuryl chloride, 82, 83
Sulphur trioxide,
  determination in oleum, 3, 4
  electrophile, 291, 297
  rôle in hydrolysis of ethers, 185
  ultraviolet absorption, 22
  vapour pressure over oleum, 15
Sulphuryl chloride, 68, 79–83

T

Taft substituent constants, 94, 95
Tetraethylammonium ion, 49
Tetrahydrofuran, 121
Tetrahydrogenosulphatoboric acid, 57, 83, 147, 148, 204
Tetrahydrothiophene, 122
Thermometric titration of oleum, 4
Thioalcohols, 120
Thioacetamide, 194
Thioamides, 110, 194
Thiochromone, 136
Thiocyanic acid, 208
Thiolactams, 110
Thiophenols, 120
Thioxanthone, 136
Tin compounds, 165
Titration in non-aqueous solvents, 60
Toluene, 264, 295–296
Toluenesulphonic acids, 293
Trialkylacetic acids, 303

Triarylethylene glycols, 242
Triarylmethanols, 150, 152–153
Triarylmethyl cations, 305
Trichloromethyl mesitylene, 160
2,4,6-Trimethoxybenzaldehyde, 261
4,6,8-Trimethylazulene-1-d, 269
Trimethylpentenes, 244
Trimethylphenylammonium ion, 284–285
2,2,6-Trimethyl-quinuclidone-7, 107
Triphenylacetic acid, 255–258
Triphenylcarbinol, see Triphenylmethanol
Triphenylcarbonium ion, 139, 299, 305
Triphenylmethanol, 145
Transition metal/carbonyl complexes, 143
Transport numbers, 72
Tropilium ion, 139

U

Urea,
  cation structure, 109
  hydrolysis, 193
  protonation, 109

Urethane,
  basicity, 108
  cation structure, 108

V

γ-Valerolactone, 132
Viscosity of sulphuric acid, 5, 7, 8, 72–73
Volumes of activation, 178, 202

W

Water activity in sulphuric acid, 15
Water titration of oleum, 3

X

Xanthen, 160
Xanthone, 136
Xanthyl-alcohols, 221
Xylenes, 295–296
p-Xylenesulphonic acid, 293

Z

Zero-point energy, 179